Preanesthetic Assessment 3

Preanesthetic Assessment 3

Edited by
Elizabeth A. M. Frost

Birkhäuser
Boston • Basel • Berlin

Elizabeth A.M. Frost, M.D.
Department of Anesthesiology
Albert Einstein College of Medicine
of Yeshiva University
Montefiore Medical Center
Bronx, NY 10461
U.S.A.

ISSN 0896-2103

Printed on acid-free paper.

The chapters in this volume are revised and updated articles that originally appeared in consecutive issues of *Anesthesiology News* from 1989 to 1991. They appear here as a service to the medical-anesthesiology community with the cooperation and kind permission of McMahon Publishing Company, New York.

ISBN 0-8176-3536-X
ISBN 3-7643-3536-X

Typeset by ARK Publications, Newton Centre, Massachusetts.
Printed and bound by Edwards Bros., Ann Arbor, Michigan.
Printed in the U.S.A.

9 8 7 6 5 4 3 2 1

All the residents and fellows who contributed these chapters and my colleagues who reviewed the topics and offered so many helpful comments deserve much applause. But without the tireless support, good humor and common sense of Belinda Fortsch, our administrator, who miraculously transcribed everyone's handwriting and minds to the computer, this third volume would not exist.

And to her, Preanesthetic Assessment 3 is dedicated.

Foreword

One of the easiest things an anesthesiologist can do is to administer an anesthetic, and therein lies one of the major dangers of anesthesia. Like in flying and diving, if the rules are followed, mishaps should be very rare. The *Art* of Anesthesiology so often depends on anticipation of the unexpected, knowing the patient and doing something intelligent when the *Science* does not give a clear cut answer to a problem. The risks to our patients can, therefore, be markedly reduced, if we have a good grasp of the difficulties and dilemmas that they are prone to, before they are anesthetized. Herein lies the great value of preanesthetic assessment.

Preanesthetic Assessment has now reached its third volume in the space of five years. Twenty-four new areas are covered with more than enough detailed information to satisfy the careful practitioner and the Board Examiner. Dr. Elizabeth Frost is to be congratulated on presenting a breadth of material that contains something useful and informative for everyone in training or practice. Subjects vary from the demands of the athlete to those of the Jehovah's Witness, congenital disorders to pediatric and obstetrical quandaries. Other problem areas include drug abuse, endocrinology, the patient undergoing neuroradiology and a mixed bag of diseases that provide the unprepared anesthesiologist with problems. Each chapter is a revised version of a continuing medical education article appearing each month in *Anesthesiology News*. These articles were part of a very successful educational program started in the Department of Anesthesiology, Albert Einstein College of Medicine/Montefiore Medical Center. This has been taken by thousands of anesthesiologists, who were awarded CEU credit.

The authors vary from recognized experts to young anesthesiologists, either completing their training or newly graduated from their residency or fellowship program. Dr. Frost has reedited and updated each article so that new references and information is included. *Preanesthetic Assessment* truly represents state-of-the-art and science of Anesthesiology. This has not usually been achieved in the larger and more comprehensive textbooks, dealing with what the anesthesiologists should know about various disease states and patient situations, and how they affect their practice.

The twenty-four new areas covered in this volume make up a very readable and informative refresher course for those who have become a bit rusty, as well as giving an excellent insight into anesthetic practice and patient management for those learning about a subject for the first time. The reader's attention is focused at the beginning of each theme by a case study. This brings immediate clinical relevance and identifies the major problems at the outset. Each chapter is a complete entity, so the reader does not have to refer elsewhere in the text for associated information. Even in areas of contention, if the readers find some disagreement with other authorities, they will always be left with a sensible, rational and justifiable plan of action. The wide range of topics in this series leaves few subjects uncovered for practicing anesthesiologists either meeting unusual cases, or for those that they rarely see.

Keeping up-to-date is becoming a major issue, not only within the Speciality, but also with those outside who feel the patient receiving anesthesia services deserve to be served safely and well. Continuing education has been promoted as a necessity for some years, and has even become a requirement for licensure in many jurisdictions. A recredentialling process is getting underway. This will be designed to ascertain whether boarded anesthesiologists have kept up with the Speciality sufficiently to be deemed competent. *Preanesthetic Assessment* makes a valuable contribution to dealing with this newer issue, while providing a comprehensive set of tutorials on the management of both the routine and extraordinary patient.

Christopher W. Bryan-Brown, M.D.
Vice-Chairman, Clinical Affairs Anesthesiology/Critical Care Medicine
Professor of Anesthesiology, Albert Einstein College of Medicine
Montefiore Medical Center, Bronx, New York

Contents

Contributors

KENNETH J. ABRAMS, M.D.
Instructor in Anesthesiology, Albert Einstein College of Medicine/Montefiore Medical Center, Bronx, New York 10461

ROBERT ADAM, M.D.
Attending in Anesthesiology, Windham Community Hospital, Willimantic, CT 06226-2082

MARK J. BADACH, M.D.
Assistant Professor of Anesthesiology, Seton Hall University, School of Graduate Medical Education, Attending St. Joseph's Hospital & Medical Center, Paterson, New Jersey 07503-2691

NIRMALA BALAN, M.D.
Attending in Anesthesiology, Loma Linda University, Loma Linda, CA 92350

SHARDE DAVE, M.D.
Attending in Anesthesiology, Mount Vernon Hospital, Mount Vernon, NY 10550-2026

JILL FONG, M.D.
Assistant Professor of Anesthesiology, New York Hospital-Cornell Medical Center, New York, New York 10021

ELIZABETH A.M. FROST, M.D.
Professor of Anesthesiology/Director of Neuroanesthesia, Albert Einstein College of Medicine/Montefiore Medical Center, Bronx, New York 10461

DARAN HABER, M.D.
Fellow in Anesthesiology, Yale-New Haven Hospital, New Haven, CT 06504

INGRID HOLLINGER, M.D.
Professor of Anesthesiology, Assistant Professor of Pediatrics/Director, Division of Pediatric Anesthesia, Montefiore Medical Center/Albert Einstein College of Medicine, Bronx, New York 10467

SETH LANDA, M.D.
Director of Obstetric Anesthesia, St. Joseph's Hospital Medical Center, Paterson, New Jersey 07503

YA-TSENG WILLIAM LU, M.D.
Resident in Anesthesia, Albert Einstein College of Medicine/Montefiore Medical Center, Bronx, New York 10461

LLOY ANDERSON M.D.
Attending in Anesthesiology, Suburban General Hospital, Norristown, Pennsylvania, 19401-1849

JAMES B. MUELLER, M.D.
Staff Anesthesiologist, Norfolk General Hospital, Norfolk, Virginia

MICHAEL PECK, M.D.
Attending in Anesthesiology, Georgetown University Hospital, 3800 Reservoir Road, N.W., Washington, D.C. 20037-2327

LAUREN A. PLANTE, M.D.
Instructor in Anesthesiology, Albert Einstein College of Medicine/Montefiore Medical Center, Bronx, New York 10461

JON D. SAMUELS, M.D.
Instructor in Anesthesiology, Good Samaritan Hospital, Suffern, NY, 10901

STEVEN S. SCHWALBE, M.D.
Assistant Professor, Director of Obstetric Anesthesia, Department of Anesthesiology, Albert Einstein College of Medicine/Montefiore Medical Center, Bronx, NY 10461

MICHAEL R. SEIDEL, M.D.
Resident in Anesthesia, Albert Einstein College of Medicine/Montefiore Medical Center, Bronx, New York 10461

DAVID R. SOFAIR, M.D.
Attending in Anesthesiology, Westchester Square Hospital, Bronx, NY, 10461-3198

STEPHEN G. SOLOMON, M.D.
Department of Anesthesiology, Horton Memorial Hospital, Middletown, NY 10940-4199

JAMES E. TOBIN, M.D.
Staff Anesthesiologist, Anesthesia Dept. (66), Portsmouth Naval Hospital, Portsmouth, Virginia 23708-5000

SCOTT IRA WINIKOFF, M.D.
Attending in Anesthesiology, Bridgeport Hospital, Bridgeport, CT, 06610-2875

Introduction

It is difficult to realize that 7 years have now passed since the first meeting on City Island in the Bronx with Mr. Ward Byrne, the editor of *Anesthesiology News*, when we discussed the development of an educational series for clinical anesthesiologists. The articles, although complete in themselves, were to be bound by the common theme of Preanesthetic Assessment. In the beginnning, I had many doubts that we would be able to complete enough articles on time for even one year. However, we passed the first hurdle. Practitioners across the United States expressed interest both by writing to *Anesthesiology News* and by completing the self assessment quizzes for Continuing Medical Education credits.

We soon had enough material to present in book form. Volume 1 contained the first 26 lessons. It was enthusiastically received both in this country and abroad as a convenient, practical reference source and as an aid in studying for Board examinations. Almost immediately, we received inquiries as to when Volume 2 would be ready and if we could establish a biannual project. The next 24 articles, each one reviewing a different topic, were assembled and we proudly presented Volume 2. Now 24 articles later, Volume 3 is completed, and as of writing, we are already on our way to Volume 4. Slowly, the idea of a biannual project is becoming a reality.

As with the early lessons, I have continued to involve our residents and junior faculty as much as possible. I am delighted that increasingly anesthesiologists from other programs have asked to contribute. Few of these young physicians have realized at the outset how much research and hard work are involved before a lesson is completed. There have been many discussions, some quite heated, before a manuscript has been finally sent to press. But I am proud to congratulate them and share their joy in their achievements and I am delighted at the understanding and awareness that has continued to grow between the residents and the attending physicians who have acted as reviewers. As the series has become more widely known, some of our authors have even found placement interviews much easier.

As was done in Volumes 1 and 2, each lesson has been updated although the material is essentially the same as that published in *Anesthesiology News*. All of the 76 case scenarios presented so far are unique—to date there have been no duplications, even though our understanding and treatment protocols for some disease states have been modified. Questions have been included at the end as a self-assessment test—answers are also provided.

Again, I express by gratitude to the staff of *Anesthesiology News,* Ward Byrne, Elizabeth Douglas, Tatiana Chillrud and the most recent member of the team, Damian McNamara, for their willing cooperation and continued support. I thank the staff of Birkhäuser for continued confidence in this project, and most or all my sincere appreciation to the residents and staff who have helped me make the deadlines on time.

Elizabeth A.M. Frost, M.D.

Chapter 1
The Athlete

Stephen G. Solomon

Case History. *A 21-year-old man presented to his orthopedic surgeon with a knee injury warranting arthroscopic examination. He had no significant past medical or surgical history, was taking no medications, and had no known allergies. He neither smoked nor drank apart from beer occasionally. He was in training for the New York marathon and ran at least 5 miles a day, 5 days a week. In addition, he regularly engaged in many sports and represented his college in football. Physical examination showed a healthy-looking, tall young man, weighing about 205lb; blood pressure, 120/80mmHg; pulse, 55/min; respiratory rate, 16/min; temperature, normal.*

The patient came to the ambulatory unit on the day of surgery. Laboratory studies, performed a few days earlier, were reviewed at that time by the anesthesiologist. They revealed the following: hematocrit, 29%; hemoglobinuria; ECG demonstrating first-degree heart block; heart rate, 54/min; left ventricular hypertrophy; and ST elevation in V_2 *to* V_4.

The orthopedic surgeon, concemed about the passing time, was anxious to operate immediately. The anesthesiologist felt that further cardiac investigation was indicated.

Introduction

The increased participation of the population in running and other endurance events has brought new challenges to the anesthesiologist. Athletes often require surgery for sports-related injuries. An apparently healthy person, paradoxically, can be a diagnostic dilemma because of abnormal laboratory and ECG data, as demonstrated by our patient. Not infrequently, the surgeon, anxious to proceed, notes that all athletes

Reviewed by Dr. Martin I. Levy, Director of the Center for Sports Trauma and Associate Professor of Orthopedic Surgery, Montefiore Medical Center and the Jack D. Weiler Hospital of Albert Einstein College of Medicine.

demonstrate changes in laboratory profiles that are not significant and do not contribute to adverse outcomes. On the other hand, most anesthesiologists have had experiences in treating patients with abnormal ECG patterns, especially those suggestive of ischemia, that belie the encouraging optimism of the surgeon.

It is well recognized that congenital heart problems have been successfully treated allowing young people who may still have some cardiac disability reach teenage years and thus engage in major athletic activities.

Moreover, drug abuse that significantly interacts with anesthetic management and adversely affects most body systems has been frequently implicated in high performance athletics.

In the setting of the ambulatory center extensive testing and other consultations are not always readily available. Time is an important factor. Most important, concern has been raised in the patient's mind that he may have myocardial disease and be about to suffer a heart attack, an event not unheard of in athletes. What constitutes appropriate further evaluation? What warrants delay? What may be safely ignored?

Several changes commonly found in athletes are referable to the cardiovascular, hematologic, and endocrine systems. Additionally, narcotic or steroid abuse may have occurred.

Cardiovascular System

Cardiovascular changes are adaptations to meet the stresses of exercise. The type of exercise plays a role in determining the nature of change. The isotonic athlete an endurance runner, for example) has a left ventricle (LV) with increased diameter, a proportional increase in LV wall thickness, an increased end-diastolic diameter, and a low heart rate.

The isometric athlete (eg, power weight lifter) has a hypertrophied LV without the proportionate increase in chamber size. This is a result of the pumping of the heart against aortic pressures as great as 300mmHg (and is similar to the LV hypertrophy seen in hypertension, where the heart must pump against high systemic pressures). Increases in the size of the interventricular wall and LV posterior wall occur. The left atrium also can increase in size. Stroke volume (SV) increases, but the ejection fraction remains the same.[1,2]

The ECG of the athlete is often a source of confusion, as it can be suggestive of disease (see Table 1). Resting bradycardia is very common and almost the norm in the athlete and is the result of two factors. There is an increase in vagal tone and a decrease in resting and sympathetic tone, with a reduction in heart rate to maintain cardiac output in the normal range and compensate for the increased stroke volume.[1,2]

TABLE 1. Common Abnormal ECG Patterns in Athletes

Bradycardia
First-degree atrioventricular block
Wandering atrial pacemaker
Wenckebach phenomenon
$ST_{12}T$ wave changes
Left and right ventricular hypertrophy

Stroke volume is increased in the athlete. Sinus dysrhythmias are found in 35% to 60% of athletes, as opposed to 2.4% of the general population. Among these abnormalities are wandering atrial pacemaker (present in 7.4% to 19% of athletes and not found in the less-athletic population) and first-degree atrioventricular block (10% to 33% of athletes, 0.65% of the general population).

In a study[2] of 126 marathon runners, a Wenckebach phenomenon that disappeared with exertion was found in 2.4% of the runners. The incidence of this dysrhythmia was related to the intensity of training and disappeared with cessation of the activity. Junctional rhythms are found in 0.31% to 7% of athletes, compared to 0.06% of the general population. These dysrhythmias are thought to be related to autonomic changes. Common repolarization changes include ST elevation (found in a widely varying percentage of the athlete population, depending on the study), ST depression, peaked T waves, and inverted T waves. ST elevations are most prominent in leads V_2–V_4, with reciprocal changes in the inferior leads.

There also can be J-point elevations, which are thought to be nonpathologic, as they frequently revert to normal during exertion. One possible explanation for these changes is that they are evidence of nonhomogeneous repolarization of the ventricle, with initial repolarization of the epicardium. Another explanation is that training-induced decreases in resting sympathetic tone uncover an inherent asymmetry of repolarization. With exercise, ST segments become isometric as increased sympathetic tone overrides resting asymmetric repolarization.

Often ST and T-wave abnormalities normalize with deconditioning. However, reversion may be delayed, as occurred in one case[2] with nonpathologic T-wave inversion that persisted 9 years after the cessation of endurance training. Also, athletes with T-wave inversion have been compared to controls by use of thallium treadmill studies.[2] Results showed no changes between the groups and did not reveal evidence of ischemic areas in the myocardiums of the athletes. Voltage changes between the groups did not reveal evidence of ischemic areas in the myocardiums of the athletes.

Voltage changes suggestive of left and right ventricular hypertrophy (LVH, RVH) are often found in the ECGs of athletes. Of 1000 Olympic athletes and marathon runners, 14% to 85% had LVH changes, compared to 5 of 122,000 healthy men (criterion was 35mm summation in precordial voltage).[2] RVH was found in 18% to 69% (voltage greater than 10.5mm). Increased T-wave amplitude often occurs. The axis of the heart can become more vertical.

Clinicians should be cautioned that although most changes are physiologic, not pathologic, deep T-wave inversions accompanied by a symmetric l-wave contour, ST depression, a prolonged QT interval, or absence of normal septal Q waves warrant further investigation. These changes are not normally found and may be associated with a pathologic state.[2]

Sudden cardiac death has been described in athletes. Probably because of their apparent paradoxic nature, such tragedies in young, apparently extremely healthy persons have been afforded considerable publicity. The most common cause of sudden death is hypertrophic cardiomyopathy,[2,3] and athletic training is probably not a causative factor. As mentioned earlier, increases in the size of the interventricular wall and LV posterior wall occur with conditioning. Ratios of 2:1, suggestive of hypertrophic cardiomyopathy, can occur. However, in contrast to hypertrophic cardiomyopathy, there is no such reduction in LV cavity size in the athlete as is seen in the disease state.[2]

Those with hypertrophic cardiomyopathy should be discouraged from taking part in endurance sports, but the healthy athlete need not be fearful that exercise will cause this disease.[4] Likewise, the anesthesiologist does not have to be concerned that his or her athletic patient will respond like one with hypertrophic cardiomyopathy.

Other causes of sudden death include congenital coronary artery abnormalities, ruptured aorta (secondary to cystic medial necrosis), and idiopathic LVH with coronary artery atherosclerosis. Less common causes include myocarditis, mitral valve prolapse, aortic stenosis, and sarcoidosis.[3] If persons with such conditions were to lead lives devoid of endurance training, they probably would not experience problems referable to their hearts. However, congenital abnormalities may be manifest first as a result of athletic activities.

During the preanesthetic assessment the anesthesiologist should question the patient (and the parents if possible) for any suggestion of congenital abnormalities and any indication from childhood visits to the pediatrician that a cardiac murmur was detected. It is unlikely in the ambulatory setting that the patient has a prior ECG available, but if available this should be obtained.

Athletes can suffer from symptomatic dysrhythmias, including paroxysmal atrial tachycardia, supraventricular tachycardia, paroxysmal ventricular tachycardia (sustained and nonsustained), and ventricular fibrillation. These dysrhythmias are usually associated with abnormal substrates, as would be present in their nonathletic counterparts. Exercise could uncover these changes.[5] It is possible that the stress of surgery could also induce dysrhythmias.

Knowing that these physiologic adaptations exist, what, if anything, should be done about ECG changes in the athlete? In one report,[6] a large group of intercollegiate athletes was studied, with history, physical examination, and 12-lead ECG as initial screening tools. If any of the tests showed positive results, additional studies were done, including an echocardiogram or stress ECG. These tests were found to be extremely expensive, with too low a yield to be practical. Also, the false-positive ST-segment response of the stress ECG was considered too high to be helpful. In more than 500 athletes no cardiovascular disease was found.

Another study[7] has confirmed that the yield is too low to justify the cost. Thus, to cancel a case and demand further expensive evaluation is probably not warranted in the absence of a positive cardiac history, certain ECG patterns as outlined above, or a symptomatic dysrhythmia. If abnormal changes in the ECG are observed, the pattern should be documented and appropriate notation made. The patient should be assured that although a pattern has been demonstrated that is slightly different from what we have rather arbitrarily described as normal, this is due to a physiologic adaptation by the heart rather than the result of pathologic determination. Sudden death must not be an all-consuming concern.

Hematologic System

Anemia has a wide differential diagnosis, ranging from the relatively benign (eg, secondary to menstruation) to more serious disorders such as aplastic anemia. Because there is a high incidence of anemia in athletes, it is important to understand the physiology of "athlete's anemia." Several theories have been suggested: First, the red blood cell mass is actually normal or increased, but the plasma volume is increased, resulting in a dilutional fall in the hemoglobin concentration. Second, there is hemolysis of cells. Hemolysis occurs mostly in impact sports, such as running; it also can occur in nonimpact sports, such as swimming, but to a lesser extent. The hemolysis from nonimpact sports may be due to hemolysis of older red cells (from compression of vigorously contracting muscles) and changes in intraerythrocyte and plasma osmolarity.[8] Eating disorders, especially anorexia nervosa, are also seen in athletes and may contribute to anemia.

The finding of depleted iron stores has led some to believe that iron depletion may play a role in anemia. However, measured iron stores are lower because, during excessive hemolysis (as evidenced by decreased haptoglobin), catabolism of red cells takes place in the hepatocytes as well as in the reticuloendothelial system (RES). Ferritin in serum mainly reflects iron stores in the RES.[9]

If a patient has a marginally low hematocrit and the procedure entails little anticipated blood loss, further evaluation and delay of the case are probably not warranted. If the patient is to remain in the hospital, it may be prudent to seek a hematologic consultation. Blood is often found in the urine specimens of athletes, usually as a result of excretion of the hemolyzed cells. It resolves after a few days of abstinence from athletic activities.

A large survey performed by the Armed Forces Institute of Pathology[10] revealed an association between the finding of sickle-cell trait and sudden death in physical training. The mortality rate with sickle-cell trait was at least 27 times greater, during exercise, than that of recruits without sickle-cell trait. Although it is generally held that patients with the sickle-cell trait incur a minimal increase in risk compared to patients without sickle-cell trait, these data suggest that certain stresses that may occur in the athlete or under anesthesia may prove hazardous to this select group.

Endocrine Disturbances

Although the athlete is not, for the most part, more prone to endocrinologic disorders than the nonathlete, some situations related to the athlete are relevant to the physician.

Regular exercise can produce changes in hypothalamic pituitary function, particular reducing pulsatile luteinizing hormone secretion. However, this change probably does not cause the menstrual dysfunction that is not uncommon in female athletes (such abnormalities are more likely related to hypothalamic, pituitary, or ovarian dysfunction associated with elevated levels of antireproductive hormones, including beta endorphins, dopamine, prolactin, and catechol estrogens, induced by exercise).[11]

It was thought that there would be a relationship between body fat percentage and menstrual patterns, but no correlation has been demonstrated.[12] Thus, in the preanesthetic assessment of the athlete, a history of amenorrhea or decreased libido may be obtained. Most likely, the cause is benign. However, the possibility of a serious disease state such as pituitary tumor should not be overlooked. Questioning should be aimed at excluding the occurrence of headaches and visual difficulties.

A secondary issue pertaining to menstrual dysfunction is not anesthesia-related but involves general patient care. Those with menstrual dysfunction have a higher incidence of osteoporosis, endometrial hypoplasia, and adenocarcinoma.[11] Referral to a gynecologist or endocrinologist for further care may be indicated.

Drug ingestion also causes endocrine abnormalities. In spite of claims that anabolic steroids do not enhance athletic performance, their use has become widespread. Anabolic steroids act to create a positive nitrogen balance and build muscle if caloric intake is sufficient. Users of these drugs may present with a history, physical findings, and laboratory abnormalities suggestive of several diseases. They often have changes in libido, increased aggressiveness, muscle spasms, and mood swings.

With anabolic steroid ingestion, the release of follicle-stimulating hormone (FSH) and luteinizing hormone LH) is inhibited, leading to testicular atrophy and gynecomastia. In women, masculinizing effects can be seen, including hirsutism and a deepened voice.[13] Laboratory abnormalities include a decrease in thyroxin-binding globulin; decreased total T_4 cell counts, abnormalities of liver function, including elevated enzymes (SGOT, SGPT, alkaline phosphatase, bilirubin, and LDL (as well as increased LDL/HDL ratio); decreased endogenously produced testosterone; and as already mentioned, decreased FSH and LH.[13,14]

If these findings are present in otherwise healthy person, they probably can be attributed to the use of steroids. Because of the legal ramifications of the use of steroids, truthful responses to direct questioning may be difficult to obtain. If abnormal laboratory findings are uncovered, their significance should be explained to the patient and his or her cooperation sought. (i.e., the possibility of severe hypotension intraoperatively that can be averted by pre-operative administration of additional steriods, or otherwise unexplained hyperglycemia that may be prudently treated with insulin).

Longterm use of steroids causes cholestatic hepatitis, peliosis hepatitis, liver tumors (both benign and malignant), increased incidence of atherosclerotic heart disease (secondary to the increased LDL/HDL ratio), and in pregnant women an increased incidence of pseudohermaphroditism and intrauterine death.[13]

Street Drugs

The use of such illegal drugs as cocaine has received much publicity because of the deaths of well-known athletes such as Len Bias. Highly trained athletes, especially sprinter-type athletes, are at higher risk of death from cocaine exposure, probably because of their muscle composition. Sprinter-type athletes have a very high percentage of white (as opposed to red) muscle. This type of muscle allows great bursts of speed for a short period of time and can produce great amounts of lactic acid.

Cocaine increases adrenergic activity, which increases glycolysis, glycogenolysis, lactic acid production, and heat generation. This last effect is compounded by the vasoconstrictive effect of cocaine, which compromises the body's ability to dissipate heat. Cocaine also can induce seizures. A cocaine-induced seizure, in an athlete whose muscles already possess the ability to produce large amounts of lactic acid and heat, may prove lethal. (See Chapter 16)

The brain is spared the vasoconstrictive effects of cocaine, and an increased brain lactic acid concentration results. Hypertension (especially associated with crack abuse) may disrupt autoregulation with cerebral edema as a result. A vicious cycle of hypoxia, lactic acidosis, and edema is established.[15] Topical cocaine is often used by anesthesiologists as well as by dentists and otolaryngologists. In light of the potential for disaster associated with cocaine use in the elite athlete, it should be used cautiously if at all.

Contact sports often involve painful injuries. To continue playing, athletes may take or be given painkillers, including narcotics. Also, the glamour and the pressure of financial rewards of spectator activities sometimes make athletes more prone to indulge in illicit drug ingestion. The problems of the anesthetic management of the drug addict were considered in Volume 1 Chapter 3 of this series; as they pertain to the athlete, they include liver function abnormalities and an inappropriate intolerance of pain.[16] The diastolic blood pressure may be low because of decreased sympathetic tone, which also decreases systemic vascular resistance.[17]

Anesthetic Plan

The fear of pain and unpleasant awareness of needles may limit the use of regional anesthesia in the athlete. Also, a very large muscular physique may make placement of the spinal needle difficult. There is, however, no contraindication to the use of either spinalor epidural block. Indeed, spinal anesthesia with appropriate intravenous sedation including fentanyl may be the anesthetic of choice for arthroscopic examinations. Instillation of small amounts of Duramorph®or fentanyl to the subarachnoid space can eliminate the need for post-operative analgesics and allow the athlete to resume exercise within 24 hours of surgery.

In choosing a general anesthetic, agents and neuromuscular blocking drugs that are metabolized by the liver (eg, opioids, pancuronium, halothane) should be avoided if narcotic-induced hepatic abnormalities exist. Induction with thiopental is probably preferable to the use of etomidate or ketamine. Etomidate has been implicated in suppression of adrenal cortical function, which may compound the effects of previous steroid ingestion. Ketamine is associated with postoperative hallucinations and may make the care and nursing of a large muscular man very difficult.

Di-isopropyl phenol (propofol), a recently introduced short acting intravenous agent may offer many advantages for ambulatory patients undergoing relatively short procedures. In a dosage of 1–2 mg/Kg, followed by bolus injections of 10–20 mg/10 minutes or by infusion of 1–2 ug/min adequate anesthesia may be achieved in conjunction with N_2O/O_2 inhalation with prompt arousal allowing early discharge from an outpatient facility.

For more extensive procedures, isoflurane is the maintenance drug of choice because less than 1% is metabolized by the liver, even in the presence of hepatic enzyme induction. Atracurium, which depends on Hoffmann elimination and plasma hydrolysis, is useful.

Routine monitoring includes pulse oximetry, capnography, ECG (including V_5), pulse, blood pressure, respiratory rate, and temperature. Sugar-containing solutions should be administered cautiously because of the risk that hyperglycemia will develop and will aggravate the detrimental effects of lactic acidosis. Occasional determinations of blood glucose levels by fingerstick are indicated.

In summary, the anesthetic management of the athlete does not differ much from that of the nonathlete. The anesthesiologist must differentiate adaptive changes of exercise from disease states. If the definition of "normal findings" is widened, the anesthesiologist should be able to make reasoned decisions with regard to the anesthetic preparation of the athletic patient requiring surgery. The need for open discussion with and reassurance of the patient is underscored.

These patients are not cardiac cripples.

References

1. Colan S.D., Sanders S.P., Borow K.M.: Physiologic hypertrophy: Effects on left ventricular systolic mechanics in athletes. *J Am Coll Cardiol* 1987, 9:776–83.
2. Huston T.P., Puffer J.C., Rodney W.M.: The athletic heart syndrome. *N Engl J Med* 1985, 313:24–32.
3. Maron B.J., Epstein S.E., Roberts W.C.: Causes of sudden death in competitive athletes. *J Am Coll Cardiol* 1986, 7:204–14.
4. Maron B.J.: Structural features of the athlete heart as defined by echocardiography. *J Am Coll Cardiol* 1986, 7:190–203.

5. Coelho A., Palileo E., Ashley W., et al.: Tachyarrhythmias in young athletes. *J Am Coll Cardiol* 1986, 7:237–43.
6. Maron B.J., Bodison S.A., Wesley Y.E., et al.: Results of screening a large group of intercollegiate competitive athletes for cardiovascular disease. *J Am Coll Cardiol* 1987, 10:1214–21.
7. Epstein S.E., Maron B.J.: Sudden death and the competitive athlete: Perspectives on preparticipation screening studies. *J Am Coll Cardiol* 1986, 7:220–30.
8. Selby G.B., Eichner E.R.: Endurance swimming: Intravascular hemolysis, anemia, and iron depletion. New perspective on athlete's anemia. *Am J Med* 1986, 81:791–4.
9. Magnusson B., Hallberg L., Rossander L., Swolin B.: Iron metabolism and "sports anemia": II. A hematological comparison of elite runners and control subjects. *Acta med Scand* 1984, 216:157–64.
10. Sullivan L.W.: The risks of sickle cell trait: caution and common sense. *New Engl J Med* 1987, 317:830–1.
11. Noakes T.D., van Gend M.: Menstrual dysfunction in female athletes: A review for clinicians. *S Afr Med J* 1988, 73:350–5.
12. Ouellette M.D., MacVicar M.G., Harlan J.: Relationship between percent body fat and menstrual patterns in athletes and nonathletes. *Nurs Res* 1986, 35:330–3.
13. Bierly J.R.: Use of anabolic steroids by athletes. Do risks outweigh the benefits? *Postgrad Med* 1987, 82:67–8, 1987, 71–2, 74.
14. Allen M., Rahkila P., Reinila M., Vihko R.: Androgenic-anabolic steroid effects on serum thyroid, pituitary and steroid hormones in athletes. *Am J Sports Med* 1987, 15:357–61.
15. Glammarco R.A.: The athlete, cocaine, and lactic acidosis: a hypothesis. *Am J Med Sci* 1987, 294:42–4
16. Gevirtz C.: The intravenous drug abuse patient, in Frost E. (ed): *Preanesthetic Assessment 1*. Boston: Birkhäuser, 1988, pp 21–29.
17. Prystan G.H.: Autonomic responsivity to sensory stimulation in drug addicts *Psychophysiol* 1975, 12:170–8.

CHAPTER 2

The Patient With Chronic Spinal Cord Injury

James B. Mueller

Case History. *A 28-year-old man with a 1-year history of a T2–3 spinal gunshot wound was admitted for a basket extraction of a left ureteral stone. Past medical history was significant for recurrent urinary tract infections, periodic decubitus ulcers, and a stress fracture of the right femur. During his last surgery 4 months ago, the patient stated, the doctors had "had problems with my blood pressure being too high." The patient denied any allergies.*

Physical examination revealed a thin man in a wheelchair. Weight was 145 lb, blood pressure 110/65mmHg, pulse 72/min and regular, respiratory rate 32/min. Breath sounds were equal bilaterally with scattered rhonchi and crackles in both lung bases. There was a well-healed tracheostomy scar. The findings on cardiovascular examination were normal. Neurologic examination revealed a paraplegic patient with a sensory level at T3. Motor function below T3 level was absent, and arm strength was essentially normal. Skin examination revealed several decubitus ulcers on his right buttock and right posterior iliac crest.

Laboratory data included: hemoglobin 9.8gml/dl, hematocrit 28%, white blood cell count 9700/mm^3*, calcium 8.8mg/dl, potassium 4.0mEq/L, sodium 145mEq/L.*

Introduction

In the United States traumatic injury continues to be a major problem. Approximately 1.5 million people per year sustain substantial injury requiring hospitalization. Spinal cord injury (SCI) occurs in nearly 11,000 of these patients. The most common causes of SCI include motor vehicle

---

Reviewed by Dr. Alan Kantrowitz, Assistant Professor, Department of Neurosurgery, Albert Einstein College of Medicine/Montefiore Medical Center.

accidents, 48%; falls, 21%; violence (gunshots/stabbing), 15%; and sports, 15%. Men are more frequently involved than women (4:1). Spinal cord injury occurs most frequently in persons between the ages of 15 and 20 years, the most frequent age being 19. Initial mortality from SCI approaches 48%. For the remaining 52% of these patients the postinjury course is long and complicated.[1–3]

Pathophysiology

Depending on the level of the injury, all bodily systems are affected to varying degrees. Those of particular concern to the anesthesiologist are reviewed below. (See Table 1.)

Respiratory

In acute SCI, respiratory disturbances are very common after high cervical and thoracic spine injuries (Figure 1 and 2). In fact, respiratory compromise is the leading cause of morbidity and mortality in the acute phase of SCI.[2]

TABLE 1. Systemic Effects of Thoracic and Cervical Injury.

System	Level	Complication
Respiration	Above C3	Apnea
	Above C6	Intercostal paralysis
		Diaphragmatic respiration
Cardiovascular	Above T4	Bradycardia
		Hypotension
	Above T6	Autonomic hyperreflexia
	All levels	Hypovolemia
Renal	All levels	Recurrent infections
		Uremia
		Amyloidosis
Musculoskeletal	All levels	Decubitus
		contractures
Hematopoietic	All levels, especially cervical	Deep-vein thrombosis
		Anemia
Cerebral function	All levels, especially cervical	Hallucinations
Gastrointestinal	All levels	Bleeding
		Ileus

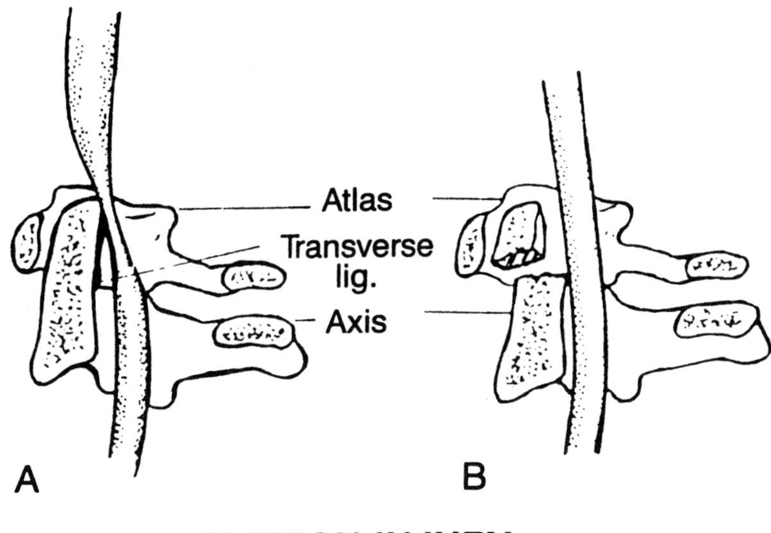

FIGURE 1. The flexion with vertical force causes: *A*, rupture of transverse ligament causing backward displacement of the odontoid with compression of cord; and *B*, the fractured odontoid process, but intact transverse ligament with displacement of axis compressing the spinal cord.

The amount of respiratory impairment depends on the level of injury. The phrenic nerve originates chiefly from the fourth cervical nerve but receives additional input from the third and fifth cervical nerves. Therefore, lesions above the level of C5 require total ventilatory support because of the loss of diaphragmatic, intercostal, and abdominal muscle action. Lesions at or below the C6 level leave the function of the diaphragm intact but cause varying degrees of intercostal and abdominal muscle paralysis. Obviously, the lower the level of injury, the less respiratory impairment.

Respiratory compromise secondary to SCI is classified as restrictive lung disease. There is generalized decrease in total lung capacity (TLC), vital capacity (VC), tidal volume (TV), forced expiratory volume in 1 second (FEV_1), and inspiratory and expiratory reserve volumes. Associated with the loss in lung volumes, there is a loss of lung compliance. As a result, the patient is unable to generate an adequate cough, has difficulty clearing secretions, develops lung collapse and atelectasis, and is susceptible to pneumonia.[2,4,5]

Following SCI, the work of breathing is increased as the patient tries to expand the noncompliant lungs. A pattern of rapid, shallow breathing develops in an attempt to minimize the work. It is not uncommon to find a slight degree of hypocapnia secondary to alveolar hyperventilation.

Troyar and Heilporn[5] studied respiratory mechanics in quadriplegic patients. They concluded that longstanding paralysis of intercostal muscles contributed to the loss in compliance and reduced the passive (outward) recoil of the chest wall. Patients without intercostal activity have reduced functional reserve capacity (FRC), lower transpulmonary pressure at FRC, and reduced static expiratory compliance. The resulting pressure-volume curves resemble those of patients with generalized respiratory muscle weakness.

Associated with the intercostal muscles in maintaining adequate respiratory exchange are the abdominal muscles. The abdominal muscles are innervated via the nerves of T6–L1. The chief respiratory function of

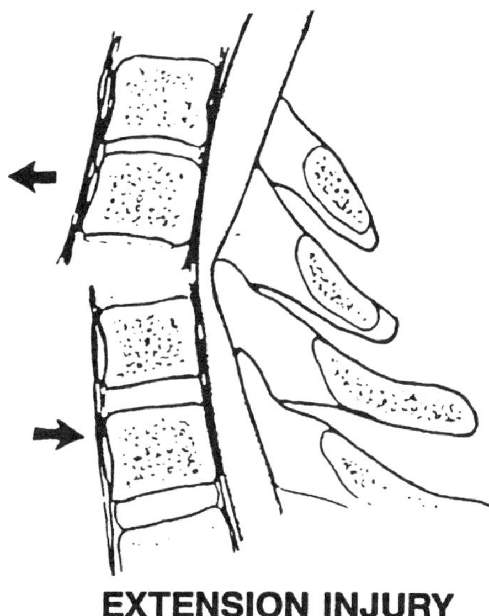

FIGURE 2. Extension injury of lower cervical spine causing displacement of body of cervical vertebra due to rupture of anterior and posterior longitudinal ligaments.

the abdominal muscles is that of forced expiration and coughing. Loss of action of these muscles is associated with a decrease in vital capacity.[4]

Since the SCI patient has varying degrees of respiratory compromise, a careful preoperative assessment of the respiratory system is mandatory. As with any patient, preoperative "tuning" helps to prevent postoperative complications. Therefore, the SCI patient should undergo a thorough, preoperative respiratory assessment (see Table 2.).

Cardiovascular

The cardiovascular system is labile in both the acute and chronic phases of SCI.[6–8]

The sudden loss of sympathetic nervous system function and of the control exerted by the brain stem contributes to spinal shock, which may last from days to many weeks. The sudden, complete cord transection/injury causes interruption of communication between the higher cerebral centers and peripheral tissues. Interruption of the sympathetic nervous system results in the loss of normal cardiovascular compensatory reflexes along with all other spinal reflexes below the level of injury. Interruption of sympathetic outflow from T1 to causes vasodilation and peripheral pooling of blood, resulting in orthostatic hypotension .

Although the sympathetic nervous system is damaged, the parasympathic fibers remain intact. Bradycardia is commonly seen as the ninth and tenth cranial nerves carry signals from the carotid sinus and aortic arch to the sinoatrial node. Also, if the level of injury is high enough the function of the cardiac accelerator fibers (Tl–T4) is lost.

TABLE 2. Preoperative Pulmonary Evaluation.
Analyze arterial blood gas (ABG)
Check ventilatory reserve (PFTs)
Check for signs of infection
Institute pulmonary therapy:
Breathing exercises
Chest PT/postural drainage
Bronchodilators
Check for signs of deep-vein thrombosis or pulmonary embolism; consider prophylactic therapy

Spinal shock is characterized by flaccid paralysis below the level of injury (2) paralytic ileus, (3) loss of visceral and somatic sensation, (4) loss of vascular tone, and (5) loss of vasopressor reflexes. The chronic phase of SCI starts when spinal cord reflexes begin to return. It is characterized by sympathetic hyperactivity and involuntary muscle spasm. SCI patients have low red cell mass and venous pooling, and thus relative hypovolemia. The level of hypovolemia is determined by the level of injury as well as the level of activity. Interrupting the sympathetic nervous system's control reduces orthostatic capacity. Patients with lesions above the level of T6 have insufficient capacity for orthostatic regulation. With lesions below the level of T6, patients are able to react normally (ie, have stable blood pressure with an increase in heart rate in response to position change on a tilt table).[6]

Autonomic hyperreflexia (AH) is a syndrome that commonly occurs in patients with chronic SCI (Figure 3). It is triggered by cutaneous or visceral stimulation below the level of the SCI and is reported to occur in up to 85% of patients if the leslon is at the T6 level or higher.[7,8]

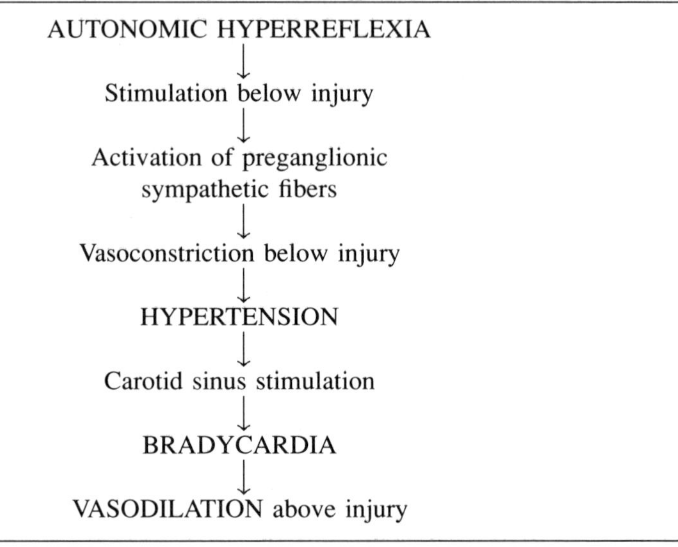

FIGURE 3. Schematic representation of the sequence of events that causes the clinical symptoms of the mass reflex.

Autonomic hyperreflexia also occurs during labor in up to two thirds of pregnant women with cord lesions above T6.[9]

Reflex activity in isolated spinal cord segments via afferent nerve endings causes a massive stimulation of the splanchnic nerves. The resultant sympathetic outflow from the splanchnic bed results in an explosive onset of symptoms, including sweating, flushing, piloerection, hypertension, bradycardia, and severe headache.

In response to the sudden rise in blood pressure, the body attempts to lower the pressure via the aortic arch and carotid slnus pressure receptors. A reflex bradycardia occurs via the vagus nerve. Also, in response to the increased circulating blood volume, there is vasodilation above the level of injury. The reflex vasodilation usually does not adequately lower the blood pressure because of insufficient area to handle the excess fluid volume. The severe changes in blood pressure have been linked to myocardial infarction, cerebral and retinal artery hemorrhages, and cardiac dysrhythmias.

Cardiac output and stroke volume are usually increased during this time.[7,8,10] The usual low level of circulating serum catecholamines is significantly elevated during episodes of AH as a result of sympathetic overactivity rather than adrenal medulla secretion. The catecholamine predominantly found during AH is norepinephrine, the neurotransmitter for the sympathetic nervous system, not epinephrine, which is secreted by the adrenal medulla.[11]

Electrolyte Imbalances

After SCI, calcium levels are frequently increased. The normal bone metabolism is changed after injury. Osteoclastic activity is increased, resulting in bone resorption. The increased resorption of bone causes osteoporosis and hyper calciuria. The onset of hypercalciuria is seen within 7 to 14 days after injury and continues for several months.[12] It occurs if resorption exceeds excretion via the kidney. Significant hypercalcemia has been demonstrated in adolescents, usually male.

Immobilization hypercalcemia has been reported with serum calcium levels ranging from 11 to 16mg/dl. Symptoms of hypercalcemia include nausea, vomiting, polydipsia, polyuria, and lethargy. Hypercalcemia may be associated with ventricular dysrhythmias. Shortened QT intervals may be observed on the ECG.[13]

Treatment of hypercalcemia consists of hydration with a balanced salt solution and diuresis. Because excretion of calcium parallels that of sodium, increasing sodium excretion enhances calcium excretion. Furosemide and ethacrynic acid also increase calcium excretion by blocking the tubular reabsorption of sodium and calcium. Supplemental therapies may include steroids, phosphates, and mithramycin.[11]

Serum electrolytes are usually within normal limits unless major organ system damage has occurred. Serum potassium levels have been shown to change drastically with the administration of the depolarizing muscle relaxant succinylcholine. Serum potassium levels as high as 14mEq/L have been reported.[3] This unusual sensitivity to succinylcholine is noticed within several days of the injury and lasts for an indefinite period.

In the SCI patient entire muscle membranes behave as giant endplates. When exposed to a depolarizing muscle relaxant, the muscles release large quantities of potassium into the circulation. The acute and often large rise in serum potassium has been associated with ventricular dysrhythmias and cardiac arrest.

The degree of hyperkalemia is proportional to the amount of

muscle involved, not to the amount of succinylcholine given. Also. pretreatment with a nondepolarizing muscle relaxant does not reliably prevent the massive depolarization. Because of the dangers associated with its use in SCI patients and the availability of shortacting nondepolarizing muscle relaxants, the administration of succinylcholine is probably contraindicated.[14,15]

The treatment of symptomatic hyperkalemia includes the administration of calcium, glucose and sodium bicarbonate, and hyperventilation.

Renal

Renal deterioration is common in longterm survivors of SCI and is the major cause of death in 40% of patients. Factors predictive of renal failure include vesicoureteral reflux, kidney stones, blunting of the caliceal pattern, and development of multiple decubitus ulcers. Urinary tract infections are a major factor in determining the rate at which these problems occur.[16] They are characterized by culture of a mixed flora and by lack of symptoms such as fever, leukocytosis, and pain. Decubitus ulcers are common and are usually infected. Cross-contamination between the ulcer and the urinary tract is common. Septicemia resulting from these infections is frequently the cause of death.[17,18]

Amyloidosis of the kidney is an associated disease of paraplegics who survive more than 12 years. Impaired renal function causes alterations in electrolyte balance and intravascular volume.

Other Organ Systems

The incidence of gastrointestinal disease in patients with SCI is increased. Peptic ulcers, cholelithiasis, and secondary amyloidosis are common.[19] Secondary amyloidosis has been reported to be as high as 56% and is related to the high rate of infection. Tanaka et al.[20] endoscopically studied the upper gastrointestinal tract in 40 patients with chronic SCI. Of these patients, 51.4% had abnormal findings, including gastric erosions, ulcers, hyperemic changes, and duodenal ulcers.

Thermal regulation may be problematic in patients who have lost sympathetic control. Loss of central regulation from the hypothalamus produces temperature disturbances, and patients tend to be poikilothermic but with more sensitive response to heat. During long operations particularly, safeguards must be taken to avoid hypothermia, including use of a temperature-controlled mattress, high ambient temperature, warmed irrigating and intravenous solutions, and heated humidified gases.

Preanesthetic Plan

The potential complications and disease processes associated with SCI demand a thorough preoperative evaluation. Included in this evaluation are chart review, knowledge of planned procedure, complete history and physical examination, and review of previous anesthetic experiences (type and responses), as well as chest x-rays, ECG, and laboratory tests, including CBC, SMA6 and creatinine, calcium clotting studies, urinalysis, and blood sample for type and crossmatch if applicable.

Recent studies have suggested improved spinal cord survival after high dose steroid therapy. Patients who have been so treated should receive supplemental steroids pre- and intraoperatively to avoid hypotension and compensate for the stress of surgery and anesthesia.

Additional tests may be included and largely depend on the patient's preoperative status. (PFTs), arterial blood gas analyses, and sputum samples may be included. Cardiovascular evaluation should be considered if there is a history of high SCI, prolonged bed rest or inactivity, or a history of cardiovascular dysfunction.

Explanation of procedures and of type of anesthesia, as well as a discussion of the associated complications, is mandatory, as denial of symptoms or even frank hallucinations may occur in patients with high SCI. It is important to emphasize that the proposed operation will not restore neurologic function. A well-informed patient is usually more relaxed, and in the case of AH earlier diagnosis and prevention or treatment of the autonomic instability may be possible.

Evaluation

Alteration in pulmonary function is a major concern. Because of the restrictive defect, recurrent infections, and loss of pulmonary function associated with anesthesia and surgery, knowledge of preoperative PFTs is required. Maximizing preoperative pulmonary function minimizes postoperative complications. Any evidence of infection should be aggressively treated with antibiotics and pulmonary toilet in conjunction with postponing the surgical procedures if possible until the infective process has resolved. Most of the surgical procedures performed on the SCI patient are peripheral, involving, for example, the lower extremities or lower urinary tract, and have minimal effect on pulmonary function.

If the surgical procedures are more invasive and ascend toward the diaphragm, respiratory function is further impaired. Following high SCI, the diaphragm is the major source of respiratory function. Therefore, respiratory function should be carefully observed and assessed during the intra- and postoperative periods. Deterioration of pulmonary function leads to retained secretions and development of atelectasis.

Loss of the sympathetic nervous system along with inactivity promotes a low circulating blood volume. However, chronic infections and blood loss from the urinary or gastrointestinal tract or from decubitus ulcers can cause marked anemia. These factors combine to produce an unpredictable response to anesthesia, surgery, and fluid loss or administration. Careful evaluation of fluid status and cardiovascular performance is essential. Placement of indwelling arterial, central venous, and pulmonary artery catheters can help in assessing cardiovascular status.[7]

Anesthetic Plan

Goals include provision of adequate analgesia and patient comfort, prevention of hypothermia, avoidance of pathologic fractures, prevention or control of AH, and maintenance of electrolyte homeostasis.

The type of anesthesia used for SCI patients depends on the site of surgery, patient status, the skill of the anesthesiologist, and patient acceptance.

If surgery is planned below the level of the injury, the patient may already have no sensation. However, even if no anesthetic is to be administered, full monitoring—including ECG, blood pressure, and pulse oximetry—is required because of the cardiovascular changes, especially AH, that may occur.

Although both regional and general anesthesia have been used, the indications are not clearly defined. One study found that 39% did not require anesthesia.[21] In another study 20% required no anesthesia, 70% were given general anesthesia, and 10% received regional anesthesia. The highest incidence of hypertension occurred in the general-anesthesia group.[22] In a third series 33% received general anesthesia[8]; blood pressure was maintained and no dysrhythmias were observed after inhalation anesthesia or spinal anesthesia, but 2 of 9 patients receiving nitrous oxide/narcotic anesthesia developed AH.

Moderate hypertension was reported intraoperatively and postoperatively with general anesthesia and postoperatively with spinal anesthesia. The incidence was higher with general anesthesia. A 3% incidence of bradycardia occurred during spinal anesthesia, perhaps because of a decrease in the pulse associated with a decrease in right atrial pressure (Bainbridge reflex). Both regional and general anesthesia provide adequate analgesia and optimize the surgical field.[8,10,23]

In the early stages of high SCI, muscle spasms may impair operating conditions. Spinal and epidural anesthesia are effective in blocking muscle spasticity and preventing AH. Visceral afferent and efferent pathways are blocked, and a mass reflex is prevented without myocardial depression.

Problems associated with regional anesthesia include (1) spinal deformities and osteoporosis causing technical difficulties, (2) spasticity leading to positioning problems, (3) decreased intravascular volumes potentiating hypotension, and (4) loss of sensation, making assessment of effectiveness of blockade difficult.

The problems associated with the use of general anesthesia are related to the decreased intravascular volume and myocardial depression caused by inhalational agents. Patients may be unable to respond to hypotension because sympathetic nervous system damage prevents tachycardia and vasoconstriction.[8] Probably any form of general anesthesia, inhalational or balanced, can be effective in preventing AH, provided there is an adequate depth of anesthesia before a triggering stimulus is applied.[8,24]

If AH should occur during surgery or postoperatively, lowering the blood pressure is of prime importance (Figure 4).[25] Table 3 shows the steps that may be helpful in treating AH.

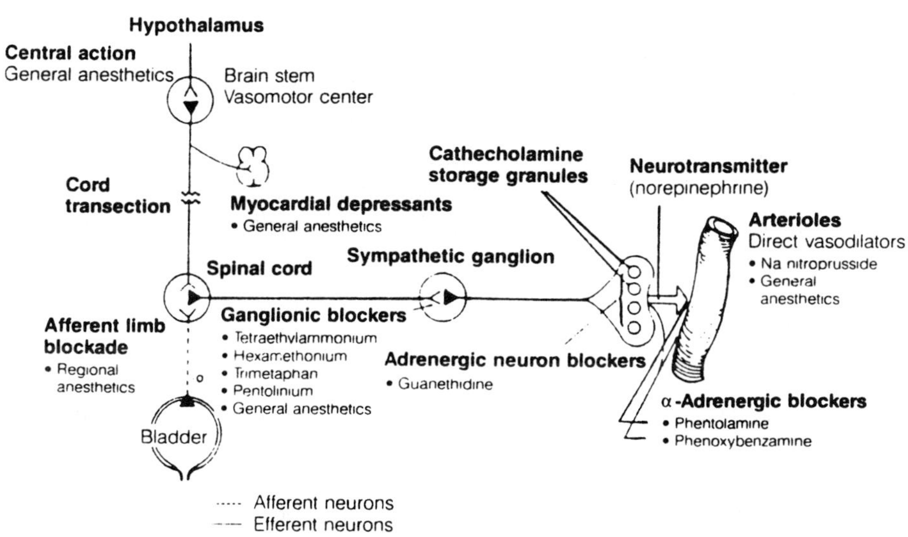

FIGURE 4. The sites of action of the agents used for control of the hypertension of autonomic hyperreflexia are shown. (From Schonwald G, Fish KJ, and Perkash I: Anesthesiology 1981, 55:550. With permission.)

TABLE 3. Procedure for Treating AH.

Remove the triggering stimulus
Stop surgery
Empty bladder or rectum
Deepen anesthesia
Use drug therapy
Direct-acting vasodilators
sodium nitroprusside
nitroglycerin
Ganglionic blocking drugs
trimethaphan
Alpha-adrenergic blocking drugs
phenoxybenzamine
phentolamine

During the postoperative period there should be close monitoring for the occurrence of AH, which can occur in the recovery room secondary to a distended bladder or rectum. Prompt evaluation and treatment of the cause of AH usually controls the fluctuating blood pressure.

Hypothermia should be aggressivelv treated and, if at all possible, prevented. Warming attempts must be continued throughout the preoperative period. Covering exposed extremities with blankets and towels retards heat loss.

Patients with SCI have a large degree of mineral loss and develop osteoporosis; they also have decreased muscle mass, muscle flaccidity, and minimal protection from hard surfaces. Attention to patient positioning and adequate padding of extremities prevents pathologic fractures and skin breakdown. During long procedures it is advisable to check systematically the position, padding, and blood flow to the extremities to help prevent iatrogenic injury.

References

1. Kopaniky D.R.: Pathophysiology and management of spinal cord trauma, in Frost E.A.M. (ed): *Clinical Anesthesia in Neurosurgery.* Boston: Butterworth, Heinemann 1990, pp 445–484.
2. Bendo A.A., Giffin J.P., Cottrell J.E.: Anesthetic and surgical management of acute and chronic spinal cord injury, in Cottrell J.E., Turndorf H. (eds): *Anesthesia and Neurosurgery*, 2nd ed., St. Louis: C.V. Mosby, 1986, 392–404.
3. Stover S.L., Fine D.R.: The epidemiology and economics of spinal cord injury. *Parapiegia* 1987, 25:225–8.

4. Ohry A., Molhom, Rozin R.: Alterations of pulmonary function in spinal cord injured patients. *Paraplegia* 1975, 13:101–8.
5. Troyer A.D., Heilporn A.: Respiratory mechanics in quadriplegia: The respiratory function of the intercostal muscles. *Am Rev Respir Dis* 1980, 122:591–600.
6. Engle P., Hildebrandt G.: Long term studies about orthostatic training after high spinal cord injury. *Paraplegia* 1976, 14:159–64.
7. Troll G.F., Dohrmann G.J.: Anesthesia of the spinal cord injured patient: Cardiovascular problems and their management. *Paraplegia* 1975, 13:162–71.
8. Schonwald G., Fish K.J., Perkash I.: Cardiovascular complications during anesthesia in chronic spinal cord injured patients. *Anesthesiology* 1981, 55:550–8.
9. McGregor J.A., Meevwsen J.: Autonomic hyperreflexia: A mortal danger for spinal cord-damaged women in labor. *Am J Obstet Gynecol* 1985, 151:330–3.
10. Barker I., Alderson J., Lydon M., Franks C.L.: Cardiovascular effects of spinal subarachnoid anesthesia. *Anesthesia* 1985, 40:553–6.
11. Debarge O., Christensen N.J., Cabett J.L., et al: Plasma catecholamines in tetraplegias. *Paraplegia* 1974, 12:44–9.
12. Bern R.G., Walsh J.J.: Urinary calcium and kidney stones in paraplegia: Report of an attempted prospective study. *Paraplegia* 1974, 12:38–43.
13. Maynard F.M., Imai K.: Immobilization hypercalcemia in spinal cord injury. *Arch Phys Med Rehabil* 1974, 589:16–23.
14. Gronert G.A., Theye R.A.: Pathophysiology of hyperkalemia induced by succinylcholine. *Anesthesiology* 1975, 43:89–99.
15. Snow J.C., Kripke B.J., Sessions G.P., Finck A.J.: Cardiovascular collapse following succinylcholine in a paraplegic patient. *Paraplegia* 1973, 11:199–204.
16. Merritt J.L.: Urinary tract infections, causes and management with particular reference to the patient with spinal cord injury: A review. *Arch Phys Med Rehabil* 1976, 57:365–70.
17. Vaziri N.D., Cesario T., Mooto K., et al: Bacterial infections in patients with chronic renal failure: Occurrence with spinal cord injury. *Arch Intern Med* 1982, 142:1273–6.
18. Hackler R.H.: A 25-year prospective mortality study in the spinal cord injured patient: Comparison with the long-term living paraplegic. *J Urol* 1977, 117:486–8.
19. Meshkinpoor H., Vaziri N., Gordan S.: Gastrointestinal pathology in patients with chronic renal failure associated with spinal cord injury. *Am J Gastroenterol* 1982, 77:562–561.
20. Tanaka M., Uchiyama M., Kitano M.: Gastroduodenal disease in chronic spinal cord injuries: An endoscopic study. *Arch Surg* 1979, 114:185–7.
21. Rocco A.G., Vandam L.D.: Problems in anesthesia for paraplegics. *Anesthesiology* 1959, 20:348–53.

22. Patel C., Miller S.M., Chalon, J., et al: Analgesia and spinal cord lesions. *Bull NY Acad Med* 1978, 54: 924–30.
23. Watson D.W., Downey G.O.: Epidural anesthesia for labor and delivery of twins of a paraplegic mother. *Anesthesiology* 1980, 52: 259–61.
24. Alderson J.D., Thomas D.G.: The use of halothane anesthesia to control autonomic hyperreflexia during transurethral surgery in spinal cord injury patients. *Paraplegia* 1975, 13: 183–8.
25. Basta J.W., Niejadlik K., Pallares V.: Autonomic hyperreflexia: Intraoperative control with pentolinium tartrate. *Br J Anaesth* 1977, 49: 1087–90.

Chapter 3

The Child With Congenital Heart Disease for Noncardiac Surgery

Ingrid Hollinger, M.D.

Case History. *A 10-month-old child was brought by his mother to the day surgery unit for bilateral removal of supernumerary digits.*

The family had recently relocated from central Europe, and the history was sparse. However, pregnancy and delivery had been normal. Birth weight was 6lb. The child had reached developmental milestones rather slowly. He could sit up but required some support. His mother noted that he had had several respiratory infections during his life. During those times, she had noted wheezing and a slightly bluish tinge to his lips.

On examination, the patient's weight was 15lb, and the heart rate was 130/min. A pansystolic murmur was heard. Respiratory sounds were clear. Hematocrit was 28%; white blood cell count, 9700/mm^3.

The surgeon was anxious to proceed, arguing that the child appeared healthy and that as the procedure was relatively short, intubation was not even necessary. The anesthesiologist, concerned about the significance of the murmur, anemia, and low weight, requested a delay in the operation for further evaluation. He suspected a ventricular septal defect.

Introduction

Congenital cardiac lesions are present in approximately 6-8/1000 live births and are the most frequent (90%) cause of heart disease in infancy and childhood. Half of the infants with such lesions require some sort of medical or surgical therapy within the first year of life, generally because of severe or life-threatening abnormalities. However, 80% of all

---

Reviewed by Dr. Henry Issenberg, Director of Diagnostic Services, Division of Pediatric Cardiology, Montefiore Medical Center, and Assistant Professor of Pediatrics, Albert Einstein College of Medicine.

congenital heart disease is represented by only eight lesions.[1] In order of decreasing frequency, they are ventricular septal defect, atrial septal defect, pulmonary stenosis, patent ductus arteriosus, tetralogy of Fallot, aortic stenosis, coarctation of the aorta, and transposition of the great vessels.

With advances in medical, surgical, and anesthetic management in recent years, early correction or palliation of the most severe anomalies takes place, and fewer children present for noncardiac surgery with their lesions uncorrected or unpalliated.

To plan a rational anesthetic management for the child with congenital heart disease presenting for noncardiac surgery, a thorough understanding of the cardiac anatomy and of the pathophysiologic changes caused either by the cardiac lesion or as a result of corrective or palliative surgery is necessary. No child should present for surgery with an undiagnosed lesion, and catheterization and/or echocardiographic studies should be reviewed by the anesthesiologist, together with the pediatric cardiologist, to plan appropriate anesthetic management.

General Considerations

Congenital cardiac anomalies appear in a great variety of lesions. However, they result in only a very limited number of major pathophysiologic derangements and can be grouped accordingly. The classifications most useful for the anesthesiologist are those that are related to pulmonary blood flow; to the presence, size, and location of abnormal intra- or extracardiac communications between the systemic and pulmonary circulations; and to the presence of obstruction to the output of either ventricle (Table 1).

Approximately 25% of children with congenital heart disease have associated defects. Most commonly invoived are the genitourinary, musculoskeletal, and central nervous systems.[2] Chromosomal anomalies and many congenital syndromes are associated with congenital heart disease, often with a characteristic defect such as trisomy 21 (Down's syndrome), with its endocardial cushion defect; Turner's syndrome, with coarctation; Noonan's syndrome, with pulmonic stenosis; and congenital rubella, with peripheral pulmonic stenosis. Trisomy 18 and 13 and DiGeorge and Williams syndromes demonstrate associated cardiac anomalies in nearly all patients.[1]

The problems of the primary anomaly may further complicate several aspects of anesthetic management, eg, establishment and maintenance of the airway, difficulties in intubation, and the presence of hyperreactive airway disease, metabolic derangement, and mental retardation.

TABLE 1. Classification of Congenital Lesions According to Major Pathophysiology.

Left-to-right shunting with increased pulmonary blood flow
- Ventricular septal defect (VSD)
- Atrial septal defect
- Endocardial cushion defect
- Patent ductus arteriosus
- Anomalous origin of the coronary arteries
- Partial anomalous pulmonary venous return

Right-to-left shunting with increased pulmonary blood flow
- D-transposition (with or without VSD)
- Double-outlet right ventricle
- Single ventricle
- Truncus arteriosus
- Total anomalous pulmonary venous return
- Aortic atresia/hypoplastic left heart syndrome

Right-to-left shunting with decreased pulmonary flow
- Tetralogy of Fallot
- Pulmonary atresia with ventricular septal defect
- All forms of large intraventricular communication with pulmonic stenosis
- Ebstein's anomaly
- Tricuspid atresia
- Pulmonary atresia with intact ventricular septum
- "Critical" pulmonic stenosis

Obstruction to ventricular output
- Aortic stenosis (sub-, supra, valvular)
- Coarctation of the aorta
- Idiopathic hypertrophic subaortic stenosis
- Pulmonic stenosis (valvular, infundibular, peripheral)

In the first few years of life, even the normal infant has an immature cardiovascular system. It is characterized by reduced cardiac contractility, with the heart muscle exhibiting reduced compliance.[3] Cardiac output, however, is high. As a result of reduced contractility and compliance, stroke volume is limited; therefore, cardiac output is rate-dependent. Heart rates below 90 beats are associated with hypoperfusion.

The sympathetic nervous system is immature, and baroceptor reflexes are less active. Less catecholamines are released in response to stress, and exogenous catecholamines are less effective. The pulmonary vascular bed is highly reactive to normal stimuli in the first few months of life and remains so in patients with large left-to-right shunts.

The most potent regulator of pulmonary vascular tone is blood pH. Hypercarbia, hypoxia, acidosis, hyperinflation, atelectasis, sympathetic stimulation, and polycythemia all will increase pulmonary vascular resistance, whereas hyperoxia, hypocarbia, alkalosis, normoventilation at normal functional residual capacity, sedation, and hemodilution all decrease pulmonary vascular resistance.[4]

Prevention of Air Embolism

All patients with shunt lesions are at risk for air embolisms in the systemic circulation irrespective of their usual shunting pattern. Upon sudden obstruction by air embolism of right ventricular output, a left-to-right shunt may convert to a right-to-left shunt. The I.V. line should be meticulously debubbled and rechecked after warming of the OR, because the warming may cause nitrogen to come out of solution in the I.V. fluid, forming additional hazardous bubbles.

All I.V. lines should be connected while free-flowing. All syringes should be cleared of air; and before injection into an I.V. line, a small amount of fluid should be aspirated into the syringe to clear the needle and injection port of air.[5] A recommended method is to dilute the drug to be given in a 10-ml syringe so that the calculated dose of medication is contained in 1ml. Aspiration of I.V. fluid into the syringe will therefore not significantly change the concentration. This is in contrast to aspirating fluid into a small volume of undiluted drug.

These precautions are important for any patient in whom a communication exists between the systemic and pulmonary circulations, regardless of the presence or absence of pulmonary outflow obstruction. Shunts are often bidirectional, and the earlier relaxation of the left ventricle, compared to the right ventricle, may transiently reverse a left-to-right shunt during this portion of the cardiac cycle.

TABLE 2. Endocarditis Prophylaxis: Recommended Antibiotic Regimen

For dental, otolaryngologic (eg, tonsillectomy, adenoidectomy), and bronchoscopic procedures:

A. **Standard regimen**	B. **Special regimen** (recommended for all patients with) prosthetic heart valves	C. **Oral regimen** for minor or repeated procedures in low-risk cases
1. *Oral*— Penicillin V(2.0gm) orally 60 minutes prior to the procedure and then 1.0 gm orally 6 hours later. Children less than 60 lb receive half the above doses. 2. *Parenteral*—Aqueous penicillin G (50,000U/kg) intravenous or intramuscular 30–60 minutes before the procedure followed by aqueous penicillin G (25,000U/kg) 6 hours later. 3. *For patients allergic to penicillin* a. Erythromycin (20mg/kg) orally 1 hour prior to procedure, followed by 10 mg/kg orally 6 hours later. b. Vancomycin (20mg/kg) slowly intravenously over 60 minutes prior to the procedure. Because of the long half-life of vancomycin, a repeat dose is not necessary.	1. Ampicillin 50 mg/kg IM or IV *plus* gentamicin 1.5mg/kg or 1V 30 minutes before the procedure followed by penicillin V 1.0gm (500 mg if less than 60lb) orally 6 hours later or by initial parental dose once 8 hours later. 2. For patients allergic to penicillin—Erythromycin (20mg/kg) orally 1 hour prior to procedure, followed by10 mg/kg orally 6 hours later.	Amoxicillin 50mg/kg orally 1 hour prior to procedure followed by 25mg/kg 6 hours later.

For genitourinary or gastrointestinal surgery or instrumentation or any surgery of infected or contaminated tissue):

A. Standard regimen	**B. Patients allergic to penicillin**
Ampicillin 50 mg/kg IM or 1V plus gentamicin 2.0 mg/kg IM or IV 30–60 minutes prior to the procedure. One follow-up dose may be given 8 hours later.	1. *Oral*— Erythromycin 20 mg/kg orally 1 hour before the procedure followed by half the dose 6 hours later. 2. *Parenteral*—Vancomycin 20mg/kg slowly IV over 1 hour plus gentamicin 2.0mg/kg IV or IM 1 hour prior to the procedure. May be repeated 8–12 hours later

Maximal dose of drugs: aqueous penicillin G, 2 million units; amoxicillin, 3 gm; vancomycin, 1 gm; erythromycin, 1 gm; ampicillin, 1 gm; gentamicin, 80 mg; vancomycin, maximal 44mg/kg/24 hr. In patients with compromised renal function the dose of antiobiotics may need modification.

Endocarditis Prophylaxis

All patients with congenital heart disease require antibiotic coverage for the prevention of endocarditis before any surgical procedure having the potential for bacteremia, including endoscopy and nasotracheal intubation. The only lesions exempt are secundum atrial septal defects, before surgery and after 6 months following closure without a patch, and a ligated patent ductus arteriosus repaired more than 6 months before the surgery. Presently recommended dosage regimens are listed in Table 2.[6]

Preoperative Assessment

A thorough history and physical examination are essential to assess the significance of the congenital heart disease and how well it is controlled.

Two major sequelae of significant congenital heart disease are congestive heart failure and cyanosis. In addition, most congenital cardiac anomalies manifest a pathologic murmur. Criteria that define pathologic murmurs are listed in Table 3.[7]

TABLE 3. Characteristics of Pathologic Murmurs.
Diastolic murmur
Pansystolic murmur
Late systolic murmur
Very loud murmur
Continuous murmur
Associated cardiac anomaly

The clinical signs and symptoms of congestive heart failure are listed in Table 4.

TABLE 4. Signs and Symptoms of Congestive Heart Failure.
Tachycardia (with or without gallop)
Cardiomegaly
Poor peripheral perfusion and pulses
Pulsus alternans
Cyanosis (relieved by O_2)
Fatigue and sweating
Growth and developmental retardation
Wheezing (bronchospasm)
Respiratory distress
Hepatomegaly/splenomegaly

Heart failure should be controlled with digitalis, diuretics, and/or an after-load-reducing agent such as hydralazine prior to any elective intervention.[8]

Cardiac assessment requires consultation with a pediatric cardiologist. An echocardiogram must be obtained preoperatively. Drug therapy should be maintained perioperatively; however, if the patient is on digitalis, care should be taken to maintain adequate serum potassium levels and prevent hypocarbia to avoid digitalis toxicity. Control of congestive heart failure and pulmonary edema will improve pulmonary function and reduce the possibility of perioperative hypoxemia or respiratory failure.

Cyanosis is a feature of all cardiac lesions with right-to-left shunting because of limited pulmonary blood flow and/or venous admixture to the systemic circulation. Severe hypoxemia results in polycythemia with an increase in blood volume and viscosity, neovascularization, alveolar hyperventilation to maintain arterial normocarbia, and a poorly defined coagulopathy. Digital clubbing (osteoarthropathy) of the distal phalanges of fingers and toes is indicative of longstanding arterial desaturation.

Increased blood viscosity increases cardiac work by increasing peripheral vascular resistance. Cerebral and/or renal thrombosis may occur with high hematocrits, particularly in the presence of dehydration. At a hematocrit greater than 60%–65%, viscosity dramatically increases. Oxygen transport is not improved, and the frequency of serious thrombotic complications and coagulopathies increases. To improve organ perfusion and reduce cardiac workload, the hematocrit should be controlled below these levels—if necessary, by hemodilution or erythrophoresis.[9]

In the presence of insufficient dietary intake of iron, however, children with cyanotic congenital heart disease may exhibit normal hematocrits and low hemoglobin levels despite severe arterial desaturation. These children are at risk for inadequate oxygen supply with depression of cardiac output or further reduction of hemoglobin levels. The preoperative hemograms will demonstrate microcytosis and hypochromia, and iron therapy should be instituted prior to elective major surgery.[10] In most cases, increasing hematocrit is an indication for surgery, either to improve pulmonary blood flow or to correct the lesion. Because of the danger of hemoconcentration with prolonged fasting, preoperative fasting (NPO) times should be held to a minimum in cyanotic children, or the patient should be intravenously hydrated.

Special Considerations

Left-to-Right Shunting

Intracardiac lesions with left-to-right shunting include all forms of atrial and ventricular septal defect, patent ductus, and other large aortopulmonary connections. The magnitude and direction of the shunt depend on the difference between the outflow resistance of the right and left ventricles and the size of the defect, except for lesions with a communication between the left ventricle and the right atrium. Under these circumstances obligatory shunting occurs because of the large difference in pressure between these two cardiac chambers.

In atrial septal defects, shunting occurs between low-pressure areas, creating a right ventricular volume overload. Pulmonary pressures remain normal despite markedly increased pulmonary blood flows. Over time, pulmonary artery pressure increases but rarely progresses to an Eisenmenger situation (reversal of left-to-right shunt due to pulmonary vascular obstructive disease) prior to the fourth decade. Patients are generally asymptomatic throughout childhood and adolescence. Of main concern in anesthesia planning is the possibility of embolization of air or debris to the systemic circulation if right-sided pressures are transiently increased (airway obstruction, massive air embolism).

Shunting in ventricular septal defects not only depends on the relative outflow resistances of the two ventricles but also on the size of the communication. Since the majority are located in the muscular or perimembranous septae and may close spontaneously in up to 50% of patients, they are treated conservatively, and patients may present for noncardiac surgery with the defect open. If it is a hemodynamically small defect, shunting is limited, and patients generally are without signs of congestive failure (restrictive defect). They are, however, at risk for endocarditis and systemic embolization.

Patients with large ventricular septal defects (nonrestrictive defect) often present with congestive heart failure in early infancy once pulmonary vascular resistance, which is high at birth, has decreased. The combination of high flow and pulmonary hypertension at systemic levels that occurs in this lesion will lead to pulmonary vascular obstructive disease. With time this will result in a reduction in left-to-right shunt with improvement in congestive failure. With progression of the pulmonary vascular disease, resistance to right ventricular outflow eventually increases to such a degree that right-to-left shunting (shunt reversal) occurs (ie, Eisenmenger's syndrome).

In the early development of an increased pulmonary vascular resistance the increases are dynamic and can be influenced by pulmonary vasodilators most notably high concentrations of oxygen. The decrease in pulmonary vascular resistance produced by ventilation with oxygen may result in such large increases in left-to-right shunting that congestive heart failure ensues.

Patients with large ventricular septal defects generally undergo repair within the first 2 years of life to prevent the irreversible arteriolar plexiform changes of pulmonary vascular obstructive disease. If these children require surgery prior to the closure of their defect, they present several problems. They usually have dynamically increased pulmonary vascular resistance that, in the presence of high oxygen concentrations, may decrease so much as to cause failure. Hypoxia, on the other hand, may lead to such increases in pulmonary vascular resistance as to cause small-airway disease with airway closure and atelectasis, and careful attention has to be paid to ventilatory management during and following anesthesia.[11]

Right-to-Left Shunting

Intracardiac lesions with right-to-left shunting and obstruction to pulmonary blood flow comprise all forms of tetralogy of Fallot and ventricular septal defects with pulmonic stenosis. The physiology of Eisenmenger's complex follows the same flow patterns. Shunting in these lesions and the severity of cyanosis depend largely on the degree of pulmonary obstruction and systemic vascular resistance.

In the presence of mixed or dynamic muscular obstruction to right ventricular outflow (infundibular stenosis), hypercyanotic spells may occur as a result of acute diminution of pulmonary blood flow. Triggering events include infundibular spasms, a fall in venous return, a fall in peripheral vascular resistance, and a sympathetic stimulation. Hyperventilation, decrease in mental status, and acidosis occur. The attack may be reversed by squatting in a knee-to-chest position, which increases venous return and peripheral vascular resistance, and by administration of oxygen and morphine.

In infants with severe pulmonary arterial hypoplasia, an aortopulmonary shunt is constructed to ameliorate cyanosis and stimulate angiogenesis, allowing for total correction at a later time. In these patients and in all patients with pulmonary atresia and a ventricular septal defect, right-to-left shunting depends solely on the loading conditions of the heart and peripheral vascular resistance. Pulmonary blood flow is dependent on systemic blood pressure and the size of the aortopulmonary connection. Hypotension will invariably result in reduced pulmonary perfusion and worsening cyanosis.

Mixing Lesions

Lesions with right-to-left shunting and no pulmonary obstruction (mixing lesion) include the D-transposition, common ventricle, and double-outlet right ventricle. These lesions combine the problems of increased pulmonary blood flow with those of moderate to severe cyanosis.

In transposition, the aorta arises from the right ventricle and the pulmonary artery from the left ventricle, with the result that the systemic and pulmonary circulations run in parallel instead of in series. Survival depends on shunting/mixing between these two parallel circulations. Patients with intact interatrial and interventricular septa are cyanotic within hours of birth and require creation of an atrial septal defect by balloon atrioseptostomy.

The presence of a ventricular septal defect allows for shunting but often results in severe congestive heart failure within the first few weeks of life. The high pulmonary artery pressures and flow, together with polycythemia, lead to the early development of irreversible pulmonary vascular obstructive disease in such infants, and corrective surgery is performed within the first year of life.

The most commonly performed of the so-called physiologically corrective procedures is the atrial inversion, Mustard or Senning, in which a baffle is constructed within the atria, resulting in diversion of pulmonary venous return to the right ventricle and systemic venous return to the left ventricle. In newborns or patients with large ventricular septal defects switching of the great arteries with coronary reimplantation (arterial switch) resulting in anatomic correction, is currently favored.

Anesthetic Plan

Left-to-Right Shunt

Uptake of inhaled anesthetic agents is only minimally influenced by a left-to-right shunt unless peripheral perfusion is poorly maintained, in which case induction may be accelerated.[12] (See Table 5.) Induction with intravenous agents, however, is markedly delayed since much of the injected agent is recirculated through the lungs. Increasing the amount injected to achieve a rapid effect may result in an overdose because eventually all drugs will reach the systemic circulation.

Potent inhaled anesthetic agents may be poorly tolerated in the presence of congestive heart failure because of their myocardial depressant effect. Narcotics or ketamine maintain cardiovascular stability and have minimal influence on pulmonary vascular resistance.[13] Nitrous oxide has little influence on the pulmonary circulation and can be used to potentiate the effects of other anesthetic agents while at the same time contributing to limiting inspired oxygen concentration in patients with labile pulmonary hypertension.

TABLE 5. Principles of Anesthetic Management for Patients with Large Left-to-Right Shunts.

- **Monitor Oxygenation**
  Excessive O_2 → PVR* ↓
    PBF ↑→ pulmonary edema,
    and peripheral collapse
- **Maintain Blood Volumes/ and BP**
- **Ventilating Management**
  closing capacity ↑
  atelectasis common
  compliance ↓
  airway resistance ↑

PVR - pulmonary vascular resistance
PBF - pulmonary blood flow

Right-to-Left Shunt

Patients should arrive in the operating room well sedated but normotensive. Preoperative fluid restriction should be minimized and maintenance fluid given intravenously to prevent hemoconcentration and hypotension. Smooth induction is important to prevent increases in oxygen demand or hypercyanotic spells. The agents used should have minimal peripheral vasodilating effects. Mild myocardial depression may relieve infundibular obstruction. If intravenous agents are used, they should be carefully titrated to prevent relative overdose. Intravenous barbiturate requirements may be reduced by half.

The airway should be secured by endotracheal intubation in all procedures except those lasting only a few minutes. Excessive airway pressures should be avoided. Adequate ventilation should be ascertained by arterial blood gas measurements. Low-dose halothane or narcotic relaxant techniques may be used. Oxygenation is frequently improved under general anesthesia, probably by relaxation of the infundibular muscle, pulmonary vasodilation from higher oxygen concentrations, and reduced peripheral oxygen demands.[14]

Monitoring blood pressure may become problematic in patients with previous shunting procedures using the subclavian arteries (eg, following a right Blalock-Taussig shunt, arteries in the right upper limb are not cannulated). For major surgery, intraarterial and intravenous pressures should be measured directly. This will also allow blood sampling for blood gas and acid-base measurements. Because the major stress in these patients rests on the right ventricle, central venous pressures can be used to assess cardiac performance.

The presence of a right-to-left shunt prolongs induction with poorly soluble inhaled anesthetics. This may be offset by the presence of a surgically created systemic-to-pulmonary arterial shunt. Induction time with highly soluble agents may be nearly normal because the patients usually hyperventilate to maintain a normal arterial PCO_2. Intravenous induction is rapid, but conventional doses may produce undesirable side effects because of high systemic concentrations, since the drugs enter the systemic circulation without passage through the lung.

With a fixed obstruction to pulmonary outflow, right-to-left shunting is influenced by changes in peripheral vascular resistance. A decline in systemic vascular resistance may lead to increasing cyanosis, acidosis, and myocardial depression, creating a vicious cycle. In patients with dynamic infundibular stenosis, severe infundibular spasm may be triggered intraoperatively by hyperventilation, hypovolemia, relative anemia, manipulation of intracardiac monitoring lines, and inotropic agents. Prompt treatment is essential.

Intravenous morphine (0.1–0.2mg/kg) or propranolol (0.05–0.1mg/kg) should be used. Oxygen may have little effect in improving systemic oxygenation. Acidosis should be corrected and venous return improved by administration of fluids. If peripheral vasodilation was a contributing factor, peripheral vascular resistance should be increased by a peripherally acting vasoconstrictor such as phenylephrine, 5–20 μg/kg I.V. Maintenance of cardiac output (rate-dependent) is essential because the oxygen content of the arterial blood is low; therefore, bradycardia is very poorly tolerated.

In patients with systemic-to-pulmonary arterial shunts, adequate systemic blood pressure is necessary to maintain pulmonary perfusion. All patients with right-to-left shunts are at increased risk of systemic embolization of air or blood clots from I.V. lines. In severe polycythemia, hemodilution to a hematocrit of between 55 and 60 should be performed prior to elective surgery. Since nitrous oxide will lead to an increase in the volume of accidentally introduced air bubbles, it is generally avoided in patients with right-to-left shunts.

Endocarditis prophylaxis is necessary for all surgery associated with bacteremia. Patients with systemic-to-pulmonary arterial shunts are at particularly high risk.

Patients with cyanosis have a blunted chemoreceptor response to hypoxia, which persists after correction of the underlying lesion. In patients with reduced pulmonary blood flow, marked ventilation/perfusion inequalities exist. Positive pressure ventilation may worsen this problem, leading to an increase in dead-space ventilation and increased arterial PCO_2.[15]

Patients who have undergone correction of their cardiac lesion(s) present less of an anesthetic problem. They may, however, have impairment of right and left ventricular function and dysrhythmias. Ventricular extrasystoles should be treated aggressively because they have been implicated in sudden death. Pacemaker wires and generators should be available. Patients require endocarditis prophylaxis for the remainder of their lives. The presence of residual defects should be determined prior to elective surgery in postcorrection patients.

Mixing Lesions

Other than patients with severe pulmonic obstruction who are awaiting a Rastelli procedure, it is rare that a patient with transposition will be encountered who requires noncardiac surgery prior to atrial or arterial repair. No experience with anesthesia for such patients who need noncardiac operations has been reported. These patients are at very high risk for systemic embolization.

Patients with only balloon septostomy have a 10%–20% incidence of cerebrovascular accidents, which are thought to be due to stasis and thrombosis. Because of the low effective pulmonary blood flow, induction of anesthesia with inhaled agents is prolonged, and adverse side effects may linger after the anesthetic is discontinued.

Intravenous agents must be used with caution; they have a rapid onset of action because most of the venous return enters the aorta. Chronic congestive failure is often present, with congestion of the lung and low ventilatory volumes, which predispose the patient to atelectasis. Myocardial depression is poorly tolerated, and bradycardia is usually associated with acidosis. Because of the compromised cardiac and ventilatory function, postoperative mechanical ventilation may be advisable.

Anesthetic agents should have minimal cardiovascular side effects. Narcotic-relaxant techniques, particularly with fentanyl and/or ketamine, are appropriate choices.[13] Low concentrations (0.5%–0.75%) of isoflurane may be used to supplement relaxant techniques. In patients with limited pulmonary blood flow, increase in pulmonary vascular resistance through coughing and struggling should be avoided. In the presence of systemic-to-pulmonary arterial shunt, maintenance of blood pressure is essential to allow pulmonary perfusion. An ever-increasing number of children who have survived reparative procedures present for noncardiac surgery in later years. Although they appear clinically normal, without evidence of cyanosis or failure, they may not tolerate anesthesia and surgery well.[16]

Atrial dysrhythmias are common after intraatrial repairs and may become more frequent with time. The most serious of these is sick sinus syndrome, which may necessitate the placement of a temporary pacemaker (if it is not already in place). There may be systemic venous obstruction at the baffle site, leading to signs of venous congestion. Significant baffle obstruction usually requires reoperation. Left-sided baffle obstruction (more common after the Senning procedure) may lead to pulmonary edema and hypertension. Increase in peripheral vascular resistance will worsen the condition of the patient under these circumstances.

Since the right ventricle is the systemic ventricle in these patients, myocardial failure may occur early, with massive fluid shifts or myocardial depression. Patients who have undergone an arterial switch operation may show signs of coronary insufficiency, supravalvular aortic and/or pulmonary stenosis, or aortic insufficiency.

Maintenance of myocardial oxygen balance and prevention of tachycardia and marked falls in systemic vascular resistance are important for anesthetic management in patients with ventricular inflow or outflow obstruction and possibly impaired coronary reserve. Anesthetic management of all other mixing lesions is similar to that of the patient with a large ventricular septal defect; congestion, heart failure, and pulmonary hyperperfusion are the pathophysiologic disturbances in these patients. Cyanosis is always present, however, and intensifies in the presence of increased pulmonary vascular resistance.

Obstruction to Ventricular Output

These include all patients with pulmonic or aortic stenosis and coarctation. The indication for correction of valvular pulmonary stenosis in childhood is a pressure gradient across the pulmonary valve greater than 50mmHg or a right ventricular pressure greater than 65mmHg in the presence of a normal cardiac output. In general, only valvotomy is performed, leaving the patients with varying degrees of residual gradient. Anesthetic management is similar to that of adult patients with valvular stenosis (Table 6).

TABLE 6. Principles of Anesthetic Management for Patients with Ventricular Outflow Obstruction.

- □ *Avoid* tachycardia
- □ *Avoid* peripheral vasodilation
- □ *Maintain* myocardial contractility

Aortic valvotomy is performed for a peak systolic ejection gradient of 50–75mmHg with a normal cardiac output, depending on age, valve morphology, and presence of coexisting aortic insufficiency. Since cardiac output is limited by the pressure gradient across the obstruction and time of ejection, tachycardia is poorly tolerated. The hypertrophied myocardium will become ischemic, and cardiac output will fall because of decreased stroke volume. Falls in peripheral vascular resistance will lead to cardiovascular collapse because stroke volume cannot be increased. Myocardial dysfunction persists in many patients after valvotomy, and anesthetic agents that cause myocardial depression or irritability should be used cautiously.

Patients with coarctation present either as infants with congestive heart failure or in childhood with systemic hypertension and a systolic murmur. The defect in symptomatic newborns and infants is repaired as soon as they are stabilized with anticongestives and prostaglandin E_1 (PGE_1).

In older children, repair should be performed when the diagnosis is established to prevent persistent systemic arterial hypertension. These patients have an increased incidence of cerebral aneurysms. Repair of a coarctation in infancy frequently results in loss or distortion of the left subclavian artery. Blood pressure monitoring in these patients should always be done on the right arm. Because they may show exaggerated blood pressure response to sympathetic stimulation, an adequate depth of anesthesia is essential.

Unusual Problems

Surgical correction of single-ventricle or triscupid atresia is currently performed by using a modified Fontan technique. This procedure results in diversion of all systemic venous return from the right atrium directly to the pulmonary arteries. Systemic venous pressure therefore becomes the determinant of pulmonary perfusion. Hypovolemia and/or increases in intrathoracic or pulmonary arterial pressure result in cardiovascular collapse. Maintenance of an adequate heart rate and prevention plus prompt treatment of dysrhythmias are essential to avoid high systemic venous pressures and their sequelae (ascites, effusion, cirrhosis). Anesthetic management should be tailored to these physiologic considerations.

Conclusion

An understanding of the pathophysiologic mechanism of various congenital cardiac lesions and of the results of palliative or corrective surgery allows for the rational planning of anesthetic management. The demands of the proposed surgery, the effects of anesthetic agents on cardiovascular physiology, and the patient's medical status have to be balanced to achieve the goal of safe anesthesia with no deterioration—or even an improvement—in the patient's hemodynamic and respiratory status.

References

1. Hollinger I.: Diseases of the cardiovascular system, in Katz R., Steward D. (eds.): *Anesthesia and Uncommon Pediatric Diseases.* Philadelphia: WB Saunders Co., 1987, pp 93–154.
2. Noonan J.A.: Association of congenital heart disease with syndromes or other defects. *Pediatr Clin North Am* 1978, 25:797–816.
3. Friedman W.F.: The intrinsic physiologic properties of the developing heart. *Prog Cardiovasc Dis* 1972 15:7–111.
4. Nelson N.M.: Respiration and circulation after birth, in Smith C.A., Nelson N.M. (eds.): *The Physiology of the Newborn Infant*, 4th ed. Springfield, Ill: Charles C. Thomas, 1976, pp 117–262.
5. Campbell F.W., Schwartz A.J.: Problems in anesthetic management of children with congenital heart disease for non-cardiac surgery, in Kirby R.R., Brown D.L., (eds.): *Problems in Anesthesia (Vol. 1).* Philadelphia: JB Lippincott, 1987, pp 411–33.
6. Shulman S.T., Amrin D., Risno A.: Prevention of bacterial endocarditis. *Circulation* 1984, 70:1123A–27A.
7. Rosenthal A.: How to distinguish between innocent and pathologic murmur in childhood. *Pediatr Clin North Am* 1984, 31:1229–40.
8. Gersony W.M., Steeg C.N.: Congestive heart failure, in Dickerman J.D., Lucey J. (eds.): *The Critically Ill Child*, 3rd ed. Philadelphia: WB Saunders Co., 1985, pp 320–36.
9. Rosenthal A., Nathan D.G., Marty A.T., et al.: Acute hemodynamic effects of red cell volume reduction in polycythemia of cyanotic congenital heart disease. *Circulation* 1970, 42:297.
10. Nadas A.S., Fyler D.C.: *Pediatric Cardiology*, 3rd ed. Philadelphia: WB Saunders Co., 1972, pp 567–8.
11. Duncan P.G.: Anaesthesia for patients with congenital heart disease. *Can Anaesth Soc J* 1983, 30:520–6.
12. Tanner G.E., Angers D.G., Barash P.G.: Effect of left-to-right, mixed left-to-right, and right-to-left shunts on inhalational anesthetic induction in children. *Anesth Anal* 1985, 64:101–7.
13. Hickey P.R.: Anesthesia for children with cardiac disease. *IARS Review Course Lecture*, 1988, pp 86–90.
14. Laishley R.S., Burrows F.A., Lerman J., et al.: Effect of anesthetic induction regimens on oxygen saturation in cyanotic congenital heart disease. *Anesthesiology* 1986, 65:673–7.
15. Lister G., Pitt B.R.: Cardiopulmonary interactions in the infant with congenital heart disease. *Clin Chest Med* 1983, 4:219–32.
16. Hickey P.R., Hansen D.D., Norwood W.I., Castaneda A.R.: Anesthetic complications in surgery for congenital heart disease. *Anesth Analg* 1984, 63:657–64.

CHAPTER 4

The Patient With Cor Pulmonale

James E. Tobin, M.D.

Case History. *A 47-year-old man presented to the operating room from the ambulatory care unit for repair of bilateral inguinal hernias. The patient's history was significant for a 30-year history of cigarette abuse with resultant severe chronic obstructive pulmonary disease (COPD), including increasing dyspnea, ankle edema, and ascites over the past year. He admitted to drinking about 1–2 pints of whiskey daily. His medications included theophylline, hydrocortisone, digoxin, furosemide, and isoetharine by inhalant. Vital signs were blood pressure 110/70mmHg, heart rate 125/min, respiratory rate 24/min, and temperature 37.0°C.*

Physical examination of the patient revealed a cachectic, extremely anxious white man. Jugular venous distention without collapse with inspiration was evident. A parasternal heave was noted on palpation of the chest. On auscultation of the chest, an S_3 gallop accentuated by inspiration and a mild expiratory wheeze with diffuse rhonchi were heard. The liver was palpable 3cm below the right costal margin and was mildly tender. Moderate pedal edema was present. Roentgenograms of the chest showed marked pulmonary hyperinflation, flattened diaphragms, and a normal heart size. The ECG showed sinus rhythm, P pulmonale, right axis deviation, R:S ratio in V_6 of 1.0, dominant R wave in V_1, and inverted T waves in the right precordial leads.

Introduction

The term *cor pulmonale* refers to hypertrophy and/or dilatation of the right ventricle secondary to pulmonary hypertension from lung disease. Pulmonary hypertension that results from disease of the left side of the heart or congenital heart disease must be excluded to make an accurate diagnosis of cor pulmonale. Acute cor pulmonale is the term used to

Reviewed by Dr. James Mueller, Department of Anesthesiology, Norfolk General Hospital, Norfolk, Virginia.

describe the right ventricular dysfunction that results from an acute pulmonary embolism. The patient presented in this case has chronic cor pulmonale, which is less specific in terms of describing the etiology of the pulmonary hypertension.

The incidence of cor pulmonale is difficult to quantify because of the difficulty of detecting asymptomatic pulmonary hypertension. The diagnosis of cor pulmonale is usually made only after right ventricular failure has occurred. Approximately 12 million people in the United States have chronic obstructive pulmonary disease (COPD), and 3 million of those are chronically hypoxic with $PaO_2 < 60$mmHg.[1] Chronic hypoxia leads to pulmonary hypertension. With prolonged pulmonary hypertension, right ventricular hypertrophy and cor pulmonale occur.

The prognosis of cor pulmonale in patients with COPD is poor; 3-year mortality is estimated at 60%.[2] Autopsy studies indicate that the cause of 10% to 30% of hospital admissions for congestive heart failure may be related to cor pulmonale.[3] In urban areas, where smoke pollution is heavy (eg, in some northern European cities), the percentage may be much higher. Thus, cor pulmonale is a common but often unrecognized problem.

Etiology

Several pulmonary disorders can lead to pulmonary hypertension.[4] They can be divided into three major categories: (a) diseases affecting the airways and alveoli; (b) diseases affecting the movement of the chest bellows, including the central hypoventilation syndromes; and (c) diseases affecting the pulmonary vasculature, including primary pulmonary hypertension (see Table 1). The most common etiology is COPD.

Pathophysiology

To create a better understanding of the pathophysiology of cor pulmonale, a brief review of the development and physiology of the pulmonary circulation and right ventricular function follows.

The prenatal right heart is slightly larger than the left heart, presumably because right ventricular afterload is greater than left ventricular afterload in utero.[4] At birth the lungs are inflated and the pulmonary vascular resistance falls with a resultant increase in pulmonary blood flow. Normal infants living at sea level retain this right ventricular hypertrophy for the first 3 months of life. However, a hypertrophic state can persist throughout the first decade of life in high altitude dwellers, presumably secondary to the pulmonary hypertension caused by hypoxia from breathing air with low inspired concentrations of oxygen. Thus, the right ventricle is clearly capable of adaptive changes to certain pathologic conditions. This is demonstrated not only by right ventricular hypertrophy in high-altitude dwellers but also by the pulmonary hypertension and right ventricular hypertrophy that develop in patients with lung disease such as COPD.

TABLE 1. Etiology of Cor Pulmonale

Diseases affecting airways and alveoli
COPD
Cystic fibrosis
Infiltrative or granulomatous disease
Upper airway obstruction
Pulmonary resection
High-altitude disease
Disease affecting thoracic cage movement
Kyphoscoliosis
Thoracoplasty
Pleural fibrosis
Neuromuscular weakness
Sleep apnea syndromes
Idiopathic hypoventilation
Diseases affecting the pulmonary vasculature
Primary diseases of the arterial wall
Primary pulmonary hypertension
Toxin/drug-induced pulmonary hypertension
Granulomatous pulmonary arteritis
Chronic liver disease
Peripheral pulmonic stenosis
Thrombotic disorders
Sickle cell disease
Pulmonary microthrombi
Embolic disorders
Thromboembolism
Tumor, amniotic fluid, or air embolism
Parasitic disease (schistosomiasis)

Adapted from Braunwald E: heart Disease. Philadelphia, W.B. Saunders, 1988, p 1603.

The right ventricle is a compliant, thin-walled, crescent-shaped chamber that functions well as a volume pump. Historically, it had been thought that the right ventricle was little more than a passive conduit through which blood flowed from the systemic venous circulation to the lungs. This conclusion was reached after experiments in which the free wall of the right ventricle was ablated without effect in previously healthy dogs.

More recently the importance of right ventricular function has been emphasized.[5] The right ventricle normally has a low afterload because it must pump only against the low-resistance pulmonary circulation. The pulmonary vascular resistance changes that can occur with cor pulmonale represent up to a 400% increase over baseline values. The magnitude and rate of development of pulmonary hypertension dictate whether right heart failure will occur.

Acute increase in mean pulmonary artery pressure from a normal of 20mmHg to only 30mmHg—as, for example, after an acute pulmonary embolus—can be associated with right heart failure; however, patients with chronic pulmonary disease such as COPD can often tolerate mean pulmonary artery pressures approaching 80mmHg without manifesting right heart failure. Thus, the right ventricle must develop adaptive changes in order to function against the high pulmonary vascular resistance seen in cor pulmonale.

The normal pulmonary circulation is a high-flow, low-resistance circuit, in contrast to the high-resistance systemic circulation. The normal pulmonary circulation possesses an enormous adaptive capacity to maintain low pressure and resistance. During the threefold increases in blood flow that occur with exercise, pulmonary pressures normally increase very little. The pressure is kept low by a combination of vasodilation and recruitment of blood vessels, recruitment being the predominant mechanism. The supernumerary vessels involved in this process actually outnumber conventional vessels. However, their functional addition increases the cross-sectional area of the pulmonary vascular bed by only 25% to 33%.[6]

When a pulmonary disease process either destroys the pulmonary vasculature or interferes with its ability to vasodilate, pulmonary hypertension may develop. The lungs of patients with COPD not only have a reduced capacity to vasodilate but also show a destruction of the pulmonary vascular bed. Thus, the adaptive capacity of patients with COPD to compensate for increases in blood flow is impaired, and pulmonary hypertension occurs. When pulmonary hypertension is sustained, the changes seen with cor pulmonale develop.

The pulmonary hypertension associated with lung diseases such as COPD has a variety of causes (see Table 2). The most common of these causes and the pathophysiology most centrally involved in cor pulmonale is increased pulmonary vascular resistance.[7] Increased pulmonary vascular resistance may be caused by vasoconstrictor mechanisms and/or anatomic changes in the pulmonary vasculature.

TABLE 2. Factors Affecting Right Ventricular Afterload

Increased pulmonary vascular resistance
Decreased vascular bed
Hypoxia-induced vasoconstriction
Acidemia-induced vasoconstriction
Increased intrathoracic pressure
Increased pulmonary blood volume
Increased blood viscosity
Increased pulmonary venous pressure
Increased cardiac output with a noncompliant vascular bed

The most potent stimulus for pulmonary vasoconstriction is hypoxia. Acidosis also increases pulmonary vascular resistance and acts synergistically with hypoxia. However, an increase in $PC0_2$ appears to exert no direct effect on the pulmonary circulation but works indirectly through increases in the hydrogen ion concentration.[4] The localization of the vasoconstriction appears to be in pulmonary arterioles less than 200μ in diameter. The mechanism for the vasoconstriction is unclear. Theories include release of endogenous mediators such as eicosanoids or other vasoactive substances, a direct effect of hypoxia on pulmonary vascular smooth muscle, and extrapulmonic reflexes such as may be induced by the adrenergic system.

The anatomic changes in the resistance vessels of the pulmonary circulation that occur with pulmonary disease contribute greatly to the development of pulmonary hypertension and cor pulmonale. Most of the resistance of the pulmonary circulation occurs in the pulmonary arterioles. A neomuscularization of the pulmonary arterioles has been associated with chronic hypoxia.[5] This process increases the length of the muscle-containing parts of the pulmonary arterioles.

When hypoxia is prolonged over days in rats, pulmonary hypertension secondary to hypertrophy and hyperplasia of the smooth muscle of the pulmonary arterioles develops. Experimental evidence suggests that chronic hypoxia does not, however, cause destruction of the pulmonary vascular bed. Contractile cells, called pericytes, have been identified adjacent to the endothelial cells and are thought to contribute to the regulation of pulmonary blood flow. These pericytes are known to increase in size on exposure to hypoxia.

In summary, vasoconstrictive mechanisms and anatomic changes in the pulmonary arterioles lead to pulmonary hypertension and right ventricular hypertrophy, which we call cor pulmonale. The vasoconstriction may be reversible, but the anatomic changes usually are irreversible and lead to fixed pulmonary hypertension.

Lung volume also affects pulmonary vascular resistance. Pulmonary vascular resistance is minimized when functional residual capacity (FRC) is normal. Increases in FRC compress intraalveolar vessels. Decreases in FRC cause large-vessel vasoconstriction secondary to hypoxic pulmonary vasoconstriction.[8] Therefore, one goal of perioperative ventilation is to maintain a normal FRC.

Increases in blood viscosity secondary to increases in hematocrit can increase pulmonary vascular resistance.[4] Chronic hypoxemia increases red cell production through erythropoietin release, which serves as an adaptive mechanism to maintain oxygen delivery. However, high hematocrits can increase blood viscosity to the point where capillary blood flow and oxygen delivery are actually impaired. These considerations have led clinicians to use phlebotomy in some patients as therapy to reduce pulmonary vascular resistance and improve pulmonary capillary blood flow.[4]

A complex interaction occurs among the right heart, the left heart, and the pulmonary vasculature as interdependent parts of the cardiopulmonary unit. The interdependence in function of the right and left ventricles is well described.[5,9] The interventricular septum is an important component in the contractile mechanism of both ventricles.

Right ventricular dysfunction can lead to a shift of the interventricular septum within the pericardium and impair left ventricular filling. Intrathoracic and pericardial pressures also play important roles in the interaction between right and left ventricular function. Thus, right ventricular dysfunction seen with cor pulmonale may indeed lead to left ventricular dysfunction and eventually to cardiopulmonary failure.

Clinical Manifestations

The clinical manifestations of cor pulmonale are often nonspecific and are difficult to detect because a sensitive, noninvasive test for pulmonary hypertension does not exist. Patients with pulmonary hypertension and cor pulmonale often present with increasing dyspnea and easy fatigability, which are interpreted by the clinician as being simply a progression of the underlying lung disease. The diagnosis of cor pulmonale is usually not entertained until overt right heart failure has occurred, although the condition may exist and may be problematic for many years.

Physical examination of patients with pulmonary hypertension often reveals accentuation of the pulmonic component of the second heart sound and sometimes a diastolic murmur secondary to incompetence of the pulmonary valve. A third heart sound, jugular venous distention, hepatosplenomegaly, and peripheral edema may be detected when right ventricular failure occurs. Chest radiographs can demonstrate right ventricular hypertrophy by a decrease in the retrosternal space on the lateral film. Pulmonary hypertension can be demonstrated on a chest radiograph by a prominence of pulmonary artery shadows and decreased pulmonary vascular markings in the peripheral lung fields, the so-called pruning effect.

Early signs of right ventricular dysfunction are often difficult to detect with traditional ECG lead placement. Because of the location of the right ventricle, standard monitoring with a lead V_5 in combination with lead II provides limited ECG view. A newer lead placement system using a V_4R lead placed in the fourth intercostal space in the right midclavicular line is more sensitive and specific in detecting right ventricular ischemia.[10]

Other ECG changes include peaked P waves in the inferior leads suggestive of right atrial enlargement. Right axis deviation, right bundle branch block, prominent R waves in V_1, R:S ratio in V_6 of 1.0, or inverted T waves in the right precordial leads are findings suggestive of right ventricular hypertrophy.[4]

The other potential ECG findings of right ventricular hypertrophy are listed in Table 3. Echocardiography may reveal tricuspid regurgitation or right ventricular dilation. Radionuclear stress testing may also reliably demonstrate right ventricular dysfunction by indicating abnormal elevation in right ventricular, systolic, and pulmonary artery pressures.

If time permits, several bedside pulmonary function tests may be performed that indicate lung reserve and may predict the need for postoperative ventilatory support. At a minimum, baseline arterial blood as analyses should be performed. Normal values are listed in Table 4.

In a recent position paper, the American College of Physicians recommend preoperative pulmonary testing prior to lung resection. For patients scheduled for coronary artery bypass surgery, spirometry and blood gas analyses are useful if there is a history of tobacco use and dyspnea. However, there is no good evidence demonstrating that routine pulmonary screening improves identification of patients at risk for development of postoperative pulmonary complications.[11]

Our patient is a poor candidate for out-patient surgery. Preoperative pulmonary function testing can indicate the amount of damage and provide a basis or rationale for behavior modification counseling with regard to smoking and drinking. Also, cessation of smoking for as short a time as 48 hours has been shown to decrease carboxyhemoglobin levels (thus increasing hemoglobin available for oxygen transport) and shift the oxyhemoglobin dissociation curve to the right (increasing available oxygen).[12] In a patient with reduced pulmonary reserve, this effect may confer a critical benefit especially if general anesthesia, which increases pulmonary shunting, becomes necessary.

TABLE 3. ECG Changes in Cor Pulmonale

Isoelectric P waves in lead I
P-pulmonale pattern P-wave axis to the right of +70°)
Right axis deviation
Incomplete right bundle branch block
R:S ratio in $V_6 < 1$
R:S ratio in $V_1 > 1$
Low-voltage QRS complex (in patients with COPD)
Normal-voltage QRS complex (in patients without COPD)
S_1Q_3 or $S_1S_2S_3$ pattern (S wave deeper than R wave in standard leads)
Clockwise rotation of the electrical axis
T-wave changes in the right precordial leads
ST-segment depression in the inferior leads
Occasional Q wave in the inferior leads

Adapted from Braunwald E: Heart Disease Philadelphia, W.B. Saunders, 1988, p 1612; and Marriott H J L: Practical Electrocardiology, Baltimore, Williams & Wilkins, 1982, pp 55–61.

Yet another reason for advocating inpatient care in this situation is the possible need for postoperative ventilatory support, a complication that may be prevented with appropriate preoperative therapy in many cases.

Treatment

The primary management goal in cor pulmonale is to relieve the pulmonary hypertension by reducing pulmonary vascular resistance through improvements in oxygenation and correction of acidosis. Gas exchange can be improved in patients with COPD through the use of bronchodilators, which promote bronchial smooth muscle relaxation and improve mucociliary clearance of secretions. The use of intravenous aminophylline has been shown to reduce pulmonary hypertension significantly without reducing cardiac index.[13] In another study it actually improved right and left ventricular ejection fractions.[14]

TABLE 4. Normal Arterial Blood Gas and Pulmonary Function Test Values (50-year old man, 70kg, 180cm, sitting position)

Arterial blood gas	pH 7.40
PCO_2	40mmHg
PO_2	90mmHg
Base excess	0
Pulmonary function tests	
FVC (forced vital capacity)	4.74L/sec
FEV_1 (forced expiratory volume in 1 sec)	3.87L/sec
FEV_3 (forced expiratory volume in 3 sec)	4.41L/sec
Expressed as % of the FVC.	
FEV_1 = 83% (ie, 3.87L is 83% of 4.74L)	
FEV_3 = 97% (ie, 4.41L is 97% of 4.74L)	
PEFR (peak expiratory flow rate)	8.97L/sec
MMF (mid-maximal flow rate, also called FEF 0.25–0.75, measures the forced expiratory flow rate in middle of effort)	3.08L/sec
DL_{∞} SB (diffusion capacity for carbon monoxide measured during single breath)	32.08
ERV (expiratory reserve volume)	0.96L
RV (residual volume)	1.77L
TLC (total lung capacity)	6.51L
RV/TLC × 100%	27%
FRC (functional residual capacity)	2.73L
IC (inspiratory capacity)	3.78L

Adapted from Frost EAM: Recovery Room Practice, Boston, Blackwell, 1985, p 200.

Supplementary oxygen has been shown to reverse the progression of pulmonary hypertension in patients with COPD when administered over an average 31-month period; less impressive results are achieved when it is administered over a 6-month period.[15] Longterm survival is improved with chronic oxygen therapy presumably by reversal of pulmonary hypertension.[2] The criteria most frequently used to initiate chronic oxygen therapy are a $PaCO_2$ of 55mmHg or less, a hematocrit of 55% or more, P pulmonale on ECG, and peripheral edema.[4]

A caveat in oxygen therapy is to be alert for patients whose ventilation is controlled by a hypoxic drive. Such patients may hypoventilate in response to oxygen therapy, and the resultant respiratory acidosis may, in fact, worsen their pulmonary hypertension.

Antibiotics are an important component of therapy in COPD patients with acute pulmonary infections that may increase pulmonary vascular resistance. The most common infecting organisms are *Hemophilus influenzae* and pneumococci, both of which usually can be treated effectively with ampicillin or a cephalosporin.[12]

The value of digitalis in patients with cor pulmonale is controversial. Early uncontrolled studies concluded that digitalis improved right heart function and led to the recommendation of the drug to treat patients with cor pulmonale. An increase in cardiac output was seen, but this was associated with an increase in pulmonary hypertension.

One clear benefit of digitalis is in the treatment of supraventricular tachydysrhythmias, to which these patients are prone because of right atrial stretch. However, one controlled study concluded that digitalis offered no clear benefit to patients with cor pulmonale except in the presence of left heart failure or supraventricular tachydysrhythmias.[16]

Diuretics have been shown to offer significant benefit to patients with right ventricular failure from cor pulmonale.[17] Excess lung water that interferes with gas exchange and increases pulmonary vascular resistance can be reduced by systemic diuresis. Careful monitoring of serum electrolytes must be performed because hypokalemia and metabolic alkalosis are signifiant complications.

Preanesthetic incentive spirometry is useful in that it familiarizes the patient with the various devices. Thus, the patient does not have to deal with a new learning experience postoperatively.

Anesthetic Plan

The preoperative management of patients with cor pulmonale requires treatment of any reversible component of the underlying lung disease. Therapy includes optimizing ventilation-perfusion matching by treatment of any pulmonary infections, good pulmonary toilet, relief of bronchospasm, and expanding alveoli with incentive spirometry. Such measures are aimed at improving oxygenation and reducing pulmonary hypertension. Preoperative medications that may depress ventilation should be avoided. Rather, the anesthesiologist should establish good patient rapport.

The choice of a general anesthetic versus a regional anesthetic should be based on the extent of surgery and the severity of the patient's disease. For minor surgery, a limited regional anesthetic would be appropriate. For major surgery that requires an extensive regional technique with its accompanying sympathectomy, systemic hypotension may be difficult to reverse. Efforts to increase systemic vascular resistance may elevate pulmonary vascular resistance and exacerbate the hypotension through right ventricular failure. Also, if the patient has a chronic cough exacerbated by lying flat, ideal surgical conditions may be difficult to achieve under regional anesthesia. Thus, general anesthesia remains the method of choice for patients with cor pulmonale undergoing major surgery.[18]

Patient preoxygenation helps to prevent hypoxia and its adverse effects on the pulmonary circulation. Hypercarbia also should be avoided at this time to prevent pulmonary hypertension associated with respiratory acidosis.

The induction of general anesthesia with either an inhalation agent or a barbiturate may be associated with increases in pulmonary artery pressure during manipulation of the airway and placement of an endotracheal tube.[19] To minimize these changes, an adequate depth of anesthesia should be obtained before intubation of the trachea. A sufficient depth of anesthesia also will help to prevent reflex bronchospasm.

Ketamine maintains systemic hemodynamics well, although it may increase pulmonary vascular resistance. Etomidate might be a prudent choice as an induction agent because it allows an adequate depth of anesthesia while avoiding the cardiovascular-depressant effects of a barbiturate or an inhalation agent.[18]

General anesthesia can be maintained with combinations of narcotics and inhalation agents. Narcotics provide a greater depth of anesthesia while avoiding cardiovascular depression. The inhalation agents are effective bronchodilators, which may be advantageous.[20] However, the volatile anesthetics appear to exert little or no predictable effect on pulmonary vascular smooth muscle tone.[21]

Hypoxic pulmonary vasoconstriction (HPV) is an adaptive mechanism by means of which blood flow is reduced to collapsed alveoli, thus minimizing the effect of shunt. The volatile anesthetics have been shown to reduce the compensatory HPV response. Isoflurane at 1 MAC used during one-lung anesthesia in dogs was shown to cause a 21% decrease in the HPV response, increasing the shunt from 20% to 24%.[22] This effect was found to be negligible in healthy individuals at clinically used concentrations.[23]

However, even slight decreases in HPV response may be significant in patients with abnormal lung function, and therefore agents that alter HPV must be used judiciously in patients with cor pulmonale (included in this caveat are drugs such as nitroprusside to induce hypotension).[24]

The use of nitrous oxide has been associated with increases in pulmonary vascular resistance.[25,26] These increases have been shown to be exaggerated in patients with preexisting pulmonary hypertension when a narcotic-based rather than a volatile anesthetic was chosen. If nitrous oxide is used as part of the anesthetic technique, the effect on pulmonary vascular resistance must be monitored.

The choice of monitors is based on the severity of pulmonary hypertension, ventilation abnormalities, right ventricular function, and the extent of surgery. A direct arterial pressure monitor is useful for beat-to-beat monitoring of systemic blood pressure and for frequent analyses of arterial blood gases. A pulse oximeter has become an invaluable tool to detect sudden, unexpected oxygen desaturations of hemoglobin that may exacerbate pulmonary hypertension.

Monitoring via a right atrial catheter can detect increases in right atrial pressure that may be a result of right ventricular dysfunction secondary to increases in pulmonary vascular resistance. When more accurate detection of changes in pulmonary vascular resistance or right ventricular output is required, a pulmonary artery catheter must be placed. Decision to use this monitor is based on the magnitude of surgery, the potential for major volume replacement, the severity of the cor pulmonale, and the presence of left ventricular dysfunction.

Ventilation must be controlled during surgery to prevent hypercarbia and respiratory acidosis that can exacerbate pulmonary hypertension. In patients with decreased lung compliance, larger than normal inspiratory pressures must be used to maintain alveoli patency for gas exchange. Increased positive pressure in the airways and alveoli can be transmitted to the pulmonary vasculature and increase pulmonary vascular resistance.

Positive end-expiratory pressure (PEEP) at $30cmH_20$ has been shown to increase right ventricular afterload to the point of right ventricular dilation, shifting the interventricular septum and interfering with left ventricular filling.[27] Also, a decrease in right ventricular ejection fraction (RVEF) has been demonstrated postoperatively at PEEP levels greater than $15cmH_20$. RVEF is most impaired with assist-controlled ventilation and least affected by intermittent mandatory or spontaneous ventilation.[28]

Ventilation should be utilized to maintain FRC at an optimal level in order to minimize pulmonary vascular resistance. In addition, induced hypocarbia is used to treat pulmonary hypertension in children whose pulmonary hypertension has resulted from congenital lesions, and it may be effective as a pulmonary vasodilator in some patients with cor pulmonale.[18,29]

Intraoperative hypothermia should be avoided through the use of humidification and heating of inhaled gases, warming of intravenous fluids, and minimizing heat loss to the patient's environment. Hypothermia can increase pulmonary vascular resistance.[30] In addition, the hypermetabolic and shivering response that occurs secondary to hypothermia to generate heat can increase cardiac output and exacerbate pulmonary hypertension.

The treatment of systemic hypotension and right heart failure in patients with pulmonary hypertension continues to be a difficult problem. The standard therapy for right heart failure consists of volume loading to ensure adequate preload and afterload reduction to the right ventricle.[4,8]

The effects of volume expansion and the use of norepinephrine for inotropic support were compared in dogs.[31] Results of the study emphasized the right ventricular dysfunction caused by volume expansion and stressed the role of inotropic support to the right ventricle. Clinical situations indicate that when pulmonary vascular resistance is normal, volume loading increases cardiac output.[28] However, when pulmonary vascular resistance (pulmonary artery pressure) is increased, volume expansion causes right ventricular failure, documented by decreasing cardiac output and increasing right ventricular end diastolic pressure.

Increased pulmonary vascular resistance in combination with increased right ventricular volume may heighten right ventricular wall stress, leading to decreased right ventricular contractility. Isoproterenol and dobutamine are alternative choices for inotropic support and cause minimal increase in pulmonary ventricular resistance.

Afterload reduction to the right ventricle remains a difficult task. The crux of the problem is that there is as yet no selective pulmonary vasodilator that does not dilate the systemic circulation. For afterload reduction to the right heart, nitroglycerin appears to be superior to nitroprusside[32]; however, systemic vasodilation and hypotension can be a problem with either drug.

The pulmonary vasodilating effects of nitroglycerin occur at doses that produce systemic venodilation, whereas systemic arterial vasodilation requires higher doses.[33] A number of pulmonary vasodilators have been tried in the treatment of primary pulmonary hypertension,[34] but none of the commonly used agents appears to surpass nitroglycerin in pulmonary vasodilating effects.

Global agents that have been tried as pulmonary vasodilaters (Table 5)

TABLE 5. Drugs Tried as Pulmonary Vasodilators

Tolazoline	Nitroprusside
Phentolamine	Nifedipine
Prazosin	Verapamil
Acetylcholine	Captopril
Isoproterenol	Prostaglandins
Diazoxide	Prostaglandin inhibitors
Hydralazine	Amrinone
Nitroglycerin	Oxygen

One agent that may hold promise is prostaglandin E_1, which is a potent pulmonary vasodilator. This agent has been used successfully in combination with a left atrial infusion of norepinephrine to treat refractory right heart failure after mitral valve replacement in patients with severe pulmonary hypertension.[35]

Summary

Cor pulmonale is the end-stage process involved in a number of pulmonary diseases the most common of which is COPD. The disease process poses a number of perioperative problems. The most important preoperative concern is correction of any reversible component of the pulmonary hypertension through improvement in gas exchange or administration of supplementary oxygen. Intraoperative concerns center around avoiding any exacerbation in pulmonary hypertension and selecting appropriate monitors according to the severity of disease and the extent of surgery. Because of the prevalence of COPD and other pulmonary disease as well as the aging of the population cor pulmonale will continue to be a significant challenge for the anesthesiologist.

References

1. Fishman A.P.: Chronic cor Pulmonale *Am Rev Respir Dis* 1975, 114:775–94.
2. Nocturnal Oxygen Therapy Group: Continuous or nocturnal oxygen therapy in hypoxic COPD. *Ann Intern Med* 1980, 93:391–8.
3. Inter-Society Commission for Heart Disease Resourccs: Primary prevention of pulmonary heart disease. *Circulation* 1970, 41:A17–A23.
4. McFadden E.R.: Cor pulmonale, in Braunwald E. (ed.): Heart Disease: A Textbook of Cardiovascular Medicine. Philadelphia: WB Saunders, 1988, pp1597–1616.
5. Hines R., Barash P.G.: Right ventricular failure, in Kaplan J. (ed.): Cardiac Anesthesia. New York: Grune & Stratton, 1987, pp 995–1020.
6. Mevrick B., Reid L.: Pulmonary hypertension—Anatomic and physiologic correlates. *Clin Chest Med* 1983, 4:199–217.
7. Matthay R.A., Berger H.J.: Cardiovascular function in cor pulmonale. *Clin Chest Med* 1983, 4:269–95.
8. Pirlo A.F., Benumof J.L., Trousdale F.R.: Atelectatic lobe blood flow: Open vs closed chest, positive pressure vs spontaneous ventilation. *J Appl Physiol* 1981, 50:1022–6.
9. Santamore W.P., Lynch P.R., Meier G., et al: Myocardial interaction between the ventricles. *J Appl Physiol* 1976, 41:362–8.
10. Klein H.O., Tardima T., Ninio R., et al: The early recognition of right ventricular infarction: Diagnostic accuracy of the electrocardiographic V_4R lead. *Circulation* 1983, 67:558–65.

11. Zilrak J.D. for the American College of Physicians: Preoperative pulmonary function testing. *Ann Intern Med* 1990, 112:793–4.
12. Davies J.M., Latto I.O., Jones J.G.: Effects of stopping smoking for 48 hours on oxygen availability from the blood: A study on pregnant women. *Br Med J* 1979, 2:355.
13. Parker J.O., Kelkar K., West R.O.: Hemodynamic effects of aminophylline in cor pulmonale. *Circulation* 1966, 33:17–25.
14. Matthay R.A., Berger H.J., Loke J., et al: Effect of aminophylline on right and left ventricular performance in COPD. *Am J Med* 1978, 65:903–10.
15. Weitzenblum E., Santegeau A., Ehrhart M., et al: Long-term oxygen therapy can reverse the progression of pulmonary hypertension in patients with COPD. *Am Rev Respir Dis* 1985, 131:493–8.
16. Mathur P.N., Powles P., Pugsley S.O., et al: Effect of digoxin on right ventricular function in severe chronic airway obstruction. *Ann Intern Med* 1981, 95:283–8.
17. Heinemann H.O.: Right-side heart failure and the use of diuretics. *Am J Med* 1978, 64:367–70.
18. Pear R.G., Rosenthal M.H.: Anesthetic management of patients with pulmonary hypertension, in Kirby R.L., Brown D.L., Thomas S.J. (eds.): Problems in Anesthesia. Philadelphia: JB Lippincott, 1987, pp 448–62.
19. Sorenson M.B., Jacobsen E.: Pulmonary hemodynamics during induction of anesthesia. *Anesthesiology* 1977, 46:246–51.
20. Hirshman C.A., Edelstein G., Peetz S., et al: Mechanism of action of inhalation anesthesia on airways. *Anesthesiology* 1982, 56:107–11.
21. Stoelting R.K.: Inhaled anesthetics, in Pharmacology and Physiology in Anesthetic Practice, vol 2. Philadelphia: J.B. Lippincott, 1987, p 45.
22. Domino K.B., Broowec L., Alexander C.M., et al: Influence of isoflurane on hypoxic pulmonary vasoconstriction in dogs. *Anesthesiology* 1986, 64:423–9.
23. Carlsson A.J., Bindsler L., Hedenstierna G.: Hypoxia-induced pulmonary vasoconstriction in the human lung. *Anesthesiology* 1987, 66:312–6.
24. Schulte-Sasse U., Hess W., Tarnow J.: Pulmonary vascular responses to nitrous oxide in patients with normal and high pulmonary vascular resistance. *Anesthesiology* 1982, 57:9–13.
25. Hilgenberg J.C., McCammon R.L., Stoelting R.K.: Pulmonary and systemic vascular responses to nitrous oxide in patients with mitral stenosis and pulmonary hypertension. *Anesth Analg* 1980, 59:323–6.
26. Frost E.A.M., Tabaddor K., Arancibia C.U.: Hypotensive drugs and outcome in aneurysm surgery. *Anesthesiology* 1979, 51:S82.
27. Jardin F., Farcot J.C., Boisante L., et al: Influence of PEEP on left ventricular performance. *N Engl J Med* 1981, 304:387–92.
28. Hines R.: Right ventricular function and failure. ASA Annual Refresher Courses, No. 521, 1988, San Fransisco.

29. Salmenpera M., Heinonen J.: Pulmonary vascular responses to moderate changes in $PaCO_2$ after cardiopulmonary bypass. *Anesthesiology* 1986, 64:311–5.
30. Benumof J.L., Wahrenbrock E.A.: Dependency of hypoxic pulmonary vasoconstriction on temperature. *J Appl Physiol* 1977, 42:56–8.
31. Ghignone M., Girling L., Prewitt R.M.: Volume expansion versus norepinephrine in treatment of a low cardiac output complicating an acute increase in right ventricular afterload in dogs. *Anesthesiology* 1984, 60:132–5.
32. Pearl R.G., Rosenthal M.H., Ashton J.P.A.: Pulmonary vasodilator effects of nitroglycerin and sodium nitroprusside in canine oleic acid-induced pulmonary hypertension. *Anesthesiology* 1983, 58:514–8.
33. Pearl R.G., Rosenthal M.H., Schroeder J.S., et al: Acute hemodynamic effects of nitroglycerin in pulmonary hypertension. *Ann Intern Med* 1983, 99:9–13.
34. Packer M.: Vasodilator therapy for primary pulmonary hypertension. *Ann Intern Med*1985, 103:258–70.
35. D'Ambria M.N., La Raia P.J., Philbin D.M., et al: Prostaglandin E_1—a new therapy for refractory right heart failure and pulmonary hypertension after mitral valve replacement. *J Thorac Cardiovasc Surg* 1985, 89:567–72.

CHAPTER 5

The Patient with Multiple Endocrine Neoplasia

Robert Adam, M.D.

Case History. *A 15-year-old girl presented to the emergency room with a complaint of headaches, excessive sweating, and palpitations. The headaches were severe and lasted about 10 minutes. They were accompanied by profuse truncal sweating palpitations, and chest pain (at the height of the attack).*

On physical exam the girl was noted to have puffy lips and multiple mucosal neuromas on the eyelids, lips, and tongue. Her neck revealed lymphadenopathy and an enlarged thyroid gland. The mandible was prominent, and marfanoid habitus was observed. She was thin, anxious, and sweating. Her pupils were dilated, and her hands were tremulous.

Blood pressure supine was recorded at 160/100mmHg; heart rate, 120/min. It was difficult to feel the radial pulse. When the patient assumed a sitting position, blood pressure was 90/50mmHg and heart rate 135/min. During abdominal palpation the patient developed a severe headache, palpitations, and profuse sweating. No abdominal tumor was detected. She was admitted for diagnostic testing.

Laboratory studies showed elevated serum and 24-hour urinary catecholamine levels. Radioimmunoassay studies revealed increased calcitonin levels but normal parathormone levels. Blood sugar was 190mg/dl; calcium and serum phosphate levels were normal. Hematocrit was elevated to 49%. Abdominal sonogram and computed tomography were unable to identify any adrenal or extraadrenal tumors.

A diagnosis of multiple endocrine neoplasia type III was made. The patient was discharged on phenoxybenzamine and scheduled for readmission and exploratory laparotomy and bilateral adrenalectomy in 1 month.

Reviewed by Dr. Adel R. Abadir, Director, Department of Anesthesiology, Brookdale Hospital Medical Center, Brooklyn, NY.

Introduction

The multiple endocrine neoplasia (MEN) syndromes are loosely divided into three groups: MEN I, MEN II, and MEN III (or IIb). Considerable overlap exists between the syndromes, all of which are inherited as an autosomal dominant trait. Many combinations of tumors and symptom complexes may occur. Relatives of MEN patients who manifest any of the classic signs or symptoms probably should be considered as subclinically expressing the disease.

MEN Type I

Clinical Presentation

MEN I (Wermer's syndrome) involves hyperplasia or tumors of two or more of the following glands: parathyroid, pancreatic islet cells, pituitary, and, less often, thyroid and adrenal cortex.

The parathyroid tumors (or hyperplasia) may involve asymptomatic hypercalcemia or nephrolithiasis, and nephrocalcinosis may be present. Ninety percent of patients with MEN I have parathyroid involvement.

Tumors of islet cells of the pancreas may produce large amounts of insulin, causing fasting hypoglycemia. Islet cells also secrete excess gastrin, resulting in hypersecretion of gastric acid and formation of peptic ulcers. Perforation, bleeding, and intestinal obstruction (Zollinger-Ellison syndrome) may result. Hypersecretion of gastric acid causes diarrhea and steatorrhea due to inactivation of pancreatic lipase. More than one half of patients presenting with Zollinger-Ellison syndrome have other endocrine abnormalities associated with the MEN I syndrome.[1]

The islet cell tumors (non-beta-cell origin) may also produce vasoactive inhibitory polypeptide (VIP), leading to severe secretory diarrhea with fluid and electrolyte depletion (ie, hypokalemia and achlorhydria). Glucagon and ACTH (producing Cushing's syndrome) also may be produced by non-beta-cell tumors. The beta-cell tumors are occasionally malignant, with metastasis, especially to the liver. Islet cell tumors are frequently multifocal and atypically located; they occur in 80% of MEN I patients.

Pituitary tumors are found in 65% of MEN I patients,[2] and 25% of these tumors are associated with hypersecretion of growth hormone, producing acromegaly.[2] The remaining 75% previously were thought to be benign but have recently been shown to secrete prolactin.[2] Large tumors may cause visual disturbances, headache, or even panhypopituitarism.

Knowledge of preexisting disease and the systems involved aid in planning and appropriate decision making to ensure a smooth anesthetic and postoperative course. A review of the history for symptoms suggestive of peptic ulcer disease, diarrhea, nephrolithiasis, hypoglycemia, and hypopituitarism is important.

Assessment of Pituitary Function

Normal levels of pituitary hormones are shown in Table 1.

TABLE 1. Blood Levels of the Pituitary Hormones

Hormone	Level (nl)
Cortisol (morning level)	4.9μg/dl (7–18μg/dl)
Tetraiodothyronine (T4)	6.1μg/dl (4–11μg/dl)
Tri-iodothyronine (T3 uptake)	25.2% (25%–36%)
Follicle-stimulating hormone (FSH)	7.4μU/ml (1–15μU/ml)
Thyroid-stimulating hormone (TSH)	3.5μU/ml (10μU/ml)
Luteinizing hormone (LH)	3.1μU/ml (1–15μU/ml)
Prolactin (PRL)	12.7ng/ml (1–20ng/ml)
Estradiol	1.6pg/ml (0.8 – 2.4pg/ml)
Growth hormone (GH)	2–5ng/ml

Note: nl = normal range.

From Osborn I: The Patient with Pituitary Disease, in E. Frost:*Preanesthetic Assessment 2*. Boston: Birkhäuser 1989, p. 18

Hypopituitarism manifests itself with deficiencies of (1) growth hormone, leading to hypoglycemia; (2) thyroid-stimulating hormone, leading to hypothyroidism; (3) adrenocorticotropic hormone, leading to hypofunction of the adrenal cortex, with hypotension and intolerance of stress or infection; (4) follicle-stimulating hormone (FSH) and luteinizing hormone (LH), leading to testicular atrophy, infertility, and a decrease in secondary sexual characteristics in men and infertility and amenorrhea in women. Visual field defects such as unilateral or bitemporal hemianopia may be present.

Physical examination should search for features of acromegaly, such as a large jaw, tongue, and epiglottis.[3] Patients must be carefully assessed for airway management as problems with intubation of the trachea are not unusual. If subglottic stenosis and edema develop intraoperatively, early extubation may not be possible. Not uncommonly, hepatomegaly, cardiomegaly, cardiomyopathies, and dyspnea coexist. Increases in growth hormone levels and lack of suppression by glucose infusion confirm the diagnosis of acromegaly.

Excess secretion of prolactin may be present, causing infertility and galactorrhea. High prolactin levels, detected by radioimmunoassay methods, and a fall of serum levels in response to L-dopa administration confirm the diagnosis of a prolactin-secreting tumor.

Assessment of Parathyroid Function

Elevated calcium and low serum phosphate levels suggest hyperparathyroidism, but the disease is usually mild and the patient asymptomatic. In more-severe cases, some hyperchloremic acidosis may be present, and parathormone levels are usually elevated. Serum calcium levels of 12 mg/100ml are usually tolerated without any symptoms and are frequently detected during routine screening examination. Blood calcium levels greater than 14mg/l00ml can lead to anorexia, nausea, vomiting, abdominal pain, constipation, polyuria, tachycardia, and dehydration.

Commonly, a history of calcium-containing renal stones is present. Subperiosteal absorption can be seen on x-rays of hands and teeth, but severe bone disease (eg, osteitis fibrosa cystica) is very rare. The signs of severe hypercalcemia include polyuria and dehydration (T interval on ECG tends to be short).

Preoperative correction requires rehydration with normal saline plus administration of furosemide, along with careful monitoring of central venous pressure and electrolytes. The cytotoxic drug mithromycin may be used in the emergency treatment of severe hypercalcemia.[5] Peritoneal dialysis or hemodialysis is indicated in patients with renal failure.

Assessment of Gastrointestinal Function

Symptoms of peptic ulcer disease and diarrhea suggest a gastrinoma. Perforation, bleeding, and obstruction are frequent and can be life-threatening. The diagnosis of gastrinoma involves the presence of a duodenal or postbulbar ulcer and excessive basal gastric acid secretory rate. The diagnosis is confirmed by finding a markedly elevated serum gastrin level and an increase of gastrin with the secretin provocation test. Treatment of this feature of the disease with H_2 blockers, anticholinergics, and antacid drugs has now become the mainstay of therapy. Total gastrectomy is reserved for patients refractory to medical treatment.

Insulinomas may produce signs of hypoglycemia, with sweating, weakness, tingling sensations, blurred vision tremors, and tachycardia. In severe cases of lowered glucose levels, bizarre behavior and even convulsions may occur. These symptoms generally develop at blood glucose levels around 50mg/dl.[6] In diabetic patients (uncommon in MEN patients), symptoms develop as glucose levels fall below 80 mg/dl.

Ratios of fasting serum insulin to serum glucose should be calculated; they are consistently above 0.3 when functioning islet cell tumors are present. Partial pancreatectomy is indicated and is curative. If surgery is unsuccessful or if the tumor has become malignant, with widespread metastasis, therapy with diazoxide or streptozocin is indicated. Diazoxide specifically blocks insulin release from the pancreas. Streptozocin is a cytotoxic drug that has specific activity against the pancreatic beta cell.

MEN type II (IIa) and Type III (IIb)

Clinical Presentation

Controversy has arisen as to whether MEN III is a variant of MEN II or a separate entity; hence the dual naming system (Table 2). Both MEN II (or MEN IIa) and MEN III (or MEN IIb) involve medullary carcinoma of the thyroid, gland (MCT) with elevated production of calcitonin.

TABLE 2. Differentiation of Multiple Endocrine Neoplasia

MEN IIa	**MEN IIb**
Bilateral medullary carcinoma of the thyroid (MCT)	Bilateral carcinoma of the thyroid gland
Pheochromocytoma	Pheochromocytoma
Familial inheritance as an autosomal dominant trait	Familial inheritance as an autosomal dominant trait, proved in 50% of patients
Variable rate of MCT progression (more commonly indolent)	Generally, MCT progresses rapidly

MEN IIa comprises MCT (100% of cases), pheochromocytoma (50% of cases), and hyperparathyroidism (60% of cases.)[7] The disease has a definite genetic pattern, being inherited as a mendelian autosomal dominant trait.

MEN IIb refers to a syndrome in which MCT (100% of cases) and pheochromocytoma (50% of cases), but usually not hyperparathyroidism, are the component parts.[8] Patients have characteristic facies with prominent mandibles, puffy lips and tongue, and multiple mucosal neuromas. Additionally, they commonly have a marfanoid habitus, pes cavus, and gastrointestinal ganglioneuromatosis and diverticulosis. The disease has presented as familial, with an autosomal dominant genetic pattern; however, in 50% of reported cases, it has been sporadic.[2] It is not clear that families were thoroughly screened in all of the reported cases.[2]

Etiologically, the cells involved in MCT are parafollicular cells that originate from neural crest tissue. Also arising from neural crest tissue is the medullary portion of the adrenal gland, which gives rise to pheochromocytomas. The presence of a parathyroid abnormality in MEN IIa is less well understood, because parathyroid cells are not derived from neural crest tissue. Some investigators feel that the hyperparathyroidism is secondary to the hypocalcemic effect of high levels of calcitonin; others feel that the hyperparathyroidism is genetically determined and is not a secondary phenomenon.

In virtually all patients with MEN IIa or MEN IIb, the thyroid tumors are bilateral, whereas MCT occurring sporadically is usually unifocal. Multiple tumor foci may be found. Clinically, MCT presents early in patients with MEN IIb and seems to have a more aggressive course (relatively few patients survive beyond 30 years of age). The pheochromocytomas of MEN IIa and MEN IIb usually present in the second and third decades of life and are bilateral and multiple in approximately 70% of patients.[8] Nonfamilial pheochromocytomas are rarely bilateral or multiple.[9] Pheochromocytomas in MEN IIa and MEN IIb patients are seldom malignant and are usually confined to the adrenal gland. The parathyroid lesion in MEN IIa consists of pseudonodular but usually generalized multiglandular hyperplasia.

Thyroid Function

Medullary carcinoma of the thyroid (MCT) occurs in all patients with MEN II. Approximately 25% of patients with MCT will present with preceding or concurrent lymphadenopathy.[8] MCT is capable of great biosynthetic activity and is known to produce prostaglandins, serotonin, histamines, and ACTH in addition to calcitonin. Cushing's syndrome associated with MCT also has been reported.[8] Thirty percent of MCT patients develop diarrhea,[8] which is thought to be caused by a hormone secreted from the tumor that causes either hypermotility or increased secretions in the intestinal tract. The diarrhea also may be caused by diffuse intestinal ganglioneuromatosis, which is often present.[2]

In reviewing the family histories of patients with MEN II syndromes, it is striking to see how frequently one finds young relatives who have experienced "strokes," "heart attacks," or even sudden death prior to the time that a thyroid tumor was observed.

The patient who presents with a neck mass or lateral neck lymphadenopathy may be diagnosed as having MCT if there is an elevated plasma calcitonin level. Elevated plasma calcitonin levels, measured by radioimmunoassay technique, are present in all patients with clinically detectable MCT.[10] In evaluating family members to determine early (preclinical) stages of the disease, minimal elevation of calcitonin indicates MCT. In family members with normal basal calcitonin levels, provocative testing by infusions of pentagastrin or calcium has shown modest increase of calcitonin. At operation, such patients have been shown to have C-cell hyperplasia, indicative of a premalignant phase of MCT.

Pheochromocytoma

Pheochromocytomas are chromaffin tumors, usually found bilaterally in the adrenal medulla. However, they may be found in other chromaffin tissue, such as the organs of Zuckerkandl, or, rarely, in ectopic rests of chromaffin cells in the thorax, bladder, brain, and sympathetic chain.

MEN IIa and MEN IIb patients may present with headaches, sweating, palpitations, nausea, and vomiting (ie, signs suggestive of pheochromocytoma). The diagnosis is confirmed by detecting elevated levels of urinary catecholamines, vanillylmandelic acid (VMA), or metanephrines. Characteristically, these patients have episodic, rather than sustained, hypertension. Even if MCT is diagnosed first, it is absolutely critical that the presence of a pheochromocytoma be ruled out prior to neck exploration. Often a general anesthetic induction in an unprepared patient will provoke a hypertensive crisis that might result in severe morbidity or perhaps death.

Signs of pheochromocytoma are hypertension (paroxysmal or chronic), tachycardia, sweating, headaches, palpitations, abdominal pain, nausea, visual disturbances, irritability, weight loss, and fainting spells related to posture (Table 3).

Physical findings in addition to hypertension include postural hypotension (as in our patient), cardiomegaly, and mild elevation of basal body temperature. Laboratory evaluation reveals a modestly elevated blood sugar (due to suppression of insulin secretion by beta catecholamines), a rise in hematocrit that often reflects a diminished plasma volume, and an elevated white cell count with a shift to the left.

By far the most common form of pheochromocytoma is associated with a greater secretion of norepinephrine relative to epinephrine. Norepinephrine has predominantly alpha-adrenergic effects, causing hypertension, tachycardia, and sweating. Epinephrine exerts a mixture of alpha and beta effects; beta-adrenergic effects include vasodilation in skeletal and splanchnic vascular beds, increased heart rate and contractility, and bronchodilation.[8]

TABLE 3. Common Signs and Symptoms of Pheochromocytoma

Sign/symptom	% of cases
Hypertension	80–90
Paroxysmal hypertension	22–87
Headache	55
Tachycardia	44–50
Dyspnea	19–27
Excessive sweat	23–78
Weight loss	33
Organ abnormality	
Cardiac	50
Retinal	80–100
Cerebrovascular	14–26
Systemic	
Fever	18–78
Hyperglycemia	40–78
Paroxysmal	
Nausea	90
Vomiting	33
Thoracic Pain	12–27
Abdominal Pain	15

The diagnosis of pheochromocytoma in patients with MEN IIa or MEN IIb depends on analysis of levels of either plasma catecholamines or 24-hour urinary catecholamines and catecholamine metabolites.[11] Preoperative determination of pheochromocytoma is critical because this lesion may not initially manifest itself clinically.

Intravenous pyelography with sonography, computed tomography scanning, and retroperitoneal air insufflation are procedures used to localize adrenal lesions. Selective arteriography is an invasive procedure, used more often in the patient with nonfamilial pheochromocytoma, in which unilateral lesions may exist. In MEN IIa and IIb patients, both adrenal glands must be explored; therefore, arteriography has less value.[8]

Because of the morbidity and mortality associated with pheochromocytoma and because the associated hypertension is usually curable, surgical removal is the recommended therapy. Longterm medical therapy is indicated only in patients who are poor surgical risks or who have malignant and unresectable tumors. Anterior abdominal exploration is recommended, with a thorough evaluation of both adrenal fossae, the organs of Zuckerkandl, and the sympathetic chain.

If only one adrenal gland appears to be involved (ie, it appears hyperplastic), it should be resected, and a biopsy should be performed on the other adrenal gland. Some surgeons believe that bilateral adrenalectomies are indicated for all MEN II patients. The patient with unilateral resection should be tested at month intervals because a second adrenal tumor can frequently be diagnosed before it is clinically apparent.

The preoperative preparation of patients undergoing surgery for pheochromocytoma consists of a 2-week treatment with the alpha-receptor blocker phenoxybenzamine (40–200mg/day). Treatment with phenoxybenzamine should begin very cautiously with 10 to 20mg daily and slowly increased to 100mg twice daily if necessary until control of hypertension is achieved. Overtreatment is easy and may manifest itself as severe postural hypotension and tachycardia resulting from an expansion of the intravascular volume. Concomitant administration of isotonic saline solution intravenously helps correct the intravascular volume deficit. The intravascular volume is slowly replenished, and a fall in hematocrit commonly occurs.

If tachycardia occurs with alpha-adrenergic blockade, addition of a beta-blocking drug (propranolol 10–40mg orally four times daily) may be given. Tachycardia may be caused by circulating beta catecholamines (ie, epinephrine) or may occur as a reflex following alpha blockade. It must be emphasized that beta blockers should *never* be used before alpha blockers because hypertension may actually be accentuated in this circumstance (ie, blocking of peripheral beta receptors allows an unattenuated alpha response).

Also, should hypotension occur for any reason in patients maintained on phenoxybenzamine, pressor agents are usually ineffective. Epinephrine is contra-indicated because it stimulates both alpha and beta receptors. Since alpha receptors are blocked, the net effect of epinephrine is vasodilatoin and further hypotension. Caveats against epinephrine extend also to its intraoperative use as an adjunct to local anesthetic administration.

Intravenous infusion of levaterenol bitartrate may be used to combat severe hypotension because this drug primarily stimulates alpha receptors.

Assessment of Parathyroid Function

MEN IIa patients may have hypercalcemia but are usually asymptomatic and only occasionally develop renal stones, bone disease, or renal failure (Table 4.)

TABLE 4. Manifestations of Hypercalcemia Due to Hyperparathyroidism

Type	Manifestations
Renal	Polydipsia; polyuria; renal stones; decreased glomerular filtration rate
Cardiac	Hypertension; short Q-T interval; prolonged P-R interval
Gastrointestinal	Abdominal pain; vomiting; weight loss; peptic ulcers; pancreatitis
Skeletal	Bone pain and tenderness; skeletal demineralization; pathologic fractures; collapse of vertebral bodies
Nervous system	Somnolence; psychosis; decreased pain sensation
Neuromuscular	Skeletal muscle weakness
Articular	Periarticular calcifications; gout
Ocular	Calcifications (band keratopathy); conjunctivitis
Hematopoietic	Anemia

Some authors believe that the hyperparathyroidism develops secondary to the high serum calcitonin levels.[8] If this assumption is true, it is difficult to explain why hyperparathyroidism does not develop in patients with MEN or sporadic MCT. It is interesting to note that patients with MEN IIb (or those with sporadic MCT) with very high levels of plasma calcitonin rarely develop hypocalcemia.

Anesthetic Plan

Patients with the MEN syndromes may present for many operative procedures. A few of the more common situations are reviewed.

Parathyroidectomy

Once hyperparathyroidism has been confirmed, parathyroidectomy should be performed to avoid the formation of renal calculi and renal parenchymal calcium deposition with resultant serious impairment of renal function as well as other symptomatology of chronic hypercalcemia.[12] Because multiple adenomas are usually present, surgical management should involve sacrificing three of the glands and sometimes a part of the fourth.

No anesthetic agent is particularly indicated, but in cases of renal dysfunction the avoidance of enflurane and fixed drugs, such as narcotics and benzodiazepines, should be considered: Renal function sometimes improves after operation, and the importance of maintenance of hydration and urine output in patients with elevated plasma calcium levels should be emphasized. Central venous pressure monitoring is helpful in fluid management and if air embolism occurs with the head elevated. Head-up position provides a blood-free field, which facilitates the identification of the parathyroid glands.[12]

The response to muscle relaxants may be altered because skeletal muscle weakness exists in these patients. Muscle relaxation should be monitored with a peripheral nerve stimulator.[13] Careful positioning of the patient is necessary because of the possiblity of osteoporosis and the associated vulnerability to pathologic fractures.

When a definite pathology cannot be found in the parathyroids, the mediastinum may be explored for an ectopic gland. Control of the airway with an endotracheal tube allows tissue retraction and control of alterations in respiration in the event of a pneumothorax. Digitalis should be used with caution in the presence of hypercalcemia because both produce myocardial irritability.

Postoperatively, transient hypoparathyroidism is present in as many as 30% of patients[14] and can be demonstrated by eliciting a positive Chvostek's sign or Trousseau's sign. Neuromuscular irritability is increased with serum calcium levels of 7–8mg/dl, and generalized convulsions with laryngeal stridor may occur with levels below 7mg/dl. Symptoms usually occur within the first few days after operation and may persist up to 3 months. Emergency treatment consists of slow intravenous administration of up to 1gm of calcium gluconate until normal plasma levels of calcium are obtained.

Gastrectomy

Gastrinomas occur in 50% of MEN I patients. Surgical exploration with palpation of the pancreas and duodenum will often reveal these tumors. Total gastrectomy is indicated because there often are multiple adenomas. When these tumors are present, one must consider gastric hypersecretion and the likelihood of large gastric fluid volumes at the time of induction. Esophageal reflux is common. Depletion of intravascular fluid volume and electrolyte imbalance (hypokalemia and metabolic alkalosis) may accompany profuse diarrhea. Hypoproteinemia could alter the pharmacokinetics of injected drugs, and cimetidine can alter the rate of hepatic breakdovn. Immediate postoperative complications are usually due to anastomotic leaks in the gastrectomy.[15]

Pancreatectomy

Most insulinomas (sporadic type) are single and benign—90% of patients—and complete tumor excision is usually successful. However, in MEN I patients there is a greater tendency for multiple adenomas to be present. When tumors cannot be found by the surgeon, a blind distal pancreatic resection is warranted to remove small occult tumors. Total pancreatectomy is not indicated as an initial procedure.

The maintenance of normal blood glucose concentration is the anesthesiologist's greatest challenge. Profound hypoglycemia can occur, particularly during manipulation of the tumor.[16] Dextrostix testing should be used at frequent intervals to estimate blood sugars rapidly. An artificial pancreas (Biostater) machine, which constantly monitors venous blood glucose levels and automatically infuses insulin or glucose, has been used intraoperatively during resection of insulinomas.[17] Use of the Biostater is limited at present because of its enormous expense. Administration of glucose during the procedure may not be necessary because the stress of surgery increases production of counterinsulin hormones such as adrenocatecholamines and adrenocorticosteroids.

Signs of hypoglycemia, which may be masked by anesthesia, result from increased sympathetic outflow. They include sweating, tachycardia, hypertension, and dilated pupils. Hypoglycemia should be corrected by rapid infusion of 5% to 10% dextrose. A rise in blood sugar of at least 40mg/dl that is sustained (hyperglycemic rebound) in the first 30 minutes after removal of the tumor is a good indicator that the tumor has been removed. False-positive hyperglycemic rebound can occur. Postoperatively, high blood sugars are common but are usually transient. Postoperative hypoglycemia indicates that the tumor probably has not been completely resected.

Thyroidectomy

In patients with MEN IIa or IIb who have MCT it is absolutely essential that a total thyroidectomy be performed, leaving no remnant of thyroid tissue in the neck. This is necessary because in 100% of these patients the MCT is bilateral. In cases in which bilateral subtotal thyroidectomies were performed for MEN II patients, recurrent disease developed. The central and retrosternal nodes of the neck should also be removed.

If parathyroid hyperplasia is present, it is common practice to manage this disease by removing all parathyroid tissue and autografting a portion of one gland to the forearm musculature or another area that may be easily accessible in the future. As is the case in patients presenting for parathyroidectomy, elevated plasma calcium levels may be encountered. Maintenance of hydration and adequate urine output is important. Airway obstruction by enlarged thyroid glands is unlikely in MCT, but if present, the problem should be managed by an awake intubation of the trachea.

Careful fiberoptic and radiographic evaluation of the airway prior to surgery are recommended in all patients with stridor related to thyroid disease.[18]

Endotracheal intubation for control of the airway, as mentioned for parathyroidectomy, is indicated. No specific agent for induction or maintenance of anesthesia has been recommended. Postoperatively, there may be signs of transient or chronic hypocalcemia and airway problems secondary to damage of the parathyroid gland.

Adrenalectomy and Abdominal Exploration

Preoperative medication with oral barbiturates or benzodiazepines alone has been recommended.[19] Narcotics have been used as premedicants without ill effects, but histamine release or respiratory depression could possibly incite tumor activity.[20] Atropine is generally considered contraindicated, theoretically because it can cause tachycardia, central nervous system stimulation, and potentiation of catecholamine vasopressor activity.

A general anesthetic induction in an unprepared patient may provoke a hypertensive crisis that could result in severe morbidity or perhaps death. In case reports of sporadic pheochromocytoma, the most frequent causes of intraoperative death are cerebrovascular accidents, congestive heart failure with pulmonary embolism, and ventricular fibrillation.[21] The periods of greatest danger are during induction of anesthesia, during tracheal intubation, while positioning the patient, during surgical manipulation of the tumor, and after ligation of venous drainage from the tumor.

Monitoring of rapid changes in circulation caused by released catecholamines and antihypertensive treatment modalities is mandatory. Monitoring should include central venous pressure, pulmonary artery wedge pressure, and arterial pressure by indwelling catheter. Two large-bore peripheral catheters are required for the administration of blood intravenous fluids, or drugs.

Anesthetic Management

Induction of anesthesia with most of the commonly used anesthetic agents appears to be acceptable. More important, hypoxemia and hypercarbia must be avoided. Good muscle relaxation for endotracheal intubation and abdominal exploration is necessary. Deep general anesthesia tends to inhibit catecholamine release but aggravates hypotension after removal of the tumor. Halothane potentiates dysrhythmias but provides good inhibition of sympathetic activity. Thiopental (given slowly), succinylcholine, enflurane, isoflurane, and most nondepolarizing muscle relaxants have been used with generally good results.

Intraoperative hypertension is best treated with rapid-onset and short-duration drugs. Phentolamine, an alpha-receptor blocking agent, given intraoperatively as a 0.01% infusion plus supplemental boluses of 1–3mg at intervals of 30–45 seconds, is useful. Sodium nitroprusside as a 2% solution (in 5% dextrose), titrated to effect, allows even more rapid and accurate control of blood pressure. Onset of action is within 1 minute, and recovery is within a few minutes of discontinuing the infusion. The search for multiple tumors is enhanced because the hypertension that follows manipulation of a second tumor is not masked by residual effects of long-acting antihypertensives. When the tumor source is removed, blood levels of epinephrine and norepinephrine (which may have a very short half-life) fall rapidly. Severe hypotension may occur following tumor excision.[21] Therefore, phentolamine, if used, should be discontinued 30 minutes prior to ligation of tumor venous drainage. Nitroprusside infusion should be stopped several minutes prior to ligation.

A norepinephrine infusion can be used to correct hypotension and should be given via a central venous, rather than a peripheral venous, line to avoid sloughing of tissues. Other vasopressors, such as phenylephrine (Neo-Synephrine®) and metaraminol (Aramine®), also have been used successfully for this purpose.

Intraoperative ventricular dysrhythmias are common because of the effects of high circulating catecholamines on the heart; they can be controlled by intravenous administration of 1–5mg propranolol or 50–100mg lidocaine. Shorter-acting drugs, such as esmolol (Brevibloc®), also may be used to reverse the dysrhythmogenic effects of circulating catecholamines.

A central venous catheter or pulmonary artery catheter should be left in place postoperatively to monitor volume status. Control of volume postoperatively has greatly reduced the need for vasopressor therapy. Blood pressure must be carefully monitored for 24–36 hours. Hypotension is the most frequent cause of death in the immediate postoperative period. Norepinephrine should be available. If total bilateral adrenalectomy is performed, the patient must be managed as an addisonian and given steroid replacement. Even if only one adrenal gland is removed, hydrocortisone, l00mg every 6 hours intravenously, may be used (especially if unexplained hypotension persists). Postoperatively, 24-hour urine levels of VMA or metanephrine should be measured to ensure that all tumorous tissue was removed.

References

1. Neil R.: Disorders affecting multiple endocrine systems, in Isselbacher K.J., Adams R.D., Braunwald E., Petersdorf R.G., Wilson J.D. (eds): Harrison's Principles of Internal Medicine. New York: McGraw-Hill, 1980, pp 1820–4.

2. Landsberg L.: Multiple endocrine neoplasia (MEN syndrome), in Berkow R. (ed): The Merck Manual of Diagnosis and Therapy. Rahway, NJ, Merck Sharp and Dohme, 1982, pp 1029–32.
3. Arancibia C.U., Frost E.A.M.: Hypophysectomy, in Frost E. (ed.): Clinical Anesthesia in Neurosurgery. Boston: Butterworths, 1984, pp 187–201.
4. Pende J.W., Basso L.V.: Diseases of the endocrine system, in Katz J., Benumof J., Kadis L.B. (eds): Anesthesia and Uncommon Diseases. WB Saunders, 1981, pp 155–220.
5. Potts J.T. Jr.: Disorders of parathyroid glands, in Isselbacher K.J., Adams R.D., Braunwald E., Petersdorf R.G., Wilson J.D., (eds): Harrison's Principles of Internal Medicine. New York: McGraw Hill, 1980, pp 1832–43.
6. Boyle P.J., Schwartz N.S.: Plasma glucose concentrations at the onset of hypoglycemic symptoms in patients with poorly controlled diabetes and in nondiabetics. *N Engl J Med* 1988, 318:1487–92.
7. Keiser H.R.: Sipple's syndrome: Medullary thyroid carcinoma, pheochromocytoma and parathyroid disease. *Ann Intern Med* 1973, 78:561–5.
8. Wells W.A. Jr., Norton J.A.: Medullary carcinoma of the thyroid and multiple endocrine neoplasia–II syndrome, in Friesen S.R., Bolinger R.E. (eds): Surgical Endocrinology: Clinical Syndromes. Philadelphia: JB Lippincott, 1978, pp 287–90.
9. Holland O.B.: Pheochromocytoma, in Isselbacher K.J., Adams R.D., Braunwald E., Petersdorf R.G., Wilson J.D., (eds): Harrison's Principles of Internal Medicine. New York: McGraw Hill, 1980, pp 736–41.
10. Tashjian A.H. Jr.: Immunoassay of human calcitonin: Clinical measurement relations to serum calcium and studies in patients with medullary carcinoma. *N Engl J Med* 1970, 283:890–3.
11. Spergel G.: Pheochromocytoma, in Berkow R. (ed): The Merck Manual of Diagnosis and Therapy. Rahway, NJ, Merck Sharp and Dohme, 1982, pp 1025–8.
12. Edis A.J.: Prevention and management of complications associated with thyroid and parathyroid surgery. *Surg Clin North Am* 1979, 59:83–8.
13. Al-Mohaya S., Naguib M., Abdelatif M., Farag H.: Abnormal responses to muscle relaxants in a patient with primary hyperparathyroidism. *Anesthesiology* 1986, 65:554–6.
14. Schafer M., Economou S.C.: Ode to an Indian rhinoceros or the evaluation and preparation of patients for parathyroid surgery. *Surg Clin North Am* 1970, 50:227–31.
15. Stoelting R.K., Dierdorf S.F., McCammon R.L.: The gastrointestinal system, in Stoelting R.K., Dierdorf S.F., McCammon R.L. (eds): Anesthesia and Coexisting Disease. New York: Churchill Livingstone, 1988, pp 393–408.
16. Roizen M.F.: Anesthetic implication of concurrent diseases: in Miller R.D. (ed): Anesthesia. New York: Churchill Livingstone, 1986, pp 259–60.
17. Pulver J.J., Cullen B.F., Miller D.R., Valenta L.J.: Use of the artificial beta cell during anesthesia for surgical removal of an insulinoma. *Anesth Analg* 1980, 59:950–2.

18. Shaha A., Alfonsa A.: Acute airway distress due to thyroid pathology. *Surgery* 1987, 102:1068–74.
19. Cooperman L.H., Engelman K., Mann P.E.G.: Anesthetic management of pheochromocytoma employing halothane and beta adrenergic blockade. *Anesthesiology* 1967; 28:575–7.
20. Deblasi S.: The management of the patient with pheochromocytoma. *Br J Anaesth* 1966, 38:740–1.
21. Scott H.W., Riddell D.H., Brockman S.K.: Surgical management of pheochromocytoma. *Surg Gynecol Obstet* 1965, 120:707–9.
22. Stoelting R.K., Miller R.D.: Endocrine, metabolic and nutritional disorders, in Stoelting R.K., Miller R.D. (eds): Basics of Anesthesia. New York: Churchill Livingstone, 1984, pp 330–2.

Chapter 6

The Patient With Down's Syndrome

David R. Sofair, M.D.

Case History. *A 10-year-old boy with Down's syndrome was brought to the dental clinic by his mother because of poor dentition. It was deemed that extensive work was required, and he was scheduled for reconstruction under general anesthesia on an ambulatory basis.*

The mother reported that the child had an abnormal heart sound, present from birth; she had been told that it would not cause any problems. The patient was a mouth-breather and snored loudly during sleep. He had frequent upper respiratory tract infections, the most recent one three days previously, and a poor tolerance for physical activity.

Physical examination revealed a child with the stigmata of mongolism, 51in tall and weighing 35kg. In particular, his tongue was very large and protruded from his mouth. He was apparently moderately retarded and combative. Auscultation revealed a 2–3/6 midsystolic murmur, and S_2 was widely split. Lung examination revealed scattered rhonchi, which cleared with coughing.

The anesthesiologist, believing that further investigations were warranted, asked that the patient be admitted and operated on as an inpatient after cardiac consultation. The dental surgeon was rather reluctant to comply, and discussion ensued.

Introduction

Trisomy 21 (Down's syndrome, or mongolism) is one out of the most common chromosomal abnormalities in humans occurring in 1 of 600–800 live births in the United States.[1] Because the abnormality affects many organ systems, Down's syndrome patients often have concurrent disease or congenital anomalies with implications that require close preanesthetic assessment (Table 1).

Reviewed by Dr. Carlos Rosasco, Professor of Anaesthesia, University of Uruguay, Montevideo.

TABLE 1. Trisomy 21 Anesthetic Considerations

Anatomic	Short stature • Airway: small mouth, large tongue, high, arched palate, small mandible and maxilla
Musculoskeletal	Hypotonia • Joint laxity: cervical spine (unstable atlantoaxial joint), temporomandibular joint, other body joints
Cardiac	Congenital heart defects • Aortic insufficiency • Mitral valve prolapse
Pulmonary	Frequent respiratory tract infections • Pulmonary hypertension associated with CHD • Sleep apnea • Postop atelectasis, airway obstruction
Immunologic	Altered immune response • Frequent infections • Increased incidence of hepatitis B and lymphocytic leukemia
Neurologic	Mild to profound mental retardation • Higher incidence of seizure disorders • Earlier onset of presenile dementia • Perioperative agitation
Gastrointestinal	Increased risk of gastroesophageal reflux
Blood dyscrasias	Polycythemia in newborns
Endocrine	Thyroid abnormalities in adults • Decreased central and peripheral sympathetic activity

Mongolism was originally described by Langdon Down in 1866. Although a genetic basis for Down's syndrome had been postulated, the precise chromosomal basis of the syndrome was not elucidated until 1959 by Lejeune et al.[2] In approximately 95% of cases, Down's syndrome results from an additional acrocentric chromosome, number 21, in the nuclei of affected patients. There are several possible causes of this chromosomal anomaly, including radiation exposure, thyroid disease, diabetes, and viral illness. However, increased maternal age is most strongly associated. If the maternal age at parturition exceeds 45 years the incidence of Down's syndrome in the offspring is 1:80, or about 10 times the general risk.[1] There also exists an inherited or familial form of Down's syndrome that is phenotypically identical to the congenital form. It results from a translocation of a piece of chromosome 21 on to 13, 14, 15, 21, or 22 and accounts for approximately 5% of Down's syndrome cases.[3]

As mongolism is characterized by a chromosomal abnormality that usually affects every cell in the body, it follows that the clinical manifestations are widespread, extending to multiple organ systems. The classical, well-described physical features include facial stigmata, such as a small head, a small nose with a flat bridge, epicanthal eye folds, upward slanting of the palpebral fissures, strabismus, and Brushfield's spots of the iris; other findings include a small mouth with macroglossia, a short and broad neck, short extremities with small hands and feet, and a single line crease across the palm (simian crease). The last feature is not seen in all cases. Mild to moderate mental retardation is the rule. Growth retardation is common, and adults with Down's syndrome are usually obese and of short stature. Muscle hypotonia and ligamentous laxity with atlantoaxial instability occur commonly.[4] Congenital heart disease is often associated with Down's syndrome and is responsible for the high childhood mortality rate. The immunologlc system is abnormal, resulting in frequent infections, especially of the respiratory tract. Sleep apnea is more common,[5] as are gastrointestinal anomalies requiring surgery. There is an increased incidence of both leukemia and myelomeningocele.[6] Thyroid abnormalities are not uncommon in adults.

Surgery in the Patient With Down's Syndrome

As life expectancy has improved for patients afflicted with trisomy 21, the number presenting for surgery has increased. In 1958, 30% of children with Down's syndrome died by 1 month of age, 53% died in their first year of life and 70% did not survive to their 10th birthday.[7] Today 76% of those afflicted with Down's syndrome survive past the age of 10.[8] As mentioned, most of the mortality in childhood results from associated congenital heart disease.

The majority of patients presenting for surgery with Down's syndrome are children. A review of 100 consecutive patients indicated a mean age of 5.8 years at the time of operative intervention.[9] The majority of surgical procedures involve correction of congenital defects that would otherwise significantly impair the quality of life at a young age. These include cardiac procedures to correct congenital heart defects (44%) and otolaryngologic, ophthalmologic, and orthopedic procedures to correct anomalies and illnesses secondary to Down's syndrome (Table 2).

In recent years elective cosmetic surgery has become increasingly popular in an effort to normalize the patient's appearance and acceptance into society.[10] Probably an increasing number of Down's syndrome patients will present for this type of reconstructive surgery.

Because trisomy 21 is a syndrome involving abnormalities of multiple organ systems, it is not surprising that the patient with Down's syndrome presenting for surgery has congenital anomalies or concurrent disease not directly related to the planned procedure. In Kobel's study,[9] 73% of patients presenting for noncardiac surgery had concurrent abnormalities with anesthetic implications (Table 3). Of note, 20% had an unrepaired congenital cardiac defect. Concomitant respiratory, musculoskeletal, hematologic and endocrinologic disorders were also

TABLE 2. Down's Syndrome Patients Requiring General Anesthesia

Type of Procedure (no. performed) Specific procedures (no.)
Cardiac (44)
A-V canal repair (19)
ASD and VSD repair (6)
VSD repair (4)
PDA ligation (3)
Mitral valve repair (2)
Paliative surgery: Blalock-Taussig shunt (2)
PA banding (2)
Miscellaneous (4)
ENT (19)
Tonsillectomy and adenoidectomy (13)
Myringotomies and mastoidectomies (4)
Laryngobronchoscopy (2)
General surgery (16)
Genitourinary surgery (10)
Duodenoplasty (3)
Pull through (2)
Colectomy (1)
Ophthalmology (12)
Cataract needling and vitrectomy (5)
Squint repair (4)
Examination under anesthesia (3)
Orthopedic (8)
Cl–C2 fusion (2)
Tendon transfer and osteotomies (3)
Others (3)
Other (1)
VP shunt
Total (100)

From Kobel et al., see [9].

diagnosed. Identification of these abnormalities in the surgical patient with Down's syndrome is positively relevant to preanesthetic assessment and appropriate anesthetic care.

Anatomy

The anatomic considerations are of far-reaching significance for the anesthesiologist. Aside from the facial stigmata children with Down's syn-

TABLE 3. Congenital Abnormalities and Concurrent Diseases With Anesthetic Implications in 41 of 56 Down's Syndrome Patients for Noncardiac Surgery.

System Affected (no. of patients) Specific Problems (no. of patients)
Cardiac (25)
(in patients for noncardiac surgery)
Congenital defect (11)
Repaired congenital defect (5)
Pulmonary hypertension (9)
Respiratory (8)
Chronic chest infection (5)
Subglottic stenosis (2)
Tracheomaiacia (1)
Locomotor (3)
Excessive ligament laxity (2)
C1–C2 subluxation (1)
CNS (1)
Seizure disorder
Hematology (1)
Neonatal polycythemia (Hct = 75%)
Endocrine (3)
Hyperthyroidism (1)
Obesity (2)
Total (41)

From Kobel et al., see [9].

drome suffer from growth retardation, which results in low birth weight and persists into childhood, during adolescence, patients are of short stature and overweight. When anesthetizing patients with Down's syndrome, it is important to consider height, size, weight, and age. Drugs administered purely on a weight basis may be given in inappropriate doses.

The associated airway abnormalities would lead one to conclude that these patients represent difficult endotracheal intubations: large, protuberant tongue; small mouth; high, arched palate;[11] short, broad neck; abnormal dentition; small mandible and maxilla; and large tonsils. In addition, the larynx is usually small, and congenital subglottic stenosis is prevalent.[12] These adversities, however, are offset by the laxity of the temporomandibular joint and cervical spine (see below), so that intubation, in fact, is relatively easy.[10,13] Of particular note is the fact that children with Down's syndrome require endotracheal tubes 1–2mm smaller than predicted by standard formulas.[9,10] As always, a variety of endotracheal tube sizes should be immediately available.

Musculoskeletal System

Hypotonia and joint laxity are common features. Attention must be paid to positioning during anesthesia in order to avoid hyperextension, hyperflexion, and rotation movements.

Approximately 15% of patients with Down's syndrome have atlantoaxial instability, and the majority of them are asymptomatic.[4] These patients are predisposed to subluxation and cervical cord compression, especially during laryngoscopy and endotracheal intubation.[14,15]

The history, especially direct questioning of the parents, and physical examination may uncover cases of symptomatic subluxation. Common symptoms are gait abnormalities, clumsiness, and increased fatigue when walking. Physical findings may include abnormal neck motion (very mobile), upper and lower motor neuron signs such as spasticity, hyperreflexia, extensor plantar reflexes, loss of bowel and bladder control, and neck posturing (torticollis).[15] Also helpful in bedside evaluation may be the Sharp and Purser test.[16] With the patient in the sitting position and the neck flexed, backward pressure is applied against the forehead while the spinous process of the axis is palpated. A gliding motion may be felt as subluxation is reduced. This test is positive in about half of patients with atlantoaxial instability.

Some authors advocate radiographic screening for all Down's syndrome children over 4 years of age (with lateral radiographs of the neck in the flexed, extended, and neutral positions).[14] Others suggest radiographic study only when there is gait disturbance or abnormal neck motion.[9] If the patient is asymptomatic and radiographs are not done, special care is absolutely indicated to protect the cervical spine during intubation, positioning, and transport. Intubation with the head in neutral position, after stabilization with sandbags, and use of the fiberoptic laryngoscope or light wand should be considered. Any patient with symptoms, signs, or radiographic evidence suggestive of subluxation should be evaluated for posterior cervical spine fusion prior to elective surgery. Intubation should then be performed during cervical traction of 2-point pin fixation placed under local anesthesia.

Cardiac Assessment

The overall incidence of congenital heart disease (CHD) in Down's syndrome is 40%.[17] The most prevalent lesion is endocardial cushion defect (ECD), which accounts for nearly half of all congenital heart lesions in this population. Ventricular septal defect (VSD), atrial septal defect (ASD), tetralogy of Fallot (TOF) and patent ductus arteriosus (PDA) are also common (Table 4).

TABLE 4. Cardiac Lesions in 230 Down's Syndrome Patients With Congenital Heart Disease

Cardiac lesion	% Patients	% Mortality
Endocardial cushion defect	49	38
Ventricular septal defect	29	20
Tetralogy of Fallot	8	42
Patent ductus arteriosus	7	31
Atrial septal defect, secundum	2	67
Other*	5	36

* Primary myocardial disease. Ebstein's anomaly, double outlet right ventricle, total anomalous pulmonary drainage, dextrocardia, pulmonic stenosis, VSD with pulmonic stenosis, and coarctation of the aorta.

Adapted from Greenwald RD, Nadas AS: The clinical course of cardiac disease in Down's syndrome. Pediatrics 1976:58(6):893–7.

All of these lesions, with the exception of TOF, result in increased pulmonary artery hypertension; pulmonary vascular disease is a frequent complication. Down's syndrome children with CHD have higher pulmonary artery pressure and an increased incidence of postcorrective pulmonary edema when compared to their normal counterparts with CHD.[18] Adults with Down's Syndrome appear to have a higher incidence of aortic insufficiency and mitral valve prolapse (MVP) in addition to the congenital heart defects. [19]

The Down's syndrome patient with CHD offers the anesthesiologist a unique set of problems when presenting for cardiac and noncardiac surgery. The evaluation begins with the history and physical examination. A detailed history, with emphasis on feeding patterns, level of activity, fatigability, and medications, should be obtained. Determination of resting vital signs, auscultation for abnormal heart sounds, murmurs, and gallops, and pulmonary evaluation for rales, cyanosis, and clubbing should be made. Preoperative evaluation continues with complete blood work and serum electrolyte determinations, electrocardiography, and chest radiographs as warranted by the history and physical examination. Patients with previously repaired CHD may have conduction disturbances, usually left anterior hemiblock, and right bundle branch block. If there are any new findings or evidence of decompensation, consultation with a pediatric cardiologist will be helpful in determining the need for preoperative echocardiography and cardiac catheterization.

Administration of digitalis and diuretics is discontinued the night prior to surgery; beta blockers and antibiotics are continued until and during surgery.

Many patients with Down's syndrome undergoing surgery will require antimicrobial prophylaxis for the prevention of bacterial endocarditis (Table 5.)

TABLE 5. Endocarditis Prophylaxis[1]

	Dosage for adults	Dosage for children
Dental and upper respiratory procedures[2]		
Oral[3]		
Penicillin V	2gm 1 hour before procedure and 1gm 6 hours later	>60lb: adult dosage <60lb: half the adult dosage
Penicillin allergy: Erythromycin	1gm 2 hours before procedure and 500mg 6 hours later	20mg/kg 2 hours before procedure and 10mg/kg 6 hours later
Parenteral[3,4]		
Ampicillin	2gm IM or IV 30 minutes before procedure	50mg/kg IM or IV 30 minutes before procedure
plus Gentamicin	1.5mg/kg IM or IV 30 minutes before procedure	2.0mg/kg IM or IV 30 minutes before procedure
Penicillin allergy: Vancomycin	1gm IV infused *slowly over 1 hour* beginning 1 hour before procedure	20mg/kg IV infused *slowly over 1 hour* beginning 1 hour before procedure
Gastrointestinal and genitourinary procedures[2]		
Oral[3]		
Amoxicillin	3gm 1 hour before procedure and 1.5gm 6 hours later	50mg/kg 1 hour before procedure and 25mg/kg 6 hours later
Parenteral[3,4]		
Ampicillin	2gm IM or IV 30 minutes before procedure	50mg/kg IM or IV 30 minutes before procedure
plus Gentamicin	1.5mg/kg IM or IV 30 minutes before procedure	2.0mg/kg IM or IV 30 minutes before procedure
Penicillin allergy: Vancomycin	1gm IV infused *slowly over 1 hour* beginning 1 hour before procedure	20mg/kg I.V. infused *slowly over 1 hour* beginning 1 hour before procedure
plus Gentamicin	1.5mg/kg IM or IV 30 minutes before procedure	2.0mg/kg IM or IV 30 minutes before procedure

1. For patients with valvular heart disease, prosthetic heart valves, most forms of congenital heart disease (but not uncomplicated secundum atrial septal defect), idiopathic hypertrophic subaortic stenosis, and mitral valve prolapse with regurgitation.
2. Data are limited on the risk of endocarditis with a particular procedure. For a review of the risk of bacteremia with various procedures, see ED Everett and JV Hirschmann, Medicine, 56:61, 1977, and PJ Shorvon et al. Gut. 24:1078, 1983. For some useful guidelines on which procedures justify prophylaxis, see ST Shulman et al. Circulation. 70:1123A. 1984.
3. Oral regimens are more convenient and safer. Parental regimens are more likely to be effective; they are recommended especially for patients with prosthetic heart valves, those who have had endocarditis previously, or those taking continuous oral penicillin for rheumatic fever prophylaxis.
4. A single dose of parenteral drugs is probably adequate, because bacteremias after most dental and diagnostic procedures are of short duration. However, one or two followup doses may be given at 8- to 12-hour intervals in selected patients, such as hospitalized patients judged to be at higher risk.

Any patient with valvular-heart disease, unrepaired CHD (but not uncomplicated secundum ASD), idiopathic hypertrophic subaortic stenosis, and MVP with regurgitation will require such therapy.[19] Many children who have had corrective cardiac surgery—including aortic valvotomy, resection of aortic coarctation, pulmonary valvotomy, valve replacement, repair of TOF, and great vessel transposition—are included;[20] those who have had PDA ligation or closure of ASD or VSD do not require antibiotics.[9] In light of the immunologic abnormalities and often-encountered poor dentition, it is prudent to administer these drugs prior to endotracheal intubation.

Pulmonary Involvement

Patients with Down's syndrome have a high frequency of pulmonary illnesses. Thus, it is not surprising that the majority of postoperative complications are related to the pulmonary system.[9,10,18] Atelectasis, poor patency of the upper airway and respiratory depression following extubation, and pulmonary edema are most common.

There is a general predisposition toward hypoxia. Respiratory tract infections secondary to airway abnormalities, immunologic deficiencies, and institutional living are common contributing causes. Hypotonia and sleep apnea with both mechanical and central nervous system factors also contribute to hypoventilation and hypoxia. Children with CHD have left-to-right shunt (except TOF) and greater pulmonary artery hypertension than do normal children with CHD. Whether this is caused by the increased incidence of hypoxia or abnormalities in the alveoli and pulmonary vascular bed has not been established.[18]

Preoperative pulmonary function tests and arterial blood gas analyses may be needed if questions of pulmonary adequacy arise from the history and physical examination. However, because of lack of cooperation it may be difficult to obtain reliable test results. In children with a history of respiratory problems, especially sleep apnea, sedative premedication should be administered under close supervision, if at all, in the operating room holding area. Patients with a history of sleep apnea who are medicated with barbiturates are probably not candidates for same-day discharge. Although children with bronchitis and pneumonitis should be treated prior to surgery, categoric cancellation of all cases in which there is a concomitant mild upper respiratory tract infection may not be warranted.[21]

After extubation, patients should be watched for signs of airway obstruction and respiratory depression; treatment should be instituted as appropriate.

Immunologic Considerations

Patients with Down's syndrome have an altered immune response, which leaves them prone to frequent and prolonged infections. Those of the respiratory tract are most prominent. The thymus in Down's syndrome is often small and depleted of T lymphocytes. Humoral (B-cell-mediated) response to certain antigens is reduced, but serum gamma globulin levels are usually greater than normal.[22] Granulocytes are normal in number but are structurally and functionally abnormal. Meticulous asepsis is required during intravenous and arterial cannulation, and central access routes should be discontinued as soon as possible.

There is an increased incidence of hepatitis B antigen seropositivity, low-grade asymptomatic hepatitis, and chronic hepatic abnormalities in institutionalized Down's syndrome patients.[23,24] The reasons for the persistent antigenemia and progression of disease are unknown. Lymphocytic leukemia[25] and thyroid autoantibodies[21] are found in increased number in trisomy 21.

Neurologic Considerations

Mental retardation ranges from mild to profound. The development of epilepsy in adulthood and early senile dementia have been reported frequently.

Down's syndrome patients, who are normally docile and trusting, can become quite unruly when confronted with a strange environment such as the operating room. Postoperatively, the patients awaken in an agitated state and are quite difficult to console. They tend to require rather high doses of sedatives and narcotics to allay their restlessness.

Involving the parents in the induction and recovery phases should be encouraged. If anesthesia can be safely induced with the child sitting in the mother's lap, this technique is preferable. While in the postanesthesia care unit (PACU) the child who is comforted by his family may require much less sedation. The anesthesiologist should pay particular attention to gaining the child's trust during the preanesthetic visit.

Gastrointestinal Considerations

Major malformations of the gastrointestinal tract are common and often require surgical repair at a young age. Duodenal obstruction, imperforate anus, tracheoesophageal fistula, diaphragmatic hernia, and other abnormalities occur.

Gastroesophageal reflux with aspiration is also found in Down's syndrome and may contribute to the increased incidence of pulmonary infection.[26] If a history of reflux is elicited, a rapid sequence induction is warranted. Preoperative administration of cimetidine is indicated.

Blood Dyscrasias

The predisposition ot the child with Down's syndrome to have immunologic deficiencies and leukemia has already been mentioned.

Newborns with Down's syndrome often have an unexplained polycythemia (hematocrit >70%) with high blood viscosity and slow capillary circulation.[27] This may help to explain the peripheral cyanosis seen in the newborn with Down's syndrome. With polycythemia there is increased risk of thrombosis, and preoperative phlebotomy may be required. Complete blood count should be performed in all patients with Down's syndrome, irrespective of their age or the procedure to be performed.

Endocrine Considerations

Most children with Down's syndrome are euthyroid, despite their short, overweight appearance. Both hypothyroidism and hyperthyroidism appear to be more prevalent in adults with Down's syndrome. Hypothyroidism may be attributed to the late development of thyroid autoantibodies. If there is a history of weight change, lethargy, heat intolerance, or integument changes, determination of thyroid-stimulating hormone (TSH) and serum thyroxine (T4) levels and triiodothyronine (T3) and T3 resin uptake may prove helpful.

Patients of all ages with Down's syndrome have lower blood pressure than normal and mentally handicapped controls.[28] During anesthesia the heart rate and blood pressure are generally stable.[9] It has been thought that central and peripheral sympathetic activity is reduced in Down's syndrome, and several abnormalities in catecholamine metabolism have been found. There are decreased levels of resting and stress dopamine beta hydroxylase (converts dopamine to norepinephrine.[29] Excretion of epinephrine is decreased and this may signify decreased adrenal production even though plasma epinephrine levels are normal.[30]

Anesthetic Plan

The following areas must be considered in an anesthetic plan:

Premedication

Because of possibly decreased sympathetic activity, it is advisable to premedicate children with atropine. However, if the child is already excited, addition of belladonna alkaloids may worsen the situation. Atropine should be prepared to be given intravenously in small doses during induction and prior to the administration of succinylcholine if that drug is to be used. There is a known exaggerated mydriatic response suggesting altered pharmacodynamics, yet the cardiac acceleratory effects appear to be normal, and atropine premedication has been found to be safe.[31,32]

The response to sedatives in any patient with mental retardation can be quite unpredictable, and as mentioned, preoperative sedation must be used with caution in those with respiratory problems and sleep apnea. For the frightened and obstreperous patient, sedation can be given intramuscularly in the holding area under close supervision. Another approach is that of family involvement (see above).

Ambulatory Patients

The selection of patients for ambulatory surgery must be made carefully. Each organ system must be assessed. If patients are not in optimal medical condition as indicated by physical examination, further preoperative investigations and therapies are warranted, and admission to the hospital may become necessary. As in the case history presented, a vague history of cardiac disease, a recent respiratory infection, questionable sleep apnea, and physical signs indicating heart and pulmonary pathology warrant admission for identification and possible preoperative therapy.

Regional Versus General Anesthesia

With the potential for increased morbidity with general anesthesia, regional or local anesthesia may seem reasonable alternatives. However, the mental retardation, younger ages, and altered pharmacologic responses to sedatives and narcotics generally make Down's syndrome patients poor candidates for anything but general anesthesia.

Technique

Virtually all general anesthetic techniques and drugs are acceptable. For patients with CHD, the pathophysiology of the lesion should determine technique (see "Preanesthetic Assessment of the Child With Cardiac Lesions for Noncardiac Surgery," (Chapter 3).

With the suggestion that sympathetic activity is reduced in trisomy 21, it is possible that the minimum alveolar concentration (MAC) for inhalational agents is decreased. Hypotonia may suggest the need to avoid nondepolarizing muscle relaxants. It should be emphasized, however, that the pharmacology of anesthetic agents in Down's syndrome has not been systematically studied, and the true effects are not known.

Fluid administration is as usual. The propensity for pulmonary hypertension and pulmonary edema should be kept in mind in patients with cardiac disease. Monitoring must include ECG, pulse oximetry, capnography, temperature, and blood pressure.

Postoperative Care

Most complications are related to the airway and the pulmonary system, and as mentioned, monitoring for airway obstruction and respiratory depression is essential. Pulse oximetry in the PACU should be mandatory.

Use of an apnea monitor also may be indicated.

In summary, there are many pitfalls during anesthetic care of the patient with Down's syndrome. With the proper preparation and anesthetic management morbidity can be reduced to a minimum.

References

1. Hirschhorn K.: Prenatal disturbances, clinical abnormalities of the autosomes, in Behrman R.E., Vauhan V.C. III, Nelson W.E. (eds): Nelson Textbook of Pediatrics, 13th ed. Philadelphia: WB Saunders Co, 1987, pp 254–6.
2. Lejeune M.J., Gauthier M., Turpin M.R.: Les chromosomes humains en culture de tissus. *C R Seances Acad Sci (Paris)* 1959, 248:602–3.
3. Patterson D.: The causes of Down syndrome. *Sci Am* 1987, 257(2):52–60.
4. Pueschel S.M., Scola F.H.: Atlantoaxial instability in individuals with Down syndrome: Epidemiologic, radiographic, and clinical studies. *Pediatrics* 1987, 80(4): 555–60.
5. Southall D.P., Stebbens V.A., Mirza R., Lang M.H., Croft O.B., Shinebourne E.A.: Upper airway obstruction with hypoxaemia and sleep disruption in Down syndrome. *Dev Med Child Neurol* 1987, 29(6):734–42.
6. Charney E.B., Weller S.C., Sutton L.N., et al.: Management of the newborn with meningo-myelocele: Time for a decision making process. *Pediatrics* 1985, 75:58–64.
7. Carter C.O.: A life-table for mongols with the causes of death. *J Ment Defic Res* 1958, 2:64–74.
8. Baird P.A., Sadovnick A.D.: Causes of death to age 30 in Down syndrome. *Am J Hum Genet* 1988, 43(3):239–48.
9. Kobel M., Creighton R.E., Steward D.J.: Anaesthetic considerations in Down's syndrome: Experience with 100 patients and a review of the literature. *Can Anaesth Soc J* 1982, 29(6):593–8.
10. Beilin B., Kadari A., Shapira Y., Shulman D., Davidson J.T.: Anaesthetic considerations in facial reconstruction for Down's syndrome. *J R Soc Med* 1988, 81(1):23–6.
11. Samsoon G.L.T., Young J.R.B.: Difficult tracheal intubation: A retrospective study *Anaesthesia* 1987, 42:487–90.
12. Steward D.J.: Congenital abnormalities as a possible factor in the aetiology of post-intubation subglottic stenosis. *Can Anaesth Soc J* 1970, 17(4):388–90.
13. Millar W., Katz J.: Genetic and metabolic diseases, other hereditary disorders, in Katz J., Benumof J., Kadis L.B. (eds): Anesthesia and Uncommon Diseases. Philadelphia: WB Saunders Co, 1981, pp 70-1.
14. Williams J.P., Somerville G.M., Miner M.E., Reilly D.: Atlantoaxial subluxation and trisomy-21: Another perioperatlve complication. *Anesthesiology* 1987, 67(2):253–4.

15. Moore R.A., McNicholas K.W., Warran S.P.: Atlantoaxial subluxation with symptomatic cord compression in a child with Down's syndrome. *Anesth Analg* 1987, 66(1):89–90.
16. Hodgkinson M.A.: Anesthetic management of a parturient with severe juvenile rheumatoid arthritis. *Anesth Analg* 1981, 60(8):611–12.
17. Rowe R.D., Uchida I.A.: Cardiac malformation in mongolism: A prospective study of 184 mongoloid children. *Am J Med* 1961, 31:726–35.
18. Morray J.P., MacGillvray R., Duker G.: Increased perioperative risk following repair of congenital heart disease in Down's syndrome. *Anesthesiology* 1986, 65(2):221–4.
19. Prevention of bacterial endocarditis. *Med Lett Drugs Ther* 1987, 29(754): 109–10.
20. Kaplan S.: The adolescent with operated or unoperated congenital heart disease. *Postgrad Med* 1974, 56:147.
21. Tait A.R., Knight P.R.: Anesthesia and upper respiratory viral infections: A prospective cohort study. *Anesthesiology* 1985, 63(3A):A526.
22. Ablin R. I.: Immunity in Down's syndrome. *Eur J Pediatr* 1978, 127:149–52.
23. Skinhoj P., Dietrichson O., Dyggue H., Mikkelsen M., Peterson P., Stene J.: Hepatitis and hepatitis associated antigen (HAA) in Down's syndrome. *J Ment Defic Res* 1971, 15:236–43.
24. Seeff L.B., Levitsky J., Tillman P.W., Perou M.L., Simmerman H.J.: Histopathology of the liver in Down's syndrome. *Am J Dig Dis* 1967, 12:1102–13.
25. Levin S., Schlesinger M., Handzel Z., Hahn T., Altman Y., Czernobilsky B., Boss J.: Thymic deficiency in Down's syndrome. *Pediatrics* 1979, 63(1):80–7.
26. Weesner K.M., Rosenthal A.: Gastroesophageal reflux in association with congenital heart disease. *Clin Pediatr (Phila)* 1987, 22:424–6.
27. Lappalainen J., Kouvalainen K.: High hematocrits in newborns with Down's syndrome. *Clin Pediatr (Phila)* 1972, 11(8):472–4.
28. Richards B.W., Enver F.: Blood pressure in Down's syndrome. *J Ment Defic Res* 1979, 23:123–35.
29. Coleman M., Campbell M., Freedman L.S., Roffman M., Ebstein R.P., Goldstein M.: Serum dopamine beta hydroxylase levels in Down's syndrome. *Clin Genet* 1974, 312–15.
30. Keele D.K., Richards C., Brown J., Marshall J.: Catecholamine metabolism in Down's syndrome. *Am J Ment Defic* 1969, 74:125–9.
31. Mir G.H., Cumming G.R.: Response to atropine in Down's syndrome. *Arch Dis Child* 1971, 46:61–5.
32. Wark H.I., Overton J.H., Marian P.: The safety of atropine premedication in children with Downs syndrome. *Anaesthesia* 1983, 38:871–4.
33. Eger E.I. III: MAC, in Eger E.I. II (ed): Anesthetic uptake and action. Baltimore: Williams and Wilkins, 1974, pp 1–23.

Chapter 7

The Patient Who Is A Jehovah's Witness

Jon D. Samuels, M.D.

Case History. *A 39-year-old Hispanic woman, a Jehovah's Witness, was scheduled for a transplant nephrectomy and insertion of Hemo Cath®. Eleven days earlier she had received a left kidney transplant from a donor (her brother); acute rejection followed. Removal of the necrosed donor kidney and vascular access for hemodialysis were planned.*

The patient had a background of hypertension since age 15, with renal insufficiency starting about 5 years later and culminating in end-stage renal failure at 31 years of age. She had initially been hemodialyzed; but as peripheral access sites (right and left arteriovenous shunts and fistulas), as well as central access site (Hickman catheter), became thrombosed, she had received peritoneal dialysis. Six months earlier the peritoneal catheter had been removed secondary to peritonitis and a Shiley®central venous catheter had been placed.

Other past history included hypothyroidism, pericardial window placement for uremic pericarditis, and parathyroidectomy for secondary hyperparathyroidism. The patient denied other systemic illnesses. She did not smoke, drink alcohol, or abuse drugs. She reported "allergies" to vancomycin (palpitations), pork heparin (pruritus), and morphine sulfate (abdominal cramps). Medications included antilymphocyte globulin, levothyroxine sodium (Synthroid®), basic aluminum carbonate gel (Basaljel®), hydrocortisone succinate, iron sulfate, multivitamins, and docusate sodium (Colace®).

Physical examination revealed a tired-appearing, obese woman in no apparent distress, alert, oriented, and anxious; height 150cm, weight 78kg. Vital signs: temperature 37.6°C, respiratory rate 20/min, pulse 90/min, and blood pressure 95/60mmHg. A double-lumen Shiley catheter was in place in the right internal jugular vein. Both upper extremities

---

Reviewed by Dr. C. Bryan-Brown, Professor of Anesthesiology, Albert Einstein College of Medicine/Montefiore Medical Center.

had prominent dilations consistent with arteriovenous fistulas and grafts. There were no external signs of infection.

Laboratory examination yielded the following data: hemoglobin 3.3gm/dl, hematocrit 10%, white blood cell count 2300/mm^3, platelets 135,000/mm^3. SMA-12 was normal except for bicarbonate 19, BUN 93, creatinine 15.2, Ca^{2+} 8.0. Coagulation study results were within normal limits.

Preoperatively, the patient signed a form documenting her refusal to accept blood or blood products.

Introduction

In the 1870s Pastor Charles Taze Russell organized an apocalyptic sect, later known as Jehovah's Witnesses (JW).[1] Since then the group has had considerable growth; worldwide, the 1988 figures showed over 3 million members, of whom 771,000 were living in the United States.[2] The administrative headquarters of JW, the Watchtower Bible and Tract Society (WBTS), is located at 25 Columbia Heights, Brooklyn, NY. Jehovah's Witnesses attempt to win converts and to educate society, via a magazine *(The Watchtower)*, regarding the impending Battle of Armageddon. They stand apart from all forms of civil society.[1] All of their teachings are based on the Bible which they regard as literally true. Considering their large numbers, it is in the interest of anesthesiologists to become acquainted with JW, their origins and beliefs, and the legal controversies surrounding them, so that a rational approach can be developed for their anesthetic management.

Refusal of Blood Products

Jehovah's Witnesses refuse transfusion because of their interpretation of the Old Testament Book of Leviticus 7:27: "And I will turn my face against anyone, whether an Israelite or a foreigner living among you, who eats blood in any form. I will excommunicate him from his people." (See also Lev. 17:14, Acts 15:29.) An article in *The Watchtower*, July 1, 1945, forbade acceptance of blood transfusions under penalty of loss of life in God's Kingdom.[1] This doctrine places JW apart from other religious groups, potentially jeopardizing the life of the believer. It may also create an adversarial relationship between the JW and the physician, for whom transfusion, when indicated, is an accepted practice.

Jehovah's Witnesses refuse whole blood, packed erythrocytes, plasma, platelets, and bone marrow transplants. Some JW accept albumin antihemophilic preparations (eg, cryoprecipitate), and immune serum globulin; others, however, noting that these substances are derived from human blood, do not. Jehovah's Witnesses accept plasma volume expanders, such as crystalloid, hetastarch (hydroxyethyl starch), and dextran. They also accept organ transplants.

The WBTS has published a policy statement titled "Jehovah's Witnesses and the Question of Blood." Individual members vary in their beliefs, however, and peer pressure may be a factor. When JW responded to a confidential questionnaire, some members were willing to accept plasma and even autotransfusion.[4] Most JW will accept autotransfusion only if the blood removed remains in physical continuity with their body.[1] Again, some believers, fearful that such a system may fail, refuse to have even their own blood returned to them.

Legal Issues

The anesthesiologist caring for a JW is faced with a series of ethical, moral, and legal dilemmas.[5–7] Legal precedents are confusing; current policy is dictated more by judicial opinions than by written law. Difficult issues should be directed to hospital legal counsel and risk management committees. If the beliefs of the physician do not allow him or her to respect the wishes of the patient until death, that physician should not enter into an agreement to provide medical services. Rather, the patient should be referred to the WBTS, where a list of compliant health care providers is maintained. In an emergency situation or in isolated areas such an option may not be available.

Also, if a physician contracts to provide services to a patient, refusal to do so then constitutes abandonment. Once a physician accepts a JW as a patient, specific beliefs should be determined (Figure 1).

Most JW readily sign the American Medical Association form relieving physicians and hospitals of liability, and many carry a Medical Alert band.[5] A properly signed and dated "Refusal to Accept Blood Products" form is a contractual agreement and is legally binding; violating such an agreement renders the physician liable to a charge of battery in most states.[8]

To date there have been very few settled claims in which plaintiffs suing physicians were awarded large amounts of money. Perhaps because JW do not believe that financial recompense will right the alleged transgression (of a spiritual nature), they have not pursued a remunerative course. Statistically, therefore, litigation is more a threat than a reality, but it must be recognized that in the case of a poor outcome the physician may be sued whether or not he or she transfused a JW.

The few cases in which money has been awarded usually relate to too little blood given too late. Judgment is based on violation of the "lifesaver theory," which holds that once a procedure is started, it must

REFUSAL TO CONSENT

1. I have been advised by Dr. SAMUELS that the following treatment should be given to me/the above named patient (please print or type): TRANSFUSION OF BLOOD PRODUCTS

2. Dr. SAMUELS has fully explained to me the nature and purpose of the proposed treatment, the possible alternatives thereto and the risks and consequences of not proceeding. I nonetheless refuse to consent to the proposed treatment.

3. I have been given an opportunity to ask questions, and all my questions have been answered fully and satisfactorily.

4. I hereby release Montefiore Hospital and Medical Center and its employees, students and medical staff from any liability for ill effects which may result from failure to perform the proposed treatment.

5. I confirm that I have read and fully understand the above and all the blank spaces have been completed prior to my signing.

Interpreter if required ______ SIGNATURE ______ PRINT NAME

Patient/Relative or Guardian* x Jane Doe SIGNATURE
JANE DOE PRINT NAME

Witness E Brown SIGNATURE (NURSE IN CHARGE)
ELIZABETH BROWN PRINT NAME

RELATIONSHIP IF SIGNED BY PERSON OTHER THAN PATIENT
OCTOBER 1st 1989 DATE

I hereby certify that I have explained the nature and purposes of, and alternatives to, the proposed treatment and the risks and consequences of not proceeding, have offered to answer any questions and have fully answered such questions. I believe that the patient/relative/guardian fully understands what I have explained and answered.

Physician Jon Samuels SIGNATURE JON SAMUELS PRINT NAME Oct 1st 1989 DATE

REMARKS:

*The signature of the patient must be obtained unless the patient is an unemancipated minor under the age of 18 or is otherwise incompetent to sign. IF A REFUSAL TO CONSENT MAY RESULT IN THE DEATH OF OR SERIOUS INJURY TO AN UNEMANCIPATED MINOR, A LEGAL OPINION SHOULD BE IMMEDIATELY SOUGHT.

FIGURE 1. A properly drawn up and signed consent form should be placed in the chart preoperatively.

be continued in a non-negligent fashion. In Randolph v. Health & Hospital Corporation, a large settlement was paid to the family of a JW woman who died of exsanguination during a cesarean section. The patient had directed that blood not be given to her. However, as she began to exsanguinate, the anesthesiologist slowly replaced two units, which were insufficient to reverse the fatal outcome. Legal opinion on this controversial subject is evolving.[9]

Patient Categories

Management decisions relating to transfusion will be influenced by the JW patient's age, sex, and competency status, and whether the patient supports a minor.

Competent adult: The patient is usually given the freedom to choose his or her own course of treatment. At the hospital of the Albert Einstein College of Medicine, Bronx, NY, a draft policy on blood transfusions and JW states: "Any adult patient who is not incapacitated has the right to refuse treatment no matter how detrimental such a refusal may be to his health."

The courts usually do not order transfusions for competent adults with no dependents, stressing the supremacy of patient choice. The New York State Court of Appeals stated that "the patient's right to determine the course of his own treatment [is] paramount.... [a] doctor cannot be held to have violated his legal or professional responsibilities when he honors the right of a competent adult to decide medical treatment."[6] However, it has been argued that general anesthesia, by rendering a patient unconscious, also makes the patient incompetent.

Incompetent adult: Patients who are unconscious (eg, trauma) or partially or totally mentally incompetent fall into this category. Relatives do not have the authority to block lifesaving therapeutic modalities. Elective procedures should be postponed, legal counsel involved, and court decision awaited.[9]

Minors: Children of parents who refuse to authorize transfusions are the cases that generate the most publicity. An in-depth review of the subject may be found in the "Guides to the Judge in Medical Orders Affecting Children" by the Council of Judges, National Council on Crime and Delinquency, United States Department of Justice. Often legal counsel petitions the case, the minor becomes a temporary ward of the court, and transfusion is authorized.

Competent pregnant adult: Although this is still a controversial area, there is a body of case law, fetal-versus-maternal law, making the mother's right of refusal subordinate to the viable fetus's right to live.[10]

Competent pregnant adult peripartum: The high frequency of obstetrical emergencies, the unpredictability of blood loss during delivery, and the potential for sudden massive hemorrhage in certain obstetrical conditions (eg, placenta previa, abruptio placentae) make this a category in which the policy should be decided early in the gestation. There are case reports in which a gravida has received a court order to accept transfusion.[8] In the postpartum period legal opinion is often influenced by whether the patient is the sole supporting adult. Also, court orders to transfuse in emergency situations to save the life of an adult who is the sole support of a minor have been sought and quickly obtained.

Preanesthetic Assessment

When an anesthesiologist assumes care of a JW, attitude is crucial. One should be nonjudgmental and prepared to listen. One cannot reasonably expect the patient's deeply felt religious beliefs to change; one should strive rather to work with the patient. Jehovah's Witness patients are keenly afraid of having their religious beliefs violated, and an understanding of this fear is warranted. However, should the physician feel unwilling or unable to care for the JW under the circumstances, the case should be transferred to a colleague.

Ideally, the initial preoperative evaluation should be conducted on the night before surgery, at the patient's bedside. It is better to conduct the interview one-to-one, because family members may increase pressure on the patient to uphold religious beliefs. A key issue is the patient's general condition in view of the operation planned. For the otherwise healthy patient (ASA I or II) scheduled for a superficial procedure (eg, facial rhytidectomy, inguinal herniorrhaphy) with estimated blood loss (EBL) <100ml, refusal to accept blood products may be purely academic.

For the moderately sick patient (ASA III) scheduled for a more invasive procedure (eg, cardiac patients with gastric disease, modified radical mastectomy) with EBL of 500–1000ml, these concerns become real. Of course, in a sick patient (ASA III or greater), as in our case presentation, or any patient scheduled for major surgery (eg, abdominal perineal resection, carotid endarterectomy), difficulties should be anticipated and options prepared.

The anesthesiologist should make a global assessment of the patient's condition and decide how much blood loss the patient will be able to tolerate perioperatively. A healthy young adult can tolerate an acute phlebotomy of 20%–30% of blood volume to a hemoglobin of 8gm/dl if adequate replacement with crystalloid is allowed. Patients with end-stage renal failure commonly have chronic severe anemias with baseline hemoglobins of 7gm/dl or less (as did our patient).

On the other hand, a 70-year-old man with symptomatic heart disease would tolerate about a 10% decrease in blood volume (to a hemoglobin of 10gm/dl). Conditions in which acute blood loss is likely to be poorly tolerated include symptomatic atherosclerotic heart disease, cardiac valvular disease (especially aortic stenosis), cerebrovascular occlusive disease, peripheral vascular disease, β blockade, and use of antihypertensive medication.

Volume Maintenance Options

The anesthesiologist caring for a JW should use every means available to limit blood loss. Consultation with the surgeon to discuss options and to select the procedure with the least expected blood loss is appropriate. It may be feasible to use a simplified surgical technique,[11] staging surgery, in addition to an expeditious technique with meticulous hemostasis.[7] Pre- and postoperatively, the patient should be placed on iron sulfate and multivitamin supplementation.

Several questions must be answered. Will the patient consent to transfusion in the event of an emergency? Occasionally, JW will tacitly acknowledge that should life be threatened and they be unable to make a decision, the physicians caring for them should do all that is possible to maintain life. If this is the case, a blood specimen should be sent to the bank and the availability of blood confirmed before commencing surgery.

The patient will require perioperative intravascular volume with a balanced-salt solution (eg, lactated Ringer's Normosol®) as crystalloid and a synthetic colloid (hetastarch, dextran). Large volumes of hetastarch, however, may precipitate a generalized coagulopathy with shortened thrombin time,[12] prolonged prothrombin (PT) and activated partial thromboplastin times (PTT), and acquired von Willebrand syndrome.[13] Dilution of coagulation may also contribute to a bleeding diathesis.

Autotransfusion is a possibility that should be discussed before surgery. A simplified version of the Cell Saver®system for autotransfusion has been designed, in which shed blood from the surgical field is concentrated, washed, and collected in 225ml bowls with a hematocrit of 60%[14] (see Figure 2.)

Most JW will accept autotransfusion, provided that there is a continuous flow of blood from suction tubing to collection bowl. The cost is similar to that of banked blood only if more than two units are transfused. Apart from the financial consideration, other problems of this technique include the need for additional personnel, including a pump technician.

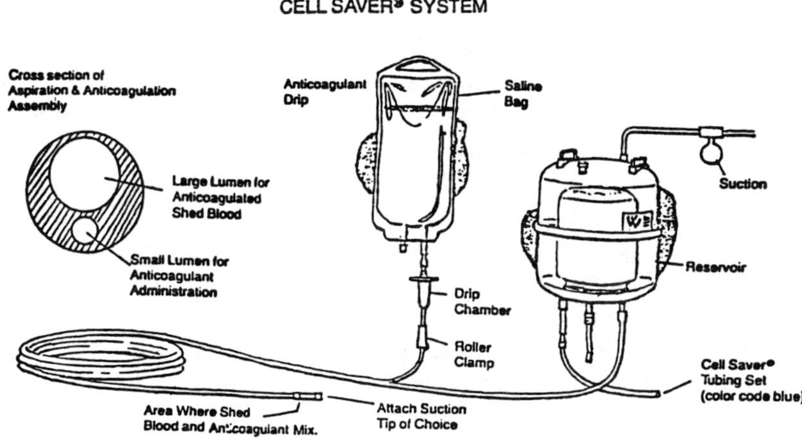

FIGURE 2. Cell Saver®System — Aspiration and Anticoagulation Assembly.

Normovolemic hemodilution anesthesia is another option, although not widely used at present. It was originally designed for major procedures in JW and has since been used for cardiac surgery and hepatic resection in children.[15,16] Whole blood is removed from the patient and simultaneously replaced with lactated Ringer's solution, 3ml per milliliter of blood loss. The blood is collected in citrate-phosphate-dextrose bags and stored at room temperature. Hypotensive anesthesia, which lowers myocardial workload and further reduces blood loss, can be achieved with inhalational agents. At the end of the procedure the blood is returned to the patient. Fluid overload caused by excessive infusion of crystalloid is prevented by administration of furosemide.

Pharmacologic substitutes for transfusion and adjuncts are available.[17] Some examples follow:

Desmopressin acetate (1-desamino-8-D-arginine vasopressin, DDAVP) is a synthetic vasopressin (antidiuretic hormone) analog. It acts by transiently increasing factor VIII and von Willebrand factor and increasing platelet aggregation.[18] The onset of action is within 30 min; maximum effect is seen in 90 min. Beneficial effect may last for several hours. It is indicated if bleeding times are prolonged, as in various thrombocytopenias (eg, as occurs in uremia), hemophilia A, type I von Willebrand's disease.

Epsilon-aminocaproic acid (EACA) is a synthetic lysine analog. It is an antifibrinolytic agent or clot stabilizer that inhibits the formation of plasmin by blocking the binding of plasminogen to fibrin. Without plasmin, peptide bonds cannot be formed in fibrin to release fibrin split products (FSP). It is used to help control severe bleeding in postpartum hemorrhage (eg, placenta accreta) and to preserve clot formation after subarachnoid hemorrhage. However, because of a high incidence of pulmonary embolism, this therapy is seldom used now.

Immune serum globulin (ISG) transiently coats reticuloendothelial sequestration sites in the spleen and elsewhere. It is, however, extremely expensive. It is used in the treatment of idiopathic thrombocytopenic purpura (ITP). Although ISG is a "blood product," some JW will accept its use.

Danazol is a synthetic androgen with slight virilizing activity. It decreases iron receptors, an action that may reduce sequestration of IgG-coated platelets by the reticuloendothelial system. However, several months may be required for response. Its principal indication is in the treatment of refractory ITP.

Conjugated estrogens, given daily for 5 days, may have an effect that lasts 2 weeks. Von Willebrand factor production is stimulated by unknown mechanisms. Although the bleeding time in uremia is shortened, therapy is not very effective.

Aprotinin is a serine protease inhibitor that inhibits plasmin, reduces fibrinolysis, and stabilizes platelet membranes. It is used mainly during open heart surgery.

Recombinant human erythropoietin stimulates erythropoiesis. Effects may not be seen for up to 6 weeks, and hypertension may be a complication. The principal indications are uremia of chronic renal failure and for patients undergoing chemotherapy who refuse blood products.[19]

20% fluosol-DA is a synthetic oxygen-carrying perfluorochemical emulsion composed of perfluorodecalin and perfluorotripropylamine. To transport oxygen in vivo, the alveolar partial pressure of oxygen must exceed 300mmHg. Thus, this therapy is effective only during general anesthesia and endotracheal intubation. One review indicates successful use of 20% fluosol-DA in JW, especially as a last resort in posthemorrhagic anoxic encephalopathy patients.[20] It is also effective in the treatment of short-term anemia. This treatment is not currently available in the United States because of reported anaphylactic reactions.

Polymerized, pyridoxylated stroma-free hemoglobin is prepared from outdated blood, polymerized, and covalently linked with pyridoxal-S-phosphate (similar to 2,3-diphosphoglycerate). Investigational use of this agent has indicated that it shifts the oxyhemoglobin dissociation curve to the right, making oxygen more available to the tissues. This is a highly experimental technique that JW probably would not accept.

Complications of Anemia

Oxygen transport is critical to maintenance of adequate oxygen delivery. Extraction ratio or utilization coefficient relates the quantitative supply of oxygen to consumption. Much work has centered on the relationship between optimal (or critical) hemoglobin content, cardiac output, heart rate, stroke volume, and blood viscosity. At a hematocrit of 45%, optimal transport of oxygen occurs only at high flow rates. Because blood is a non-newtonian fluid, as the rate of flow decreases, the blood becomes more viscous and oxygen transport capacity decreases.

With falling hematocrit and constant blood volume, oxygen transport capacity is maintained stable and even increased by three mechanisms: (1) rise in flow rate, (2) increased oxygen extraction, and (3) reduction of hemoglobin-oxygen affinity by a shift of the oxygen dissociation curve to the right.[20,21]

Limited hemodilution in dogs has shown that oxygen transport capacity rises until the hematocrit is less than 30%.[22,23] At that point oxygen transport capacity decreases, reaching the norm at a hematocrit of 20% (Figure 3).

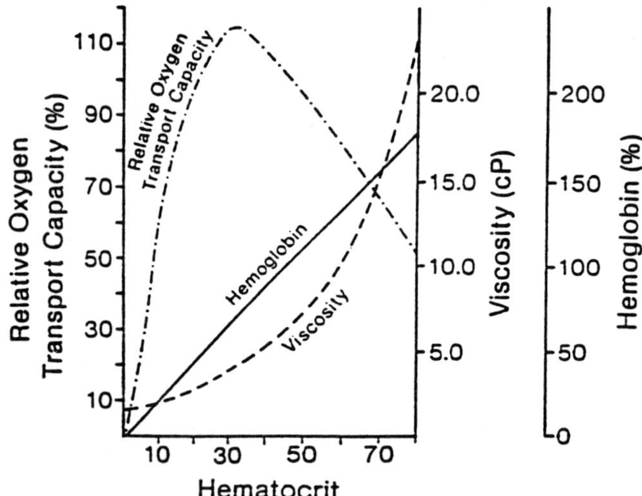

FIGURE 3. Relative oxygen transport capacity increases as hematocrit decreases to about 30–33% despite reduction in hemoglobin. As the hematocrit is reduced below 30%, relative oxygen transport capacity falls. cP=centipoise.

Modified from Hint H: The pharmacology of dextran and the physiological background for the clinical use of Rheomacrodex and Macrodex. Acta Anaesthesiol Belg 19; 119–138, 1968.)

Several studies have indicated that with intact cardiorespiratory compensatory mechanisms, anesthesia and surgery can be performed safely at hematocrit values of 25%–30%.[20,23]

A recent study of 113 operations in 107 JW patients undergoing major elective surgery concluded that mortality depends more on estimated blood loss than on preoperative hemoglobin levels. Also elective surgery may be safely performed in patients with preoperative hemoglobin levels as low as 6g/dl if the blood loss is below 500 ml.[24]

Our patient was severely anemic and might have been considered not a candidate for surgery. However, major surgery, including cardiac surgery, has been successfully performed without blood replacement.[25] The lowest recorded hemoglobin level associated with survival is 1.8gm/dl.[26] In that case the patient required ventilatory support in an intensive care setting for 3 months. He was transfused with a gelatin substance.

Low hemoglobin levels affect most body systems, as described in the following sections.

Cardiovascular System

Severe anemia is associated with high output failure. As oxygen consumption increases, systemic vascular resistance decreases. Finally, the coronary arteries fail to perfuse adequately, and myocardial ischemia results. Early ECG changes include tachycardia.

If ST and T wave changes are present, delay of surgery and administration of iron supplements or other pharmacologic agents (see above) are warranted. Emergency situations may call for supplemental oxygen during the entire perioperative period and even nitroglycerin paste may be needed.

The assessment of paleness may be very difficult in nonwhite people, and reliance should be placed more on laboratory evaluation. Also, cyanosis is present only when 5gm/dl of deoxyhemoglobin is present in the blood and therefore, despite severe respiratory difficulties, may never occur in the anemic patient.

In the cardiac assessment of the anemic patient, stress tests (including chemical stress tests) are not indicated. Muscles are easily fatigued, and as lactic acid increases after exercise, tachycardia results, causing hyperventilation and general malaise. Echocardiography may be of value in assessing previous heart disease (eg, mitral valve prolapse), which may further compromise myocardial function.

Respiratory System

As mentioned, anemia, by contributing to muscle fatigability, increases lactic acid production and causes hyperventilation. Physiologic shunting in the lungs is generally decreased, and respiratory reserve is diminished. Phasic respiratory cycling of the arterial blood pressure is commonly seen in hypovolemic patients. Again, preoperative pulmonary function tests may not be indicated if undue stress is provoked. If the patient is severely anemic, care must be taken to ensure that minimal (less than 1ml) blood is removed for blood gas analyses and that adequate pressure is applied to the punctured artery to prevent hematoma formation.

Gastrointestinal System

Because of its two circulations, the liver is relatively resistant to the effects of anemia. The intestines, on the other hand, are sensitive to hypoperfusion, and mesenteric plaques may form, causing infarction. The syndrome of intestinal angina results, and abdominal pain may add to the diagnostic dilemma.

Renal System

The kidneys are also resistant to the effects of anemia. However, in severe cases, especially if hypotension occurs, acute tubular necrosis may develop. Furosemide, which acts on the ascending loop of Henle, requires the oxygen ATP system to be effective. Thus, the anemic patient may require a higher dosage.

Central Nervous System

Anemia causes cerebral vasodilation and increases intracranial pressure. In patients with cerebrovascular disease, severe anemia may result in stroke.

Integument

Wound healing is decreased by anemia. Infection, however, is usually not problematic unless it occurs in patients with end-stage renal disease who are receiving steroids.

Pressure sores are easily acquired; thus, patient positioning and adequate padding must be undertaken very carefully.

Anesthetic Plan

Anesthetic considerations for the JW patient both in the operating and emergency rooms have been reviewed.[27,28]

Anemia per se does not appear to alter the metabolism of anesthetic drugs. However, inhalation agents will decrease cardiac output in a hypovolemic patient.

If feasible, subarachnoid or epidural block is a reasonable alternative. Central venous pressure should be monitored and maintained as close to 6–12cmH_2O as possible by crystalloid infusion. The decision to use more-invasive monitoring should be weighed against the possibility of further blood loss (eg, cannulation of the radial artery may result in loss of 5–10ml blood). In our case, given a hemoglobin of 3gm/dl and an estimated blood volume of 5L, a 30ml blood loss would reduce the hemoglobin by a further 0.1gm/dl. An additional argument for the use of regional block is that systemic vascular resistance is already very low because of peripheral arteriovenous shunting and thus will probably not decrease much further.

At very low hematocrits, the dissolved oxygen fraction becomes critical, and thus a 100% inspired oxygen concentration should be used.

The patient described in the case history received a subarachnoid block using 8mg tetracaine. Her kidney was removed in less than 1 hour. Vital signs were stable throughout. Postoperatively, hematocrit decreased from 10% to 8%. After 10 weeks of therapy with erythropoietin, the hematocrit rose to 19%. No blood products were given.

References

1. Jehovah's Witnesses and the Question of Blood. Watchtower Bible and Tract Society. New York, 1977, pp 1–35.
2. The Holy Bible. Leviticus 17:14; Acts 15:28–9.
3. Harris T.J.B., Parikh N.R., Rad Y.K., Simpson J.C., Oliver R.N.P.: Exsanguination in a Jehovah's Witness. *Anesthesia* 1983, 38:989–94.
4. Findley L.J., Redstone P.M.: Blood Transfusion in Adult Jehovah's Witness. *Arch Intern Med* 1982, 142:606–9.
5. Rothenberg D.M.: The approach to the Jehovah's Witness Patient, in Speiss, B.D. (ed): *Anesthesiology Clinics of North America*. Philadelphia: WB Saunders Co., 1990, pp. 589–607.
6. Blood — Whose Choice and Whose Conscience? *New York State J of Med* 1988, 8(9):463–94.
7. Thomas J.M.: Meeting the surgical and ethical challenge presented by Jehovah's Witnesses. *Canad Med Assoc J* 1983, 128:1153–4.
8. Sacks D.A., Kopper R.H.: Blood transfusion and Jehovah's Witnesses: Medial and legal issues in Obstetrics and Gynecology. *AM J Obstet Gynecol* 1986, 154(3):483–6.
9. Doris J.J.: Compulsory medical treatment and religious freedom: Whose law shall prevail? University of San Francisco Law Review, Vol. I, No. L, 1975.
10. The fetal patient and the unwilling mother: A Standard for Judicial Intervention. *Pacific Law J* 1983, 14:1065–85.
11. Papaioannov A.N.: Abdominal perineal resection of the rectum, preliminary experience with a simplified technique. *An J Surg* 1969, 118:417–21.
12. Howell P.J., Bamber P.A.: Severe Acute Anemia in a Jehovah's Witness. Survival Without Transfusion. *Anaesthesia* 1987, 42(1):44–8.
13. Lockwood D.N., Bullen C., Machin S.J.: A severe coagulopathy following volume replacement with hydroxyethyl starch in a Jehovah's Witness. *Anaesthesia* 1988, 43(5):391–3.
14. Kumar B.: The Haemonetic Cell Saver (letter). *Anaesthesia* 1986, 41(7): 774–5.
15. Schaller R.T., Schaller J., Furmon B.: The Advantages of Hemodilution Anesthesia for Major Liver Resection in Children. *J Ped Surg* 1984, 19(6): 705–10.
16. Martin E., Ott E.: Extreme hemodilution in the Harrington procedure, in Schmid-Schonbein H., Messmer K., Rieger H. (eds): Hemodilution and Flow Improvement. *Biblithca Haemat* 1981, 47:322–7.
17. Transfusion Medicine Topic Update. Alternatives to blood transfusion. Dept. of Laboratory Medicine. Yale University School of Medicine and American Red Cross. Vol 2, No. 2, May 1989, pp 1–35.
18. Stone D.J., Difazio C.S.: DDAVP to reduce blood loss in a Jehovah's witness (letter). *Anesthesiology* 1988, 69(6):1028.
19. Heinz R., Reisner R., Pittermann E.: Erythropoietin for chemotherapy

patient refusing blood transfusion. *Lancet* 1990, 335(8688):542–3.
20. Tremper K.K., Levine E., Friedman A., Shoemaker W.C., Katz R.: The preoperative treatment of severely anemic patients with a perfluorochemical blood substitute, Fluosol DA 20%, after massive postpartum hemorrhage. *Obstet Gynecol* 1985, 65:127–36.
21. Martin E., Hansen E., Peter K.: Acute limited normovolemic hemodilution: A method of avoiding homologous transfusion. *Wld J Surg* 1987, 11:53–9.
22. Sunder-Plassman L., Klovekorn W.P., Messmer K.: Hemodynamic and rheological changes induced by hemodilution with colloids, in Messmer K., Schmid-Schonbein H. (eds): Hemodilution; Theoretical basis and Clinical Application. Basel, S. Karger, 1976, 124–30.
23. Messmer K.: Hemodilution. *Surg Clin N Amer* 1975, 55:659–78.
24. Spence R.K., Carson J.A., Poses R., et al: Elective surgery without transfusion: Influence of preoperative hemoglobin level and blood loss on mortality. *Am J Surg* 1990, 159(3):320–4.
25. Henderson A.M., Moryniak J.K., Simpson J.C., et al: Cardiac surgery in JW. A review of 36 cases. *Anaesthesia* 1986, 41(7):784–53.
26. Newman H.S., Eikhauser M.L.: Postoperative management of a severely anemic Jehovah's Witness. *Crit Care Med* 1975, 55:142–3.
27. Benson K.T.: The Jehovah's Witness patient: Considerations for the anesthesiologist. *Anesth Analg* 1989, 69(5):647–56.
28. Fontanarosa P.B., Giorgio G.T.: The role of the emergency physician in the management of Jehovah's Witnesses. *Ann Emerg Med* 1989, 18(10):1089–95.

Chapter 8

The Patient With Wilson's Disease

Elizabeth A.M. Frost, M.D.

Case History. *A 21-year-old man came to the emergency room complaining of weakness and tremor. He reported that he had recently developed night sweats and nausea and vomiting, and that his urine was dark. There was no history of drug or alcohol abuse, and the patient denied homosexual contact. On several occasions in the past, he had been jaundiced but had not sought medical evaluation.*

Significant findings on physical examination included bilateral Kayser-Fleischer rings, hepatosplenomegaly, bilateral gynecomastia, and increased muscle tone with tremors, dysarthria, and inversion deformity of both feet. Laboratory findings were as follows: hematocrit was 41%, white blood cell count 3500/mm^3, platelets 39,400/mm^3. Liver function tests were normal, except SGOT 77 IU; SGPT 57 IU; albumin 2.6gm/dl; prothrombin time 16.1/11.3 sec. Serum ceruloplasmin was 10mg/dl (normal 20–43mg/dl); serum copper 34μg/dl (normal 81–147μg/dl); 24h copper excretion 530μg (normal 15–30μg/24h).

CT scan of the abdomen confirmed hepatosplenomegaly, and CT scan of the head demonstrated cerebral atrophy and focal atrophy of the basal ganglia. Serologic tests for hepatitis A and B were negative. A diagnosis of Wilson's disease was tentatively applied.

The patient was scheduled for hepatic biopsy. Because of his extreme anxiety and difficulty in controlling movements, anesthetic consultation and intraoperative assistance were requested.

Introduction

Wilson's disease, or hepatolenticular degeneration, is a disorder of copper metabolism, mainly affecting storage and resulting in excessive amounts of copper in several organs.

Reviewed by Dr. Steven S. Schwalbe, Assistant Professor of Anesthesiology and Director of Obstetric Anesthesia, Albert Einstein College of Medicine/Montefiore Medical Center.

Inherited as an autosomal recessive disorder, it has been attributed to a defective gene on chromosome 13.[1] The disease affects men and women equally. Approximately 1 in 200 persons is heterozygous for Wilson's disease. The incidence of the disorder is approximately 3 per 100,000 births.[2]

Although the genetic defect is present from birth, clinically manifest pathologic changes rarely appear before the age of 4 years and are sometimes delayed until the fifth decade. Fifty percent of all patients display symptoms or signs of the disease before reaching 15 years of age.

The disease is a multisystem disorder rather than simply hepatolenticular dysfunction. Onset may be insidious or sudden and may mimic acute hepatitis, infectious mononucleosis, idiopathic thrombocytopenic purpura, multiple sclerosis, parkinsonism, encephalitis, and acute gomerulonephritis among many other disorders. Thus, preanesthetic assessment requires awareness not only of the disease process but of its effect on all body systems.

Historical Background

Wilson's disease was first described by a neurologist in the late 1800s as a type of "tetanoid chorea."[3] An association with liver disease was recognized. Wilson published the first complete description of the disorder in 1912.[4] Although the etiology was unclear, he concluded that the disease "must be due to a toxin."[4] Within a year, evidence pointed to copper as the etiologic agent. However, it was not until the mid-1940s that definitive evidence confirmed this early hypothesis.

Pathophysiology

Copper, an inorganic nutrient that is essential for human life, is required in minute quantities. The normal adult body contains 100mg of copper. One third of the dietary requirement is absorbed from the gut. In this process copper is attached to a metal-binding protein (metallothionein) contained in the intestinal mucosa. This complex is transported across the intestines to the blood. Albumin strongly binds copper. The metal is also bound to various peptides and amino acids (especially histidine and threonine). The latter complexes are important in that, although small as a percentage of plasma copper, they are in enchange equilibrium with similarly bound copper and tissue copper complexes.[5]

Transported copper is metabolized mainly in the liver by incorporation into ceruloplasmin (globulin glycoprotein), in which 90% of plasma copper exists in the metalloprotein form. The liver excretes copper into bile, the main route for copper excretion. Tissue copper proteins in which copper is structurally and functionally integrated include cytochrome-c oxidase, superoxide dismutase, monoamine oxidase, and tyrosinase. Ceruloplasmin facilitates the conversion of ferrous to ferric iron and the transfer of electrons from various substrates to molecular oxygen, a critical factor in the normal biochemistry of the central nervous system.

In Wilson's disease there is an accumulation of copper, especially in the liver and brain. The rate of hepatic ceruloplasmin synthesis is diminished, apparently by blockade of incorporation of copper into the molecule perhaps because of an intrahepatic cell defect in copper metabolism.[2] There is also an inhibition of biliary excretion of copper, which may be associated with generalized damage to the liver and the subsequent development of cirrhosis. The exact mechanism is unknown.[6]

The amount of copper bound to albumin is increased. This copper is easily dissociated and excreted by the kidneys, accounting for the hypercupriuria associated with the disease. Because the reduction in ceruloplasmin is greater than the increase in albumin-bound copper, the total serum copper concentration is usually reduced.

Although it was thought that positive copper balance was the result of increased intestinal absorption, a study using radiolabeled copper 64 found no diffference in copper absorption among normal subjects, heterozygotes, and patients with Wilson's disease.[7] Nor is the uptake of copper increased by cytoplasmic proteins in hepatocytes.[7]

As hepatic copper levels become very high, liver damage progresses and copper is released into the blood. Significant erythrocyte injury can occur as copper is absorbed into the red blood cells, producing hemolysis. Clinical hepatitis may be diagnosed early. If liver injury occurs more gradually, clinical liver disease is less likely to be noted. Patients affected in this manner may have histologic evidence of low-grade chronic toxic hepatitis, which is not clinically evident. Such patients may present with neurologic or psychiatric manifestations but without evidence of cirrhosis.[2]

Diagnosis

The diagnosis of Wilson's disease may be confirmed by one or more of the following four abnormalities: (1) copper concentration of less than 20mg/100ml, (2) urinary excretion of more than 100μg of copper in 24 hours, (3) liver copper concentration of more than 250μg/gm of dried weight, and (4) serum copper level of less than 80μg/100ml.

The clinical manifestations usually begin to appear in the second or third decade of life. The following organ systems are affected (see also Table 1).

TABLE 1. Organs Primarily Affected by Increased Copper Storage

Liver	Kidney
Brain	Blood
Cornea	Bone

Liver. Necrosis of liver cells and subsequent postnecrotic cirrhosis is the most frequent mode of presentation of Wilson's disease. Hepatic involvement leads to jaundice, hepatosplenomegaly, portal hypertension, and liver failure. The abnormal laboratory findings include elevation of the hepatocellular enzymes SGOT and SGPT; prolongation of prothrombin time; and increased bromsulphalein retention. Liver biopsy specimens from patients with Wilson's disease frequently resemble those from patients with chronic active hepatitis.

Brain. Damage to the basal ganglia results in tremors and a dystonic syndrome, characterized by rigidity, ataxia, dysarthria, dysphagia, and an inability to control oral secretions. Grand mal seizures also may occur. Intellectual function is usually preserved but psychologic problems are often present.[8]

Cornea. The Kayser-Fleischer ring is a green-brown ring of pigment located at the periphery of the cornea and derived from deposition of copper in Descemet's membrane.[9] Once thought to be pathognomonic for Wilson's disease, the rings have been reported in several patients with other disorders, most notably biliary cirrhosis and intrahepatic cholestasis. Kayser-Fleischer rings are present in virtually all patients with neurologic manifestations of Wilson's disease and in most patients with hepatic manifestations.[10] Because they are somewhat difficult to visualize on routine ophthalmoscopic examination, a slit-lamp examination is recommended when Wilson's disease is suspected.

Kidney. Renal dysfunction, characterized by decreased creatinine clearance, is almost always present; usually, it is not of clinical significance. Decreased tubular reabsorption is common, and a Fanconi-like syndrome is often present.

Hematologic problems. Hemolytic anemia, which is often present, is secondary to hypercupremic inhibition of RBC glycolytic enzymatic systems. Pancytopenia, secondary to hypersplenism, cirrhosis, and portal hypertension, is commonly present. Hematologic complications also may be associated with hypersensitivity reactions to D-penicillamine therapy.

Hemolytic anemia may be the only presenting symptom in Wilson's disease. In these cases, Coombs' test is negative, and hemoglobinemia and hemoglobinuria are present.[11]

Bone. Skeletal abnormalities, seen on x-ray, are common with Wilson's disease; they include osteoarthritis, osteoporosis, and pathologic fractures.

Treatment

Therapy includes decreasing the amount of absorbable copper by eliminating foods with high copper content, ingesting copper-binding sulfides, and removal and excretion of excessive tissue copper. The first significant human decoppering drug given to patients with Wilson's disease was 2,3-dimercaptopropanol (BAL). Its drawbacks included painful intramuscular administration and frequent adverse drug reactions. A more effective oral medication, with significantly fewer side effects, is D-penicillamine (Cuprimine®Merck Sharp & Dohme). The usual dose is 1–2gm daily.

Therapy with this copper-chelating drug results in the mobilization of copper from the tissues and an increase in its excretion in the urine.[12] After prolonged therapy, the Kayser-Fleischer rings and other neurologic signs disappear; liver function abnormalities revert to normal.

The side effects of D-penicillamine therapy include (1) hypersensitivity reaction including fever, lymphadenopathy, and skin rash; (2) "toxic" hepatitis and intrahepatic cholestasis; (3) rarely, a lupus-like syndrome and glomerulopathy; (4) deficiencies of zinc, mercury, iron, and cadmium, as those metals are also chelated; (5) gastrointestinal symptoms such as nausea, vomiting, and diarrhea; (6) hematologic complications such as thrombocytopenia, agranulocytosis, and aplastic anemia; (7) neuromuscular complications, including optic neuritis and myasthenia (patients frequently take supplemental vitamin B_6 to prevent optic neuritis).

D-penicillamine therapy may be less toxic when combined with glucocorticoid and antihistamine agents.

A possibly less toxic decoppering drug, triethylene tetramine, has been used. Trientine hydrochloride (Cuprid®Merck Sharp & Dohme) is also available. Liver transplantation is used only as a last resort.[14,15]

Anesthetic Considerations

Hepatic Dysfunction

Hepatocellular dysfunction greatly increases morbidity and mortality in surgical patients.[16] Thus, there are several problems that may confront the anesthesiologist treating the patient with Wilson's disease.

Hypoalbuminemia: An unusually profound and prolonged response to drugs such as thiopental may occur because of decreased protein binding, making more pharmacologically active agent available.

Biotransformation: The liver is the primary site of drug biotransformation; prolonged action of thiopental and increased plasma half-lives for diazepam and meperidine have been demonstrated in patients with severe hepatic dysfunction.

Muscle relaxants: The duration of action of succinylcholine and ester local anesthetics may be prolonged because of decreased synthesis of pseudocholinesterase in the liver. Resistance to the effects of non-depolarizing muscle relaxants has been postulated, secondary to binding of the relaxant drugs to gamma globulins or to an increased volume of distribution in patients with liver disease, or both.[16]

Clotting factors: The production of clotting factors (prothrombin and factors V, VII, IX, and X) may be impaired, and thrombocytopenia secondary to hypersplenism (portal hypertension) may be present; preoperative evaluation of the platelet count, prothrombin time, and partial thromboplastin time are essential. The prolonged prothrombin time usually responds to a 3-day course of intramuscular vitamin K therapy unless severe hepatocellular damage interferes with prothrombin synthesis.

Fresh blood and plasma should be available. The blood should be transfused slowly to compensate for the decreased clearance of citrate by the diseased liver.

Effect of anesthetics on hepatic function: All anesthetic agents and techniques impair liver function somewhat, as evidenced by increased bromsulphalein retention postoperatively.[18] The factors involved include operative manipulation (the greatest changes occur in patients undergoing portacaval shunt procedures), preexisting liver disease, and the anesthetic agent used.

Experimentally, a progression of decreased anesthetic toxicity would seem to be as follows: chloroform → fluroxene → halothane → enflurane → isoflurane.[19]

Alterations of hepatic blood flow may be induced under anesthesia. The liver receives 30% of its blood supply from the hepatic artery and 70% from the portal vein.

All anesthetic agents and techniques decrease the hepatic blood flow (HBF). For instance, cyclopropane produces a 33% decrease in HBF secondary to increased sympathetic tone and a marked increase in splanchnic vascular resistance.[16] Halothane, 1.2%–1.6%, produces a 29% decrease in HBF, secondary to a decreased cardiac output and a consequent decrease in hepatic perfusion pressure (mean arterial or portal vein pressure minus hepatic vein pressure): the splanchnic vascular resistance is unchanged.

Methoxyflurane (rarely used today) in an end-tidal concentration of 0.2% can produce a 50% reduction in HBF, secondary to decreased perfusion pressure, combined with an increased splanchnic vascular resistance.

Isoflurane probably has little effect on HBF. Spinal anesthesia decreases HBF in proportion to the decrease in blood pressure.[20] Increased intrathoracic pressure during positive pressure ventilation increases hepatic venous pressure and therefore may decrease hepatic perfusion pressure.[21]

Reduction in HBF should be accompanied by similar reductions in the splanchnic O_2 consumption (Table 2).

TABLE 2. Ratio of Hepatic Blood Flow to Splanchnic O_2 Consumption

Drug/Technique	**Ratio**
Halothane	0.82
Cyclopropane	0.79
Spinal anesthesia	0.73
Nitrous oxide plus nondepolarizing relaxation	0.59
Methoxyflurane	0.55

As HBF decreases, splanchnic oxygen consumption is also reduced. The ratio of those factors is relatively stable.[22]

As HBF decreases, splanchnic oxygen consumption is also reduced. The ratio of those factors is relatively stable.[22] Although reduced HBF could initiate anaerobic metabolism, increased splanchnic lactate production has not been demonstrated.[22] These data are derived from healthy patients before surgery; changes in HBF in patients with preexisting liver disease have not been reported. Thus, in summary, no anesthetic agent or technique appears to be superior with regard to the effect on HBF.

Thrombocytopenia

The three components important for maintaining homeostasis include vascular integrity, platelets, and coagulation factors. Bleeding can be secondary to a defect in one or more of these three components. With regard to platelets, a deficient platelet count is more common than a disorder of platelet function. The normal platelet count is 240,000/mm^3. A count greater than 100,000/mm^3 is necessary for a normal Ivy bleeding time of less than 6 minutes. More than 25,000/mm^3 are necessary to avoid spontaneous bleeding, provided that the function of the individual platelets is adequate.

Many patients, especially those with chronic thrombocytopenia from immune destruction, may have no bleeding episodes for long periods, even though the platelet count may be as low as 5000/mm^3. The absolute minimum platelet count below which the risk of bleeding contraindicates surgery ranges from 50,000 to 75,000/mm^3, with an upper level of 100,000/mm^3.

Thrombocytopenic patients scheduled for surgery should receive preoperative platelet transfusion immediately before surgery, with an anticipated rise in platelet count, 1 hour later, of about 5000 to 10,000/mm^3 for each unit of platelet concentrate infused to a 70kg adult. Such platelets stored at 22°C can be expected to circulate for approximately 8 days, in contrast to a normal platelet lifespan of 9 to 11 days. Storage at 4°C reduces survival to 2 to 3 days.

Most of the coagulation factors, with the exception of factor VIII, are produced by hepatic protein synthetic reactions. In hepatocellular liver disease, impairment is common. Clotting variables are decreased in content and activity (prolonged PT and PTT). Vitamin K, 10mg three times a day intramuscularly, should be given preoperatively. If necessary, fresh frozen plasma can correct all deficiencies except that of fibrinogen, which requires the infusion of cryoprecipitate.[16]

If the anemia associated with Wilson's disease is of a chronic, low-grade nature, then the patient suffers only from a decreased reserve of erythrocytes and is otherwise normovolemic and hemodynamically stable. The need for packed-cell transfusion would be similar to that of any other person. The possibility of an acute hemolytic reaction during general anesthesia should be considered in the presence of unexplained hypotension and tachycardia, particularly if accompanied by hemoglobinuria.

Anesthetic Plan

Anesthetic management must be tailored individually because of the variable intensity and degree of organ involvement.

Preanesthetic Visit

The preoperative visit should be used not only to evaluate the systemic manifestations of the disease but also to assess the patient's psychologic status. Thus, an appropriate choice and dose of premedicant drugs can be made.

If the patient is mentally intact, routine premedication is appropriate. Premedication may be contraindicated or dosages modified in patients with severe liver dysfunction and impaired ability to detoxify drugs. Glycopyrrolate is preferable to atropine and scopolamine because quaternary ammonium compounds do not cross the blood-brain barrier. The patient may be receiving concomitant steroid therapy to control the hypersensitivity reaction induced by D-penicillamine, and additional glucocorticoid coverage during the preoperative period may be indicated.

Choice of Anesthetic Technique

There is no evidence that halogenated anesthetics increase the severity of hepatocellular damage in patients with preexisting liver disease.[22] However, it may be prudent to avoid halothane and halogenated hydrocarbons, as hepatic function may worsen postoperatively.

It is mandatory to assess the liver function test preoperatively. The SGOT and SGPT levels are the most sensitive indicators of hepatic dysfunction in this disorder. The dosage of muscle relaxants should be titrated with a neuromuscular blocking monitor because of the myasthenia-like syndrome that may be associated with D-penicillamine therapy. Regional anesthesia is not contraindicated and may offer a suitable alternative if the coagulation profile is normal and patient acceptability exists.

Therapy with D-penicillamine may produce dermatologic side effects, including desquamative and bullous-type lesions and increased friability. Patients on therapy should be considered at risk of injury of the skin by tight face masks, blood pressure cuffs, and electrocardiogram electrodes.

Routine monitoring includes blood pressure, pulse oximetry, capnography, and continuous ECG. Temperature monitoring is especially important, since a febrile response in not uncommon in the late stages of Wilson's disease or it may be secondary to a drug reaction to D-penicillamine.

References

1. Frydman M., Bonne-Tamir B., Farrer L.A. et al: Assignment of the gene for Wilson's disease to chromosome 13: linkage to the esterase D locus. *Proc Natl Acad Sci USA* 1985, 82:1819–21.
2. Woods S.E., Colon V.F.: Wilson's disease. *Am Fam Physician* 1989, 40(1): 171–6.
3. Idem D.: Tetanoid chorea and its association with cirrhosis of the liver. *Rev Neurol Psychiatr* 1906, 4:249–53.
4. Wilson S.: Progressive lenticular degeneration: A familial nervous disease associated with cirrhosis of the liver. *Brain* 1912, 34:295–8.
5. Katz J., Benumof J., Kadis L.E.: Genetic and metabolic disease, in *Anesthesia and Uncommon Diseases*. Philadelphia: WB Saunders Co, 1981, pp 40–5.
6. Frommer D.J.: Defective biliary excretion of copper in Wilson's disease. *Gut* 1974, 15:125–9.
7. Sternlieb I., Scheinberg I.H.: Radiocopper in diagnosing liver disease. *Semin Nucl Med* 1972, 2:176–88.

8. Scheinberg I., Sternlieb I., Richman J.: Psychiatric manifestations in patients with Wilson's disease. *Birth Defects* 1968, 4:85–7.
9. Walshe J.: The eye in Wilson's disease, in Bergshma D., Bron A.J., Cotlier E. (eds): *The Eye and Inborn Errors of Metabolism.* New York: Liss, 1976, pp 187–9.
10. Scheinberg I.H., Sternlieb I.: Wilson's Disease. Philadelphia: WB Saunders, 1984, pp 93–8.
11. McIntyre N., Clink H.M., Levi A.J., Cumings J.N., Sherlock S.: Hemolytic anemia in Wilson's disease. *N Engl J Med* 1967, 276:439–44.
12. Walshe J.: D-penicillamine, a new oral therapy for Wilson's disease. *Am J Med* 1956, 21:487.
13. Walshe J.M.: Copper chelation in patients with Wilson's disease: A comparison of penicillamine and triethylene tetramine dihydrochloride. *Q J Med* 1973, 42:441–52.
14. Sternlieb I.: Wilson's disease: Indications for liver transplants. *Hepatology* 1984, 4(1):15S–17S.
15. Scharschmidt B.F.: Human liver transplantation: Analysis of data on 540 patients from four centers. *Hepatology* 1984, 4(1):95S–101S.
16. Brown B.R.: Pre-existing liver disease: Surgical and anesthesia risk assessment, in Nunn J.F., Utting J.E., Brown B.R. (eds): *General Anesthesia.* London: Butterworths, 1989, pp 383–91.
17. Brown B.R., McLain G.E.: The role of the liver, in Smith N.T., Corbascio A.N. (eds): *Drug Interactions in Anesthesia.* Philadelphia: Lea & Febiger, 1986, pp 63–70.
18. Sear J.W.: Effect of renal and hepatic disease on pharmacokinetics of anaesthetic agents, in Prys-Roberts C., Hug C.C. (eds): *Pharmacokinetics in Anaesthesia.* Boston: Blackwell, 1986, pp 64–88.
19. Brown B.R. Jr., Dykes M.H.M.: Anesthetic Hepatoxicity. ASA Annual Refresher Course Lectures, 138A B 1–4, 1981.
20. Kennedy W.F. Jr., Everett G.B., Cobb L.A., et al: Simultaneous systemic and hepatic hemodynamic measurements during high spinal anesthesia in normal man. *Anesth Analg* 1970, 49:1016–24.
21. Heiniman H.O., Emingail C., Mijnseen J.P.: Hyperventilation and arterial hypoxemia in cirrhosis of the liver. *Am J Med* 1960, 28:234–40.
22. Stoelting R.K.: Estimation of hepatic function: effects of the anesthetic experience, Hershey S.G. (ed): Refresher Courses in Anesthesiology (vol. 4). Philadelphia: Lippincott, 1976, pp 139–50.

CHAPTER 9

The Drowning/Near-Drowning Patient

James B. Mueller, M.D.

Case History. *A 3 year old child was brought to the emergency room (ER) by the rescue squad after he was pulled from a freshwater pond. He had fallen through the ice and been submerged for approximately 5min. Initially, the child was apneic, pulseless, and cyanotic. Cardiopulmonary resuscitation (CPR) was started. Upon arrival in the ER, the patient was still receiving CPR and was being ventilated by bag mask. Initial core temperature was 30°C. Arterial blood gas values were pH 6.95,* PaO_2 *250mmHg,* $PaCO_2$ *68mmHg,* HCO_3^- *12mEq/L. Other laboratory results included Hgb 8.5gm/dl, Hct 25.4%,* Na^+ *135mmol/L,* K^+ *3.4mmol/L,* CL^- *97mmol/L.*

In the rescue attempts, both his arms had been dislocated and one leg had been crushed by a log trapped in the ice. An anesthesiologist was called to the ER for an immediate consultation.

Introduction

Each year approximately 100,000 submersion accidents occur.[1] Fewer than one tenth, or roughly 8000, of these victims die as a result of their accidents. Drowning is the second leading cause of accidental death among children, next to motor vehicle accidents. It ranks third, behind motor vehicle accidents and cancer, as a cause of death for ages 1 to 14 years.[2,3] The overall majority rate in the United States is about 3 per 100,000.[1] A recently published review of pediatric submersion victims indicates that risk of fatal outcome increases steadily with time of submersion. There is a 10% risk of severe or fatal outcome with submersion of 5 minutes or less and an 85% risk with submersions of 10 to 25 minutes.[4]

Reviewed by Dr. C. Bryan-Brown, Professor of Anesthesiology, Albert Einstein College of Medicine/Montefiore Medical Center.

A recently published review of pediatric submersion victims indicates that risk of fatal outcome increases steadily with time of submersion. There is a 10% risk of severe or fatal outcome with submersion of 5 minutes or less and an 85% risk with submersion of 10 to 25 minutes.[4]

The most common ages for drowning are 2 and 18 years. Approximately 45% of drowning victims are less than 4 years of age.[4] Male victims outnumber females by 5 to 1. Ninety percent of drowning/near-drowning incidents take place in fresh water. Alcohol is involved about 50% of the time in adolescent and older populations.[5,6]

The most common sequence of events in drowning begins with fear, panic, and struggling. As the victim struggles, varying amounts of water are swallowed. As he tires, laryngospasm develops after a small amount of liquid is aspirated. As hypoxia becomes more severe, laryngospasm subsides, and the drowning victim aspirates larger quantities of liquid. Worsening hypoxia, secondary to aspiration and prolonged immersion, leads to cardiopulmonary arrest and death.

Anesthesiologists are frequently called to assist in the resuscitation of drowning/near-drowning victims. Experience in airway management and invasive monitoring and extensive pharmacologic background can be of great benefit in this special resuscitation.

Not all victims who drown aspirate; indeed, 10% die of asphyxia during submersion without aspiration.[7] Those who survive submersion or are successfully resuscitated fall into two subgroups: near-drowning with and without aspiration. Patients who survive initially may die subsequently of pulmonary insufficiency. This sequence of events may change slightly depending on the victim's age, presence of alcohol or drugs, and degree of hypothermia.

Definitions

Several definitions have been formulated:

Drowning is death by suffocation within 24 hours of submersion in a liquid.

Near-drowning is survival past 24 hours after a submersion accident.

Secondary drowning refers to the patient who dies of complications attributed to the original submersion accident (infection/sepsis, respiratory failure, etc).

Wet drowning is defined as submersion with aspiration and is the most common form of drowning, occurring approximately 85%–90% of the time.

Dry drowning is defined as submersion without aspiration. Despite loss of consciousness and onset of hypoxia, laryngospasm continues and prevents aspiration of fluid into the lungs. The reasons for continued glottic closure and prevention of aspiration are still unclear.

Acute submersion hypothermia is the rapid development of hypothermia during drowning. Core cooling is accelerated from rapid absorption of inhaled and swallowed water.

Pathophysiologic Changes

Pulmonary System

The common problem in drowning, whether in fresh or salt water, is aspiration. Fluid aspirated by the drowning victim is frequently a mixture of stomach contents and the drowning medium. This contaminated mixture further complicates the resuscitative attempt. Aspiration causes a dramatic change in pulmonary function. Almost immediately there is increased pulmonary shunting (Qs/Qt), decreased compliance, bronchospasm, and destruction of pulmonary tissue. Loss of surfactant, atelectasis, decreased functional residual capacity (FRC), and varying degrees of hypoxemia are present.[8] The magnitude of these changes depends on the type and volume of fluid aspirated.

Although differences between drowning in fresh water and in salt water exist (see Figure 1), many of the distinct symptoms are rarely seen because the volumes are frequently small and contaminated.

Fresh water is hypotonic compared to bodily fluids. Ingestion of fresh water into the lungs or gastrointestinal tract leads to rapid absorption and a transient increase in plasma volume.[9] This increase in volume is short-lived, lasting only about 60 minutes. Large-volume ingestion (>20ml/kg) can produce hemodilution, transient hypervolemia, red blood cell hemolysis, hemoglobinuria, hyperkalemia, hyponatremia, and hypochloremia.

Salt water (3% NaCl) is hypertonic. Ingestion and/or aspiration of a hypertonic fluid creates a gradient for plasma fluid to flow into the damaged pulmonary epithelium and into the gastrointestinal tract. This loss of intravascular fluid results in hemoconcentration, hypovolemia, hypernatremia, and hyperchloremia. In such cases it is not uncommon to see varying degrees of pulmonary edema secondary to the excess fluid in the lungs. At autopsy, lungs are found to be stiff, with areas of hyperexpansion. Emphysematous changes may be the result of violent respiratory excursions against a closed glottis or obstructed airway.[7]

In principle, freshwater aspiration should be less destructive and less complicated than saltwater aspiration. However, as mentioned earlier, the aspirate is rarely "pure." Stomach acid, food particles, and dirt, as well as fluid, will lead to significant pulmonary alveolar destruction, hyaline deposits, chemical injury, and inflammation.[10] Loss of surfactant, alveolitis, noncardiogenic pulmonary edema, and ventilation/perfusion mismatching are present. The lungs are also at great risk for infection. Patients frequently develop fulminant pneumonias and adult respiratory distress syndrome (ARDS). Capillary congestion and hyaline membrane formation may cause longterm pulmonary dysfunction.[4]

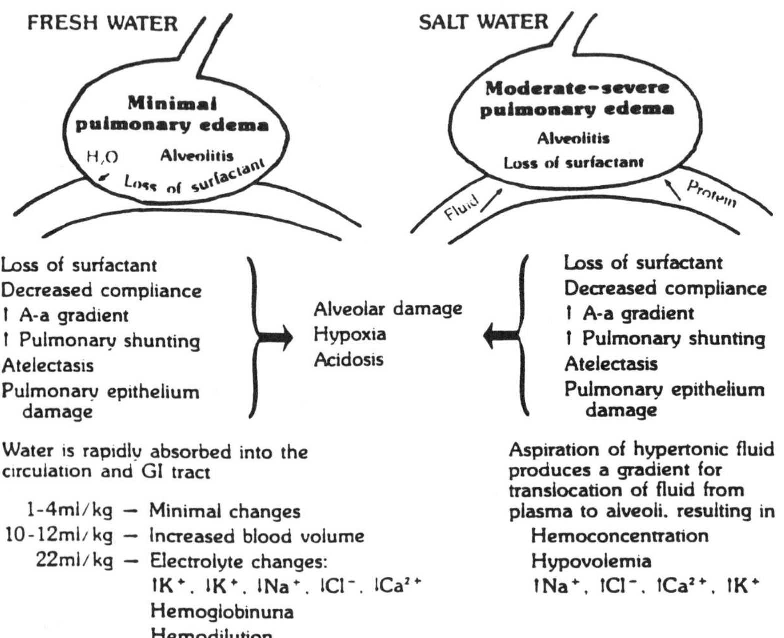

FIGURE 1. Physiologic Differences Between Drowning in Fresh Water and in Salt Water.

Initial changes in arterial blood gas status are rapid and profound. Hypoxemia occurs early. The exact level of PaO_2 at which cardiac arrest occurs is unknown but is probably about 15mmHg if the pH is above 7.2.[11]

Hyperventilation prior to submersion prolongs the time until $PaCO_2$ rises sufficiently to force inspiration. During this time, PaO_2 drops rapidly as a result of increased oxygen consumption. Thus, consciousness is quickly lost and with it the effort to suppress respiration.[12]

Cardiovascular Changes

Almost every abnormality of cardiac electrical activity has been associated with drowning or near-drowning.[7] Ventricular fibrillation has been associated with freshwater drowning.[13] However, as the basis for this dysrhythmia is probably sudden hyponatremia, and as the volume of fluid aspirated is usually small, ventricular fibrillation has rarely been reported in humans.[14] Reported blood pressure changes are variable. In dogs, cardiac output declines after aspiration of fresh water and increases if respiration is spontaneously reestablished. However, if mechanical ventilation and positive pressure are required, cardiac output decreases further, probably because of an inadequate effective blood volume.[15]

In general, if PaO_2, pH, and volume status can be restored to normal, alterations in cardiovascular function will be corrected, and pharmacologic manipulations are rarely necessary. One problem may be hypothermia. While decreased temperature may be protective to the brain, if it falls below 28°C, effective cardiac output ceases.

Neurologic Effects

Most near-drowning victims have full recoveries despite the potential for and sometimes the actuality of multiorgan system involvement. Improvements in cardiopulmonary resuscitation, ventilatory support, and pharmacologic adjuncts for therapy have decreased the mortality and morbidity associated with near-drowning. The reported incidence of severe residual neurologic deficit after resuscitation ranges from 0% to 3%.[2,16]

However, neurologic injury continues to be of major concern in patients who arrive at the ER in a state of cardiac arrest. Severe anoxic encephalopathy usually results.[17]

The central nervous system (CNS) is one of the first organ systems to suffer from sudden hypoxia and decreased cardiac output. The brain is an aerobic organ with minimal alternative energy sources. In only a matter of minutes the brain depletes oxygen supplies through a sustained metabolic demand. Unless oxygen and nutrients are quickly supplied or the metabolic requirements are decreased, cellular death follows.

As mentioned, submersion victims are frequently hypothermic. Loss of body heat is rapid in liquids because of efficient heat transfer. Hypothermia may offer some protection to the CNS and other organ systems; it reduces the basal metabolic rate (BMR) and systemic perfusion.[18] For the brain, rapid hypothermia decreases the cerebral metabolic rate of oxygen utilization ($CMRO_2$) and cerebral blood flow (CBF). These effects seem to be most marked in small children, because of their large surface-to-mass ratio and relative lack of insulation. Swallowing water further decreases core temperature by the rapid absorption of cold water into the circulation. Cerebral cooling prior to pulselessness may help delay hypoxic brain damage.[9] This form of hypothermia is commonly referred to as acute submersion hypothermia.

Reports of prolonged submersion in cold water with normal recoveries are becoming more frequent. To date, the longest submersion has been reported by Bolte et al. A 2 year-old girl survived submersion for approximately 66 minutes with a core temperature of 19°C.[19]

The mammalian diving reflex (MDR) has been credited with increasing the potential survival of the young pediatric submersion victim. The diving reflex occurs with a sudden drop in facial temperature when apnea and peripheral vasoconstriction take place. Apnea prevents aspiration, and vasoconstriction shunts blood into the central compartment to help preserve vital organ blood flow. In conjunction with the rapid core cooling, decreased metabolism and preserved vital organ blood flow may facilitate prolonged survival. Data from Hayward et al and Ramey et al raise doubt that the MDR is active or is responsible for a protective effect in humans.[20,21] Subjects tested by cold water immersion were found to have decreased breath-holding durations compared to control groups. As the water became colder, the duration of breath holding decreased, thus increasing the potential for drowning.

Gastrointestinal Function

The gastrointestinal tract frequently has decreased gastric emptying times as well as altered motility. There is also an increased incidence of gastrointestinal hemorrhage. The most common etiology is that of stress-induced bleeding or gastric erosion. Stress ulcers are more frequent in cases of multiple trauma, severe hemorrhage or shock, respiratory failure, renal failure, and sepsis.

The liver is unable to metabolize drugs effectively, leading to their accumulation in the circulation.

Renal Function

Most near-drowning victims do not have renal complications. However, the kidneys are sensitive to episodes of decreased blood pressure and perfusion. Acute renal failure (ARF) and increases in antidiuretic hormone (ADH), secretion decrease urine production, create electrolyte abnormalities, and complicate acid-base status. Renal failure also increases the risk of infection.

Other derangements include albuminuria and hemoglobinuria from hemolysis.[22]

Hematologic Changes

Hemolysis may be a direct effect of water absorption.

The near-drowning patient tends to be immunocompromised. Neutropenia is common in the hypothermic patient. Bohn et al found that neutropenia after resuscitation from near-drowning indicated a poorer prognosis for survival.[23] Hypothermia decreases the total number of circulating polymorphonuclear leukocytes by inhibiting their release from the bone marrow. The combination of multiple organ system dysfunction and depressed immune response increases morbidity and mortality.

Therapy

Once the victim has been located, prompt removal from the water is essential. Prehospital advanced cardiac life support appears to be the most important form of medical treatment and prehospital parameters are good predictors of eventual outcome. A thorough but rapid assessment includes airway patency, breathing, circulation, and other bodily injuries. Cervical/thoracic spine injuries may be present but not obvious, as well as peripheral and internal injuries secondary to rescue or blunt trauma, as in our patient. Protective immobilization is usually advisable during transport.

Pulmonary Therapy

In patients who are awake and breathing spontaneously, supplemental oxygen should be provided until the patient has been evaluated at a treatment center. In patients with absent or inadequate ventilation or oxygenation, ventilation must be started and maintained. Evacuation of the lungs is not necessary, because usually only a small amount of liquid is present in the lungs after aspiration. Mouth-to-mouth ventilation should be started until manual ventilation with supplemental oxygen can be provided. The cervical spine should be stabilized by "manual axial traction" during intubation.

Although increased inspired oxygen will increase PaO_2, the improvement is limited by the degree of intrapulmonary shunt. Further improvement in PaO_2 is achieved by increasing functional residual capacity with continuous positive airway pressure (CPAP).[15]

After seawater aspiration, CPAP with spontaneous or mechanical ventilation may be sufficient to open closed airways. However, surfactant washout in freshwater drowning may require higher inspiratory pressures to open closed alveoli. Bronchodilators may be required to treat bronchospasm.

Occasionally chest x-ray reveals collapse, and bronchoscopy is necessary to clear the airway.

If the lung damage caused by seawater or freshwater drowning can be successfully treated, the prognosis for complete recovery is excellent.[24]

Cardiovascular Support

Maintenance of the circulating blood volume provides oxygen to the hypoxic tissues. Patients without a pulse or palpable blood pressure should receive external compressions in conjunction with ventilation (ie, CPR). At best, properly performed CPR will achieve only a fraction of normal cardiac output. Detection of a pulse and blood pressure may be difficult in the hypothermic patient. Nevertheless, CPR must be continued until the patient has a pulse and adequate blood pressure, which may not occur until the core temperature is about 30°C. Hypothermic victims are prone to cardiac dysrhythmias, most commonly sinus bradycardia and atrial fibrillation. Ventricular fibrillation is common with core temperatures below 28°C.[25,26] Defibrillation may not be effective until the temperature is above 30°C.

Rewarming of the patient should be started immediately. Covering the patient with blankets and otherwise providing a warm environment will suffice until the patient reaches the hospital. Then depending on the available facilities, aggressive rewarming can take place. Heated humidified oxygen; warmed intravenous fluids; warmed irrigation fluids via the stomach, rectum, and peritoneum; heating blankets; and extracorporeal circulation are all used to increase the core body temperature. Peritoneal lavage with warm dialysate is as effective as extracorporeal circulation for central rewarming, without the need for systemic heparinization.[27] However, CPR must be continued with peritoneal lavage. Peritoneal lavage has the advantage that it can be done in any emergency facility, using minimal equipment.[18]

During rewarming the goal is to maintain perfusion and adequate oxygenation to the tissues. Hypothermia and altered cardiac performance reduce tissue perfusion. Vascular instability, as evidenced by rapid fluctuations in blood pressure, is common secondary to cold, acidotic blood returning to the heart and loss of intravascular fluid into the peripheral tissues.[18] In a hypothermic model using dogs, Moss et al showed that rewarming was best achieved by maintaining cardiac output, adequate intravascular volume, oxygenation, and pH.[27]

Essential monitoring includes electrocardiogram (ECG), invasive arterial blood pressure, urinary output, temperature (core and surface), and central venous pressures. Cardiac changes induced by cold, combined with pulmonary edema caused by aspiration, may make gauging the volume of fluid required difficult. Measurement from a pulmonary artery catheter will aid in appropriate replacement. Pulse oximetry may not be helpful if there is peripheral vasoconstriction, as the pulse may not be detected. Placing the finger inside a rubber glove may help to obtain an accurate reading at low core temperatures. Vasopressor therapy is rarely required, as fluid replacement and correction of acid-base abnormalities usually restore circulation adequately. Diuretics may be required to maintain urine output and are especially indicated in patients with free hemoglobin in plasma.

Brain Resuscitation and Neurologic Classification

Assessment by the Glasgow Coma Scale (GCS) and other systems used for trauma and encephalopathic diseases may not be as accurate for near-drowning victims.[2,28] A classification system recommended by Conn and Modell uses three basic categories for evaluation: awake (A), blunted (B), and comatose (C) (see Table 1).[9,28]

TABLE 1. Postsubmersion Neurologic Classification*

Category	Description	Glasgow Coma Scale
A (Awake)	Alert, fully conscious	15
B (Blunted)	Obtunded, stuporous but arousable; purposeful response to pain; normal respiration	10–13
C (Comatose)	Comatose, not arousable; abnormal response to pain; abnormal respirations	5
Subcategories		
C_1 (Decorticate)	Flexion response to pain; Cheyne-Stokes respirations	5
C_2 (Decerebrate)	Extensor response to pain; neurogenic hyperventilation	4
C_3 (Flaccid)	No response to pain; apneustic or "cluster" breathing	3
C_4 (Deceased)	Flaccid, apneic, no detectable circulation	

Adapted from Conn AW, Barker GA: Fresh water drowning and near-drowning—an update. *Can Anaesth Soc J* 1984: 31(3):S38–44.[3]

* Examination to take place in ER immediately following successful CPR.

Patients in category A are awake and alert, demonstrating minimal, if any, injury. They are admitted to the hospital, evaluated, and observed for 24 hours; if on reevaluation, all functions are normal, they are released.

Category B patients have had more pronounced cerebral injury; evidence of asphyxia/hypoxia is present. However, these patients still have normal pupillary reflexes and peripheral responses to pain. They are admitted to the ICU for aggressive evaluation and supportive care and are observed for a longer time for delayed pulmonary dysfunction and/or cerebral injury.

Category C patients are comatose. They demonstrate abnormal responses to pain, lack verbal response, and have irregular respiratory patterns. The depth of coma is further divided into three categories: C1, decorticate; C2, decerebrate; C3, flaccid. These patients are admitted to the ICU for aggressive and prolonged treatment.

Survival from near-drowning has been reviewed in studies by Modell et al and Conn et al. In general, they showed that patients admitted in categories A and B had approximately a 90% chance of survival and full recovery. Patients in category C had a 34% mortality rate. Of the surviving patients in category C, only 20% had no residual neurologic deficits.[22]

Allman et al studied 66 children admitted to the ICU after severe drowning and resuscitation. Each patient required full CPR and had a GCS of 3 (equivalent to C3 category). Intact survival was 24%; 26% of the patients remained in a chronic vegetative state, and the remaining 50% of the patients expired.[29]

In all of these studies, several important points were made: (1) Because of the complete recoveries demonstrated by approximately 20% of patients in the severe comatose state, all near-drowning victims should be aggressively resuscitated; determination of death should not be made during the initial resuscitative effort; (2) the presence of spontaneous respiration in the immediate post-CPR period is a reliable prognostic indicator of survival; (3) no patient who arrived in a flaccid state had a good outcome.

Several protocols have been designed to attempt to lessen cerebral damage.[30,31] For example, the HYPER regimen (HYPER-therapy) is based on the principle that after a successful resuscitation and volume expansion has taken place, patients are found to be hyperhydrated, hyperventilating, hyperpyrexic, and hyperexcitable, with hyperrigidity. The goals then are to establish mild dehydration, controlled hyperventilation, moderate hypothermia and barbiturate coma, and skeletal muscle paralysis.

There has been a great deal of debate concerning the use of the complete HYPER regimen.[32] In a comparison of the studies on outcome, hypothermia and barbiturate coma did not significantly improve survival.[2,28] In fact, the use of hypothermia and barbiturate coma was associated with problems of hemodynamic instability, decreased cerebral perfusion pressures, and potential for further immune suppression.[22] Also, in patients without severe brain damage secondary to hypoxia or trauma, intracranial pressure (ICP) is rarely increased to the point of requiring treatment. Intracranial hypertension has been shown to be a late and usually terminal event.[33] It is not clear at this time if monitoring and treating elevated ICP will increase survival. There does not appear to be any correlation between the postmortem degree of cerebral injury and raised ICP.[34,35]

Other Drug Therapies

No current evidence supports the use of steroids to improve survival or oxygenation after stomach content aspiration in freshwater drowning.[36,37]

Routine use of antibiotics is probably not useful. Rather, the preference is to monitor clinically for signs of infection and to stain and culture tracheal aspirates. Specific coverage of pathogens lessens resistant overgrowths.[7]

Summary

The near-drowning patient deserves aggressive therapy (see Table 2).

Initial effective CPR seems to be the most important factor. After successful resuscitation, the victim should be taken to an adequate treatment center for further evaluation and therapy, which includes rewarming, volume expansion, and pulmonary care. Depending on the status of the patient, further aggressive therapy should be continued in the ICU. Only after the temperature is normal and cardiorespiratory status stable can operative intervention to reset broken limbs or debride crushed areas be undertaken.

TABLE 2. Summary of Resuscitation

1. Remove from water.
2. Make rapid cardiopulmonary assessment
   - Airway patency
   - Breathing
   - Circulation
3. Begin cardiopulmonary resuscitation (CPR).
4. Look for associated injuries.
   - Cervical spine
   - Blunt trauma
   - Long-bone fractures
   - Drug ingestion
5. Start rewarming.
   - Blankets
   - Warm I.V. fluids
   - Warming lights
   - Heated/humidified gases
   - Peritoneal lavage
   - Extracorporeal circulation
6. Draw blood samples for electrolyte estimations.
7. Perform initial neurologic classification.
8. Transfer to ICU.

References

1. Modell J.H.: Drowning, in Staub N.C., Taylor A.E. (eds): *Edema*. New York: Raven Press, 1984, pp 679–94.
2. Orlowski J.P.: Drowning, near-drowning, and icewater submersions. *Pediatr Clin North Am* 1987, 34:75–92.

3. Rasch D.K., Pollaro T.G.: Near-drowning in children: Management and outcome. *NY State J Med* 1988, 88:427–33.
4. Quant L., Wentz K.R.: Outcome and predictors of outcome in pediatric submersion victims receiving prehospital care in King County, Washington. *Pediatrics* 1990, 86:586–93.
5. Redding J.S.: Drowning and near-drowning: Can the victim be saved? *Postgrad Med* 1983, 74:85–97.
6. Ornato J.P.: The resuscitation of near-drowning victims. *JAMA* 1986, 256: 75–7.
7. Modell J.H., Spoler R.W.: Drowning and near-drowning, in Nunn J.F., Utting J.E., Brown B.R., Jr. (eds): *General Anesthesia*. London: Butterworths, pp 1288–94.
8. Modell J.H.: Near-drowning. *Circulation* 1986, 74(suppl IV):27–8.
9. Conn A.W., Barker G.A.: Fresh water drowning and near-drowning—an update. *Can Anesth Soc J* 1984, 31(3):S38–44.
10. Fuller R.H.: The clinical pathology of human near-drowning. *Proc R Soc* 1963, 56:33–8.
11. Kristofferson M.B., Rattenberg C.C., Holaday D.A.: Asphyxial death: The role of acute anoxia, hypercarbia and acidosis. *Anesthesiology* 1967, 28:488–97.
12. Craig A.B. Jr.: Causes of loss of consciousness during underwater swimming. *J Appl Physiol* 1961, 61:583–6.
13. Swann H.G.: Mechanisms of circulatory failure in fresh and sea water drowning. *Circ Res* 1956, 4:241–4.
14. Modell J.H., Davis J.H., Giammona S.T., et al: Blood gas and electrolyte changes in human near-drowning victims. *JAMA* 1968, 203:337–43.
15. Bergquist R.E., Vogelhut M.M., Modell J.H., et al: Comparison of ventilatory patterns in the treatment of fresh water near-drowning in dogs. *Anesthesiology* 1980, 52:142–8.
16. Pearn J.H., Bart R.D. Jr., Yamaoka R.: Neurologic sequelae after childhood drowning: A total population study from Hawaii. *Pediatrics* 1979, 64:187–91.
17. Peterson B.: Morbidity of childhood near-drowning. *Pediatrics* 1977, 59: 364–70.
18. Bristow G.: Accidental hypothermia. *Can Anaesth Soc J* 1984, 31(3):S52–5.
19. Bolte R.G., Black P.G., Bowers R.S., Thorne J.K., Cornel H.M.: The use of extracorporeal rewarming in a child submerged for 66 minutes. *JAMA* 1988, 250:377–9.
20. Hayward J.S., Hay C., Matthews B.R., Overweel C.H., Radford D.D.: Temperature effect on the human dive response in relation to cold water near-drowning. *J Appl Physiol* 1984, 56:202–6.
21. Ramey C.A., Ramey L.D.N., Hawand J.S.: Dive response of children in relation to cold water near-drowning. *J Appl Physiol* 1987, 63:665–8.
22. Redding J.S.: Treatment of near-drowning. *Int Anesthesiol Clin* 1965, 3:255–65.

23. Bohn D.J., Biggar W.D., Smith C.R., Conn A.W., Barker G.A.: Influence of hypothermia, barbiturate therapy and intracranial pressure monitoring in morbidity and mortality after near-drowning. *Crit Care Med* 1986, 14:529–34.
24. Jenkinson S.G., George R.B.: Several pulmonary function studies in survivors of near-drowning. *Chest* 1980, 77:777–80.
25. Martin T.G.: Near-drowning and cold water immersion. *Ann Emerg Med* 1984, 13:263–73.
26. Ornato J.P.: Special resuscitation situations: Near-drowning, traumatic injury, electric shock, and hypothermia. *Circulation* 1986, 74(suppl IV):IV23–IV26.
27. Moss J.F., Harklin M., Southwick H.W., Roseman D.L.: A model for the treatment of accidental severe hypothermia. *J Trauma* 1989, 26:68–74.
28. Conn A.W., Montes J.E., Barker G.A., Edmonds J.F.: Cerebral salvage in near-drowning following neurological classification by triage. *Can Anaesth Soc J* 1980, 27:201–8.
29. Allman F.D., Nelson W.S., Pacentine G.A., McComb G.: Outcome following cardiopulmonary resuscitation in severe pediatric near-drowning. *Am J Dis Child* 1986, 140:571–5.
30. Gonzalez-Roth R.L.J.: Near-drowning: Consensus and controversies in pulmonary and cerebral resuscitation. *Heart Lung* 1987, 16:474–82.
31. Frewen T.C., Sumabat W.O., Han V.K., Amacher A.L., Del Maestro R.F., Sisbald W.J.: Cerebral resuscitation therapy in pediatric near-drowning. *J Pediatr* 106:615–7.
32. Conn A.W., Edmonds J.F., Barker G.A.: Near drowning in cold fresh water: Current treatment regimens. *Can Anaesth Soc J* 1978, 25:259–65.
33. Nussbaum E., Galant S.P.: Intracranial pressure monitoring as a guide to prognosis in the nearly drowned, severely comatose child. *J Pediatr* 1983, 102:215–8.
34. Mayer T., Walker M.L.: Emergency intracranial pressure monitoring in pediatrics. *Clin Pediatr* 1982, 211:391–6.
35. Sarnaik A.P.: Preston G., Lieh-Lai M., et al: Intracranial pressure and cerebral perfusion pressure in near-drowning. *Crit Care Med* 1985, 13:224–7.
36. Wynne J.W., Reynolds J.C., Hood C.I., et al: Steroid therapy for pneumonitis induced in rabbits by aspiration of foodstuff. *Anesthesiology* 1979, 51:11–9.
37. Calderwood H.W., Modell J.H., Ruiz B.C.: The ineffectiveness of steroid therapy for treatment of fresh water near-drowning. *Anesthesiology* 1975, 43:642–50.

CHAPTER 10

The Patient With Muscular Dystrophy

Seth Landa, M.D.

Case History. *A 13-year-old boy was scheduled for Harrington rod insertion to correct severe scoliosis. He had been diagnosed as having Duchenne muscular dystrophy at age 5 and had been confined to a wheelchair since age 10.*

Physical examination showed an obese young male patient sitting in a wheelchair with a pronounced list to the left. Weight was 190lb; height, 5'4"; blood pressure, 100/60mmHg; pulse, 106/min; respiration, 22/min; temperature, 37.2°C. Examination of the chest and heart was unremarkable. There was marked weakness of all muscle groups except the hand, forearm, and gastrocnemius muscles.

Laboratory values revealed hematocrit, platelets, and SMA-6 within normal limits. The ECG showed sinus tachycardia with a short PR interval, a tall R wave in V_1, and Q waves in leads aVL, V_5 and V_6. Pulmonary function testing demonstrated a moderate restrictive pattern with a vital capacity 35% of predicted. The chest x-ray was clear; echocardiogram showed an ejection fraction of 70%.

Introduction

"Jerry's Kids," known to the layman from the national telethon sponsored annually on their behalf, are children with muscular dystrophy. The muscular dystrophies are actually a heterogeneous group of hereditary myopathies characterized by progressive muscle degeneration and weakness. The diseases included in this group are Duchenne muscular dystrophy, facioscapulohumeral muscular dystrophy (Landouzy-Déjerine dystrophy), limb-girdle muscular dystrophy (Erb dystrophy),

Reviewed by Dr. Miguel Rosa, Staff Anesthesiologist, Department of Anesthesiology, Portsmouth Naval Hospital, Portsmouth, Virginia.

ocular muscular dystrophy, and myotonic dystrophy (Steinert's disease, Hoffman's disease, Batten-Curschmann disease). (See Table 1.)

Myopathies are primary diseases of striated muscle with neurophysiologic, biochemical, or morphologic changes occurring singly or in combination. Classification is based mainly on historic descriptions or clinical similarities. Patients suffering from muscular dystrophy may present in the operating room for a variety of procedures, often relating to their underlying condition. The anesthesiologist must be familiar with the systemic effects of these diseases as well as specific hazards associated with various anesthetic techniques and agents.

Duchenne Muscular Dystrophy

The commonest and severest form of muscular dystrophy, Duchenne muscular dystrophy (DMD), is inherited as an X-linked recessive characteristic and has a reported incidence of 3 per 10,000 births.[1] Clinical observations that served as the basis for the classification of the dystrophies were made in 1868 by Guillaume Duchenne on the basis of details of a single case described in 1861 and review of 27 other cases.[2,3]

Although defects have been described in the fetus, and a few well-documented cases have been reported in females,[4] occurrence is limited almost entirely to males, with symptoms usually becoming apparent between 2 and 6 years of age. This phenomenon has been explained by the Lyon hypothesis: only one X chromosome is active in any cell. The other X in females is inactivated early in embryonic growth. In some female heterozygotes, most cells could have the abnormal X chromosome as the active one. Also, female carriers have elevated serum enzymes and those with the highest levels have the largest proportion of abnormal cells. The enzyme increase is probably the result of increased cell membrane permeability to sarcoplasmic enzymes. Levels are highest early in the disease and decrease as atrophy progresses.

Initial complaints include waddling gait, difficulty running and climbing stairs, and frequent falling. Gower's sign, the use of the arms to "climb up the legs" and assist in standing, reflects hip girdle weakness. Lumbar lordosis develops to maintain balance. The proximal muscles atrophy first, whereas the distal muscles, especially the calves, enlarge. This pseudohypertrophy results from replacement of muscle fibers by connective and fatty tissues. The fat that replaces muscles is resistant to mobilization, even during starvation.[5] Mental retardation is present in about 60%–70%.

The disease is progressive, and the children are wheelchair-bound by their early teens. In general, the earlier the onset, the more rapid the

TABLE 1. The Muscular Dystrophies

Type	Inheritance	Age at onset	Muscle groups involved	Course
Duchenne	X-linked	Early childhood recessive	Symmetrical weakness initially pelvifemoral; weakness of shoulder girdle later and then trunk muscles; puberty; pseudohypertrophy of calves;cardiac involvement	Progressive; inability to walk by puberty death by age 20
Becker	X-linked recessive	Second decade	Milder variant of Duchenne type; cardiac involvement less common	"Benign": inability to walk into adulthood
Facioscapu-lohumeral	Autosomal dominant	Childhood to late adult life	Usually facial weakness first; scapular weakness; humeral weakness; cardiac involvement rare	"Benign": course not progressive
Limb-girdle	Autosomal recessive	Variable onset, 1st to 3rd decade	Two variants; pelvifemoral weakness, shoulder girdle weakness; cardiac involvement rare;	Variable progression; disability within 20 years
Ocular	Autosomal dominant	Variable	Group of syndromes; weakness of extraocular muscles; sometimes involvement of face, neck, limbs	Rarely progressive
Myotonic	Autosomal dominant	2nd to 3rd decade; may be seen in infancy or childhood	Facial, neck, proximal limb weakness, sternocleido-mastoid and temporalis wasting; myotonic phenomenon; cardiac involvement	Progressive; death by sixth decade

downhill course. Death usually occurs 5 to 20 years from onset as a result of respiratory failure, pneumonia, or congestive heart failure. In a subgroup of approximately 15%, there is a much slower course, with stabilization at or near puberty. In such patients contractures contribute greatly to disability,[5] although contractures and kyphoscoliosis may develop in all patients. Even after short periods of immobilization the preoperative level of motor function may not be regained, emphasizing the need for physiotherapy. Some patients become obese, others appear wasted.

Cardiac involvement occurs in up to 70% of cases but is clinically symptomatic in only 10% of cases and usually only late in the disease.[5] Electrocardiographic changes (Table 2.) are typical for DMD and are independent of the time since diagnosis or onset of symptoms.[6]

TABLE 2. ECG Abnormalities in Duchenne Muscular Dystrophy

ECG finding	Pathophysiologic correlate
Sinus tachycardia; short PR interval	Functional abnormality of autonomic nervous system with adrenergic predominance
Tall R waves in R precordial leads; Q waves in lateral precordial and limb leads	Wall motion abnormalities and fibrosis in lateral and posterobasal walls of left ventricle seen on cardiac imaging and post-mortem

Myocardial degeneration results in depressed contractility and occasionally mitral regurgitation secondary to papillary muscle dysfunction (mitral valve prolapse). Obstructive cardiomyopathy may affect right ventricle outflow and lead to right-sided heart failure. Sudden death has been reported in patients with fully compensated cardiac status.[7] The usual finding at autopsy is cardiac dilation with or without hypertrophy. Fibrosis of the myocardium is limited to the free wall of the left ventricle, unlike the fibrosis seen in other forms of muscular dystrophy.

Recently, primarily because of advances in understanding the biochemistry and immunochemistry of collagen, the connective tissue or collagen theory of the cause of DMD has been reconsidered.[8] Muscle growth failure and skeletal muscle wasting probably result from an imbalance between synthesis and degradation. Because of the proliferation of connective tissue in muscular dystrophy, Duchenne named the disorder "paralysie myosclérotique." Normal and abnormal collagen production by skin fibroblasts occurs, and collagen type III is increased. Decreased normal collagen synthesis is associated with a twofold increase in collagen degradation in DMD. There is also a consistent decrease in muscle protein synthesis in cultured muscle and skin fibroblasts. Disordered regulation of extracellular matrix metabolism is another important feature.[9]

To date no specific treatment exists other than physiotherapy and physical aids. Contractures, spinal deformities, and skin ulceration may require surgical correction. However, a promise of further help is provided by advances in recombinant DNA technology.[10] Particular hope is held out for DMD because of the presumed location of an exchange point on the short arm of the X chromosome. In studies on cells from female DMD patients, seven have been documented in whom translocation occurred between X and autosomes.[11] Band X_p21 has been identified as a potential locus tor the DMD gene. Studies indicate that the DMD gene in one patient split the block of genes encoding ribosomal RNA on the short arm of chromosome 21. These and other studies indicate rapid progress in attempts to isolate the DMD gene.

Becker Dystrophy

The syndrome is an X-linked recessive muscular dystrophy similar to DMD but with a later onset and better prognosis. It is not so common, representing fewer than 10% of all X-linked cases. Few patients are wheelchair-bound before the age of 15.[12] Life expectancy is decreased although not to the extent seen in DMD.

Facioscapulohumeral Dystrophy

Landouzy-Déjerine syndrome has a benign progress, beginning in adolescence. Patients have weak pectoral and facial muscles. Males and females are equally affected. Although cardiac involvement is rare, cases of atrial paralysis have been reported,[13] in which patients have no atrial electrical activity and cannot be paced electrically from the atrium. Severe bradycardia may be life-threatening, and ventricular pacing is necessary. A more common problem is a decrease in vital capacity and inability to cough and clear secretions.

Limb-Girdle Dystrophy

There are two subdivisions in this disorder: Erb's type, in which the shoulder girdle is primarily involved, and Leyden-Moebius type, with mainly pelvic girdle involvement. The dystrophy has a variable onset. Cardiac involvement is less common and includes sinus tachycardia and right bundle branch block. Respiratory problems again involve decrease in vital capacity.

Ocular Muscular Dystrophy

Muscular dystrophies involving weakness of the extraocular muscles and ptosis are difficult to classify. The differential diagnosis rests with ocular myasthenia gravis, although diplopia, which is very common in the latter disease, is rare. Sensitivity to muscle relaxants is often similar.[5]

Myotonic Dystrophy

Myotonia dystrophica (MyD), an autosomal dominant disease, is the most common dystrophy of adult life (2.4–5.5 per 100,000 births)[14] This dominant inherited disorder with incomplete penetrance is a multisystem condition classified under the muscular dystrophies. It is also known as Steinert's disease (after the physician who first distinguished it from other myotonic disorders), Batten-Curschmann disease, or myotonia atrophica.

Myotonia is the inability to relax a muscle after stimulation. It results from abnormal calcium metabolism as the cellular ATP system fails to return calcium to the sarcoplasmic reticulum. The calcium remains available to produce sustained skeletal muscle contraction[15] (ie, contraction with electrical silence, sustained without further stimulation by the motor nerve). When the agonist muscles contract, the antagonists also contract as the patient tries to relax. Because the defect is intracellular, the contraction is not prevented by muscle relaxants or by general or regional anesthetics. Infiltration of contracted muscles with local anesthetics may induce relaxation. Procainamide, quinine, phenytoin, tocainide, and mexiletine have been reported to alleviate myotonic contractures.[16,17]

The onset of symptoms is usually in the second or third decade. The clinical picture includes weakness and wasting of facial, cervical, and proximal limb muscles, frontal baldness, cataracts, and a low intelligence. Endocrine involvement is manifested as gonadal atrophy, diabetes mellitus, adrenal insufficiency, and hypothyroidism. Difficulty in releasing the grip after handshake may give the clue to diagnosis.

Cardiac degeneration also occurs and conduction defects are common. Sudden death may be caused by the onset of third-degree AV block. The commonest ECG abnormality is an increased PR interval unresponsive to atropine and nitroglycerin. The P wave is decreased in height, and ST elevations occur. Atrial flutter may develop. Blood pressure is usually low. Stokes-Adams seizures are not uncommon.[18]

Therapy is directed at the myotonia rather than at rehabilitating the atrophic muscle. Quinine up to toxic doses has been used but has now been replaced by procainamide. Dosage of the latter drug is 1gm four times daily. It can block myotonia after spontaneous effort but not percussion myotonia (the stimulus of surgery and electrocautery) and thus has limited effectiveness intraoperatively. It can be given intravenously in doses to 1gm at 100mg/min.

Death from cardiac failure or pneumonia usually occurs by the sixth decade. Pregnancy exacerbates MyD, and uterine atony or retained placenta may complicate vaginal delivery. Large doses of oxytocics and massive blood transfusions may be required. Fortunately, pregnancy is rare because of ovarian atrophy.

An important systemic manifestation of MyD is presenile cataract, which is characteristic of the disease. In relatives of affected patients this may be the only manifestation of the disease. It has been said, perhaps anecdotally, that cataracts occur in the first generation, muscular abnormalities in the second, and full systemic manifestations thereafter.[5]

Other Dystrophies

Two rare nonprogressive myotonic syndromes are myotonia congenita and paramyotonia. Myotonia congenita is an autosomal dominant disease characterized by general myotonia and hypertrophy of some voluntary muscles, first described by Thomsen in himself and several relatives. The myotonia is widespread, and the muscles are frequently increased in size. Cataract is not commonly associated with this syndrome. In paramyotonia, skeletal muscle contracture develops only on exposure to cold.

Preanesthetic Assessment

The preoperative evaluation of the patient with muscular dystrophy should focus on cardiac and pulmonary involvement. Because decreased cardiopulmonary reserve is often masked by impaired skeletal muscle function and limited exercise capacity, a more detailed invasive evaluation of the cardiorespiratory system may be necessary.

History and Physical Examination

As the muscular dystrophies are progressive disorders, the degree of muscle involvement and incapacity must be noted. The wheelchair-bound patient is at greater risk for postoperative cardiopulmonary complications than is a child in the early stages of the disease. Repeated chest infections and a diminished ability to cough are signs of weak respiratory muscles. Cardiac involvement may previously have been manifested as congestive heart failure or dysrhythmias.

Musculoskeletal examination determines whether weakness is limited to certain muscle groups or is diffuse, indicating advanced disease. Auscultation of the heart and lung fields may reveal a murmur, dysrhythmia, evidence of congestive heart failure, or pneumonia. Obesity and scoliosis may contribute to respiratory impairment.

Cardiac Assessment

The ECG must be carefully reviewed. If any abnormalities (as outlined in Table 2) are present, cardiac catheterization should be considered. Increased pulmonary wedge pressure and low cardiac index indicate the need for preoperative placement of a pulmonary artery catheter.[19]

Although once thought to be beneficial, longterm prophylactic digitalization has been shown to have no therapeutic value.[20] Digitalis is probably indicated only to improve ventricular contraction when there is frank congestive failure.

Episodes of unexplained hypotension suggesting adrenocortical suppression with poor response to vasopressors, may require steroid therapy.[5]

Echocardiography is indicated to document the degree of cardiac involvement and confirm the presence or absence of mitral valve prolapse (MVP). As it is performed in the resting state, it may not reflect the ability of the myocardium to respond to stress.[21] Although MVP need not constitute a significant anesthetic risk, the presence of bivalve prolapsing disease has been associated with sudden decompensation.[22] Radionuclide studies and exercise tests also have been used to evaluate cardiac involvement.

Respiratory Assessment

Three main areas must be considered in the assessment of respiratory function.

Pulmonary function tests. Vital capacity is often reduced, the major reduction being in expiratory reserve volume. Maximum breathing capacity also is decreased, as is maximal expiratory pressure. Respiratory function tests should include one-second forced expiratory volume and, if possible, minute ventilation, maximum breathing capacity, and total lung volume.

The risk of surgery varies with vital capacity. If vital capacity is greater than 45% of predicted, mechanical ventilation is not usually needed postoperatively; if it is less than 30% of predicted, serious postoperative complications ensue even with mechanical support.[23] In the patient with scoliosis the rate of deterioration of pulmonary function as assessed by vital capacity can be reduced from 20% to 5% per year by spinal fusion. Thus, although Harrington rod instrumentation constitutes major surgery, it may prove beneficial, as in our patient. Quinine has been shown to improve respiratory status by decreasing myotonia.[18] Although preoperative respiratory therapy may do little to improve the respiratory mechanics, it will serve to familiarize the patient with the equipment and allow more-efficient postoperative use.

Central nervous system effects. Somnolence and personality changes in patients with muscular dystrophy have raised the suggestion that the disease may have direct or secondary CNS manifestations. Although the CO_2 response curve is shifted to the right, $PaCO_2$ is often mildly elevated, and oxygen saturation on room air is usually low; these changes generally are considered too minor to cause cerebral abnormalities. Also, normalization of gases may not improve mental status or respiratory ability. Thus, the consensus is that muscular dystrophy does have direct CNS effects that may contribute to respiratory problems, especially postoperatively.[24] If a patient exhibits decreased mentation and somnolence preoperatively, postoperative ventilatory support may be necessary.

Oropharyngeal reflexes. Difficulty in swallowing and blunted pharyngeal reflexes are common in patients with muscular dystrophy, leading to aspiration pneumonia and lung abscesses. This complication appears to be confined to males.[5] Dysphagia is not improved by quinine, procainamide, or steroids, perhaps because the problem parallels muscle atrophy rather than myotonia. Poor gastric motility and decreased absorption of oral medication also have been noted.[5] In assessing patients preoperatively, an indication of swallowing difficulties may be obtained if the patient has a chronic cough, frequent pneumonias, emphysema, or bronchiectasis—all of which suggest recurrent aspiration.

Preoperatively, attempts should be made to treat any infections and provide respiratory therapy with bronchodilators as necessary. A combination of respiratory and swallowing difficulties may indicate the need for postoperative ventilatory support after even minor surgery.

Laboratory Data

Large elevations in serum enzymes, especially creatine phosphokinase (CPK), may be seen in all of the progressive muscular dystrophies. Aldolase, CPK, serum glutamic oxaloacetic transaminase (SGOT), lactic dehydrogenase (LDH), and serum glutamic pyruvic transaminase (SGPT) are the most commonly measured enzymes. In DMD the CPK level is highest early in the disease and then falls as muscle atrophy occurs. The cardiac muscle component is increased but correlates poorly with myocardial involvement.[1] Serum glucose may be elevated in myotonic dystrophy. The glucose tolerance test is abnormal, with delayed utilization of glucose and a slow return to normal after glucose load.

Anesthetic Plan

Premedication should be cautiously administered or avoided altogether. Anticholinergic agents may precipitate dysrhythmias. Narcotics and sedatives increase the risk of respiratory depression. Patients with MyD are extremely sensitive to the respiratory depressant effects of barbiturates, benzodiazepines, and opioids. This most likely reflects central ventilatory depression superimposed on respiratory muscle weakness.[15]

In addition to the standard monitors, arterial cannulation is useful in major procedures for rapid hemodynamic assessment and blood gas analyses. For surgery with large fluid shifts or in the presence of decreased cardiac reserve, a pulmonary artery catheter may be indicated. The anesthesiologist must weigh the risks of invasive monitoring in these circumstances (eg, inducing dysrhythmias in an irritable myocardium, pneumothorax with an already compromised respiratory status) against the benefits to be gained. The ECG leads monitored should provide information relating to rhythm disturbances (lead II) and possibly ischemic changes (V_5). Because of the unpredictable response to muscle relaxants, a peripheral nerve stimulator must be used to monitor the degree of blockade.

As mentioned, gastric hypomotility and decreased pharyngeal muscle strength increase the risk of aspiration in both DMD and MyD.[25,26] Therefore, a strict fasting period of at least 6 hours is required before elective cases. Antacid or cimetidine therapy is indicated. Post-operative gastric dilatation and the increased risk of regurgitation associated with it may be prevented by placement of a nasogastric tube.[27]

There are extensive series describing the uneventful administration of anesthesia to children with muscular dystrophy. However, there are also numerous reports of anesthesia-related complications in these patients. Although early reports indicated extreme sensitivity to thiopental, subsequent studies have shown that muscular dystrophy patients are very sensitive to all respiratory depressants.[5]

Succinylcholine administration in DMD has been associated with exaggerated potassium release, rhabdomyolysis, myoglobinuria, and even cardiac arrest.[29] Certain patients with DMD are susceptible to malignant hyperthermia, and dantrolene should be readily available,[30] although it should not be used prophylactically, because further weakness may be induced. Myocardial dysfunction may make the patient more sensitive to the depressant effects of volatile anesthetics. These problems may be minimized by the use of narcotics and nondepolarizing muscle relaxants instead of potent inhalation agents and succinylcholine.[21] Nondepolarizing relaxants may result in prolonged blockade and should be used cautiously. Hypothermia should be avoided, as shivering may be ineffective in restoring temperature postoperatively. Local or regional anesthesia may be adequate for many procedures, avoiding the risks of general anesthesia. Verapamil, often used to control dysrhythmias in otherwise normal persons, is contraindicated in DMD because it may precipitate respiratory failure and heart block.[31,32,33]

In MyD, succinylcholine injection can produce prolonged skeletal muscle contraction, making adequate ventilation difficult. An increased potassium level also may exacerbate the myotonia and may form the basis of the adverse reaction to succinylcholine.[14] In general, producing muscle relaxation in a myotonic patient is challenging. Nondepolarizing muscle relaxants block only motor nerve impulses and have no effect on percussion myotonia. A similar effect is seen with nerve blocks. Normal response to curare has been demonstrated.[31]

Reversal with neostigmine theoretically may precipitate myotonia, although it has been used without complications.[31] Inadequate reversal and even prolonged muscular weakness may be seen after neostigmine use in myotonia, possibly resulting from an acetylcholine-induced depolarization blockade.[28] Because the response to reversal is unpredictable, it is prudent to use short-acting nondepolarizing agents and continue mechanical ventilation if residual paralysis is present postoperatively.

Occasionally the jaw may be weak enough for intubation without neuromuscular blockade, and only a small dose of thiopental may suffice for induction. As in DMD, cardiomyopathy renders the patient more susceptible to myocardial depression when potent inhalation agents are used. If cardiac conduction abnormalities are present, agents such as halothane, known to delay conduction in the His-Purkinje system, should be avoided.[16] If volatile agents are used, ventilation must be controlled, as apnea will occur even at low expired concentrations postoperatively. Hypothermia must be avoided because shivering may induce myotonic contracture postoperatively.

The anesthetic management of other rare forms of muscular dystrophy is similar to that in DMD. In facioscapulohumeral dystrophy, the response to atracurium appears to be normal,[34] but in ocular muscular dystrophy small doses of curare (as low as 10% of normal) may result in complete paralysis.[35]

The main postoperative concern in all of the muscular dystrophies is adequate ventilation. Often mechanical ventilation is required. In DMD, pulmonary insufficiency may occur 5–36h postoperatively despite seemingly adequate recovery from anesthesia.[25] Use of an apnea monitor and careful observation in an intensive care setting are recommended. DMD patients are rarely justifiable candidates for ambulatory or same-day surgery.

References

1. Duncan P.: Neuromuscular diseases, in Katz J., Steward P. (eds): *Anesthesia and Uncommon Pediatric Diseases.* Philadelphia: WB Saunders, 1987, pp 516–8.

2. Duchenne G.B.A.: De l'électrisation localisée et son application à la pathologie et à la thérapeutique, 2nd ed. Paris: Baillière et fils, 1861.
3. Duchenne G.B.A.: Recherches sur la paralysie musculaire pseudohypertrophique ou paralysie myosclérotique. *Arch Gen Med* 1868, 11:5–25,179–209, 305–21, 421–3, 552–88.
4. Zundel W.S., Tyler F.H.: Muscular dystrophies. *N Engl J Med* 1965, 273: 537,596.
5. Miller J., Lee C.: Muscle diseases, in Katz J., Benumof J., Kadis L.B. (eds): *Anesthesia and Uncommon Diseases.* Philadelphia: WB Saunders, 1981, pp 530–7.
6. Hancock W.: Palpitations and abnormal ECG in muscular dystrophy. *Hosp Pract* 1989, 24:21,24.
7. Zatuchni J. et al: The heart in progressive muscular dystrophy. *Circulation* 1951, 3:846–53.
8. Fertoff B.W., Rao J.S.: Muscular dystrophy, dystrophies, in Adelman G. (ed): *Encyclopedia of Neuroscience, vol 2.* Boston: Birkhauser, 1988, pp 723–4.
9. Rowland L.P., Layzer R.B.: X-linked muscular dystrophies, in Vinken PJ, Bruyn G.W. (eds): *Handbook of Neurology.* Amsterdam: North Holland, 1979, 40:349–402.
10. Roses A.D., Pericak-Vance M.A., Yamaoka L.H., Stubblefield E., Stajiach J., Vance J.M., Roses M.J., Carter D.B.: Recombinant DNA strategies in genetic neurological diseases. *Muscle Nerve* 1983, 6:339–55.
11. Hayden M.R., Nichols J.L.: Molecular genetic approaches to the study of the nervous system. *Dev Neurosci* 1983, 6:189–214.
12. Merritt H.H., Rowland L.P.: Muscle, in Merritt H.H. (ed): *A Textbook of Neurology,* 6th ed. Philadelphia: Lea & Febiger, 1979, 576–96.
13. Baldwin B.J., Talley R.C., Johnson C., Nutter D.O.: Permanent paralysis of the atrium in a patient with facioscapulohumeral muscular dystrophy. *Am J Cardiol* 1973, 31:649–53.
14. Aldridge L.M.: Anaesthetic problems in myotonic dystrophy. *Br J Anaesth* 1985, 57:1119–30.
15. Stoelting R., Dierdorf S., McCammon R.: Anesthesia and Co-existing Disease. New York: Churchill Livingstone, 1988, pp 623–5.
16. Dierdorf S.: Rare co-existing diseases, in Barash P., Stoelting R., Cullen B. (eds): *Clinical Anesthesia.* Philadelphia: JB Lippincott, 1989, p 440.
17. Bradley W., Rebeiz J.: Progressive muscular dystrophy and chronic myopathies, in Petersdorf R. et al (eds): *Harrison's Principles of Internal Medicine.* New York: McGraw-Hill 1983, pp 2188–90.
18. Kilburn K.H. et al: Cardiopulmonary insufficiency in myotonic and progressive muscular dystrophy. *N Engl J Med* 1959, 261:1089–96.
19. Meyers M.B., Garash P.G.: Case history number 90: Cardiac decompensation during enflurane anesthesia. A patient with myotonia atrophica. *Anesth Analg* 1976, 55:433–6.

20. Fowler W.M., Pearson C.M., Egstrom G.H., et al: Ineffective treatment of muscular dystrophy with an anabolic steroid and other measures. *N Engl J Med* 1965, 272:875–872.
21. Sethna N. et al: Anesthesia-related complications in children with Duchenne muscular dystrophy. *Anesthesiology* 1988, 68:462–5.
22. Marks A.R., Choong C.Y., Sanfillipo A.J., et al: Identification of high-risk and low-risk subgroups of patients with mitral valve prolapse. *N Engl J Med* 1989, 320:1031–6.
23. Milne B., Rosales J.K.: Anesthetic considerations in patients with muscular dystrophy undergoing spinal fusion and Harrington rod insertion. *Can Anaesth Soc J* 1982, 29:250–4.
24. Tsueda K., Shibutani K., Lefkowitz M.: Postoperative ventilatory failure in an obese myopathic woman with periodic somnolence. A case report. *Anesth Analg* 1975, 54:523–6.
25. Smith C.L., Bush G.H.: Anaesthesia and progressive muscular dystrophy. *Br J Anaesth* 1985, 57:1113–18.
26. Ishizawa Y., Yamaguchi H., Dohi S., et al: A serious complication due to gastrointestinal malfunction in a patient with myotonic dystrophy. *Anesth Analg* 1986, 65:1066–8.
27. Cobham J.G., Davis H.S.: Anesthesia for muscular dystrophy patients. *Anesth Analg* 1964, 43:22–32.
28. Wislicki L.: Anaesthesia and post-operative complications in progressive muscular dystrophy. *Anaesthesia* 1962, 17:482–7.
29. Henderson W.A.V.: Succinylcholine-induced cardiac arrest in unsuspected Duchenne muscular dystrophy. *Can Anaesth Soc J* 1984, 31:444–6.
30. Kelfer H.M., Singer W.D., Reynolds R.N.: Malignant hyperthermia in a child with Duchenne muscular dystrophy. *Pediatrics* 1983, 71:118–9.
31. Mitchell M.M., Ali H.H., Savarese J.J.: Myotonia and neuromuscular blocking agents. *Anesthesiology* 1978, 49:44–8.
32. Zalman F., Perloff J.K., Durant N.N., et al: Acute respiratory failure following intravenous verapamil in Duchenne's muscular dystrophy. *Amer Heart J* 1983, 105:510–11.
33. Emerg A.E.H., Skinner R.: Double blind controlled trial of a "calcium blocker" in Duchenne muscular dystrophy. *Cardiomyology* 1983, 2:13–23.
34. Dressner D.L., Ali H.H.: Anaesthetic management of a patient with facioscapulohumeral muscular dystrophy. *Br J Anaesth* 1989, 62:331–4.
35. Jacob J.G., Varkey G.P.: Curare sensitivity in ocular myopathy. *Can Anaesth Soc J* 1966, 13:449–52.

CHAPTER 11

The Asthmatic Patient

Lauren A. Plante, M.D.

Case History *A 29-year-old woman with a 15-year history of asthma was hospitalized on the obstetrical service at about 37 weeks' gestation. During this pregnancy alone she had been admitted four times for asthma attacks; the last was a bout of status asthmaticus, less than 2 weeks previously, requiring intubation and mechanical ventilation for several days. She had subsequently been managed on theophylline 300mg po b.i.d., metaproterenol by nebulizer q4h, and prednisone 20mg daily. On physical examination she was afebrile; respiratory rate, 18/min; pulse, 100/min; blood pressure, 110/70mmHg in the lateral recumbent position. Mild diffuse expiratory wheezes were heard. Peak flow was 180L/min. Hemogram showed a leukocyte count of 10.3 without shift; hematocrit 29%; arterial blood gas on room air was reported as 7.42/30/98, and the serum theophylline level was 8.9μg/ml. The patient was proposed for cesarean section because of breech presentation.*

Pathology and Pathophysiology

Asthma is a relatively common disorder; estimates of its prevalence in the U.S. population range from 3% or 4%[1] to as high as 9% or 10%[2] (the higher figures include self-diagnosis and less stringent criteria).

A useful working definition is provided by the American Thoracic Society and the American College of Chest Physicians: asthma is a disease characterized by increased responsiveness of the airways to various stimuli, manifested by widespread airway narrowing, which changes in severity either spontaneously or as a result of therapy.[3] Various types of asthma are subsumed under this definition. The best-known classification is extrinsic versus intrinsic asthma.

Reviewed by Dr. Greg Ruskin, Assistant Professor, Department of Anesthesiology, Albert Einstein College of Medicine/Montefiore Medical Center.

In extrinsic asthma the disease is provoked by an inhaled allergen (eg, mold, insect aeroallergen, cat dander). This type of asthma is IgE-mediated and is associated with atopy, elevated serum IgE levels, and often eosinophilia of the blood or sputum. Skin testing with extract of the specific allergen is positive.

Extrinsic asthma usually begins in childhood and may disappear with age. Attacks, however severe, tend to be self-limited. Treatment of the acute episode generally suffices, with no need for chronic therapy.

Intrinsic asthma affects a much larger group, up to 50% of all people with asthma. In this type of asthma no specific immunoprovocative agent can be identified. It tends to occur in an older age group than does extrinsic asthma. The attacks are more severe, much more difficult to break, and much more often fatal. People with this type of asthma require therapy not only to treat the acute episode but on a chronic basis to prevent exacerbation. Manifestations of both intrinsic and extrinsic disease may be found in the same individual.

Additional types of asthma exist, such as those related specifically to exercise, to various agents, and to the ingestion of aspirin and other non-steroidal anti-inflammatory drugs. Regardless of etiology, the pathologic and pathophysiologic picture is constant. The patient with acute asthma presents with complaints of dyspnea and chest "tightness." Wheezing may be remarked by the patient and will be heard by the examiner. A prolonged expiratory phase is the rule. The attack may remit spontaneously, improve with therapy, or worsen over time.

During an asthma attack,[4–6] the major pathologic changes are those associated with contraction of bronchial smooth muscle. In the patient with longstanding asthma, concentric hypertrophy of muscle within bronchial walls may further reduce airway diameter. Bronchoconstriction may result from irritant receptors or from the release of inflammatory mediators, many of which have been identified, including histamine, bradykinin, leukotrienes C_4, D_4, and E_4, prostaglandins D_2, F_2, and G_2, adenosine, thromboxane A_2, and platelet-activating factors. Several of these are released with mast cell degranulation.

With the increase in vascular permeability provoked by these mediators, plasma proteins leak from capillaries into the interstitium of the lung, and interstitial edema results throughout the airways. Leukocytes are attracted, and the bronchial mucosa becomes infiltrated with eosinophils, neutrophils, plasma cells, lymphocytes, and macrophages.

Increased production and viscosity of mucus contributes further to luminal narrowing. The clearance of mucus and cellular debris is compromised by desquamation of epithelial cells, with loss of cilia, diminished function of remaining cilia, and replacement of ciliated epithelium by goblet cells.

Early in an asthmatic attack, while simple bronchoconstriction is still the major factor in airway narrowing, the attack may be aborted by appropriate pharmacologic intervention with bronchodilators. As the episode progresses and more airflow obstruction occurs because of mucus plugging, inflammation, and airway edema, such pharmacotherapy is markedly less useful.

Airway narrowing and the resultant resistance to airflow have profound effects on pulmonary function.[5,7] The obstruction is not uniform, so ventilation-perfusion mismatch ensues, with an increase in wasted ventilation and a decrease in PaO_2. Closing volume is increased because high negative intrapleural pressures must be generated to force expiration past a narrowed lumen, hence, airways close prematurely during expiration. Air-trapping with subsequent hyperinflation of the lungs results.

All static lung volumes are increased, especially functional residual lung capacity (FRC) and total lung capacity (TLC). Indeed, this may be so marked in cases of severe bronchospasm that FRC approximates TLC, leaving minimal inspiratory reserve.

The increase in FRC, which also is thought to contribute to the increased ventilatory drive seen in acute asthma, represents a means of maintaining airway patency that is achieved at great cost. A much higher transpulmonary pressure gradient must be generated, with a pronounced increase in negative intrapleural pressure. Pulmonary hypertension can result as the right ventricle is pressure-and volume-loaded. The left ventricle is relatively underfilled, with increase in left ventricular afterload.

The more closely FRC approaches TLC, the more elastic work must be done by lung and chest wall to maintain simple tidal breathing. The end-expiratory length of the inspiratory muscles—the diaphragm and the intercostals—is increased with hyperinflation of the lungs, but this impairs their ability to contract properly with inspiration. The length-tension characteristics of these muscles are such that they function most efficiently at shorter resting (end-expiratory) lengths.

Compensation is attempted by recruiting accessory muscles such as the scalene and sternocleidomastoid so that normal tidal volume can be maintained at higher FRC. Metabolic demand is increased, and less efficiently functioning muscles are required to perform increased elastic and resistive work. With the diminishing of elastic recoil and the loss of diaphragmatic contractility, expiration can no longer be passive.[8] Instead, intercostals, abdominal muscles, and the diaphragm itself must all actively contract during expiration, further increasing metabolic demand.

This increase in the work of breathing can be dramatic. Whereas a person with normal lungs can increase ventilation by an additional 1L/min at an additional oxygen cost of 1ml/min, during an attack the asthmatic patient must increase oxygen consumption by 10ml/min to achieve that same increase in ventilation. The increased oxygen demand, accompanied by the increased CO_2 production that typifies this added work of breathing, comes at a time when ventilation-perfusion mismatch produces arterial hypoxemia. Any attempt to compensate by further increasing ventilation only aggravates the problem, and respiratory failure results.

Diagnosis of Asthma

Most often the anesthesiologist will be presented with a patient who has already been diagnosed as asthmatic. Criteria for the diagnosis are signs or symptoms of bronchial hyperreactivity plus evidence of reversibility. The presence of wheezing is neither necessary nor sufficient for diagnosis. For example, dyspnea without wheezing, particularly after exercise, may represent asthma.

In other cases, chronic or recurrent cough may be the sole presenting manifestation of asthma.[9] The anesthesiologist should be mindful of these possibilities when eliciting a patient's history and may wish to consider referral for formal pulmonary function testing. In certain cases, bronchoprovocative testing with inhaled methacholine may be the only way to confirm or rule out the presence of asthma.[10]

Wheezing may represent disorders other than asthma.[11] The differential diagnosis includes emphysema, chronic bronchitis, cystic fibrosis, bronchiolitis (an infectious process in children, of varying etiologies in adults), sarcoidosis, hypersensitivity pneumonitis, pulmonary embolism, aspiration, central airway obstruction, "cardiac asthma" secondary to left ventricular failure, and carcinoid syndrome. The importance of differentiating any of these conditions from true asthma is readily apparent. In addition many of these disorders respond poorly to standard antiasthmatic therapy although they respond well to treatment of the specific disease.

Treatment of Asthma

A wide variety of drugs is available to prevent and to treat asthmatic attacks (see Table 1).

Several excellent recent reviews are available for those seeking in-depth knowledge of the pharmacology, appropriate choice, and actions of those agents.[12–16]

Preanesthetic Measures

Several tests are available to assess pulmonary function.[17–18] Simple bedside spirometry is generally sufficient in the asthmatic, and various

TABLE 1. Drugs Useful in the Prevention and Treatment of Asthma Attacks

Type	Drug	Route and Dose (Adult)
Adrenergic agents Activate adenyl cyclase, cAMP		
Mixed α and β	Ephedrine	Oral:25mg q4–6h
	Epinephrine	Subcutaneous: 0.3mg Inhalation: 0.25–0.6mg q2–3h
Mixed β_1 and β_2	Isoproterenol	Inhalation: 0.25–0.6mg q1–2h
β_2-selective	Isoetharine	Inhalation: 0.25–1.0mg q4–6h
	Metaproterenol	Oral: 20mg q6h Inhalation: 1.30–1.95mg q4–6h (MDI), 10–15mg q4–6h (NB)
	Albuterol	Oral: 2–4mg q6–8h Inhalation 180–270μg q6–8h (MDI), 1.0–2.5mg q6–8h (NB)
	Terbutaline	Oral: 2.5–5mg q6h Subcutaneous: 0.25mg q4–6h Inhalation 0.2–0.4mg q4–6h
	Fenoterol*	Oral: 50010mg q6h Inhalation: 0.1–0.3mg q6–8h
Methylxanthines Mechanism unclear; inhibit phosphodiesterase with resulting increase in cAMP; inhibit intracellular calcium activity; inhibit prostaglandin actions; direct antagonism of the adenosine receptor	Aminophylline	Oral: 13–16mg/kg/d titrated to clinical effect and serum level Intravenous: loading dose 5–6mg/kg, maintenance 0.2–0.9 mg/kg/h
Anticholinergic agents Inhibit generation of cyclic GMP, causing bronchodilation, and inhibit tracheobronchial secretions	Atropine	Oral: 0.4–1mg q6h Inhalation: 25–75μg/kg q3–6h
	Glycopyrrolate	Intravenous, subcutaneous, or inhalation: 4–5μg/kg q2–8h
	Ipratropium	Inhalation: 20–40μg q4–8h
Cromolyn sodium Stabilizes mast cells; useful only in prophylaxis	Cromolyn sodium	Inhalation: 20mg q6h
Corticosteroids Mechanisms ill-defined; potentiate response to aminophylline and sympathomimetics; reduce inflammation and airway edema		Oral: drugs and doses extremely variable Inhalation: beclomethasone 80–160μg q6h

* not currently available in U.S. MDI=metered dose inhaler NB=nebulizer

hand-held devices are available. Many asthmatics will be familiar with such devices; some will even know their own best peak flow.

Spirometry is based on a vital capacity or forced vital capacity maneuver. The subject inhales as deeply as possible, then exhales as fully and forcefully as he can. Volume is recorded as a function of time. The maneuver is thus designed to assess flow rates.

The functional vital capacity (FVC) is divided into several time intervals (see Figure below).

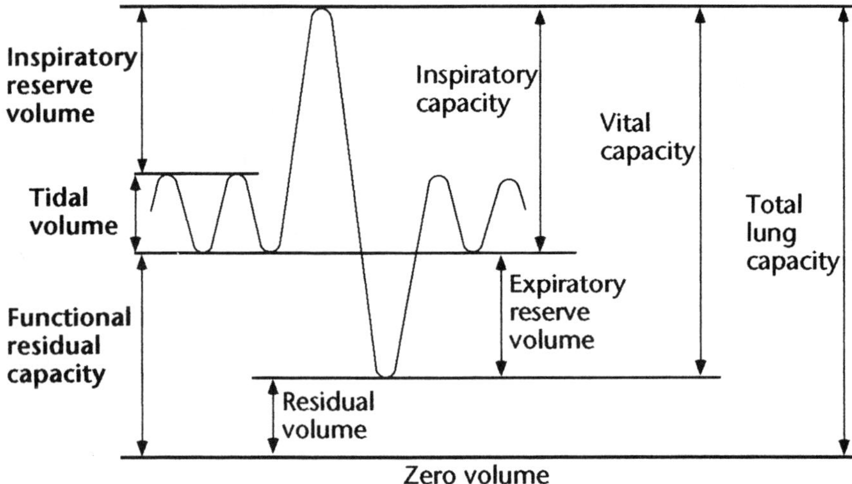

Adapted from Nunn J.F.: Applied Respiratory Physiology 3rd ed. London: Butterworths, 1987, p.39.

The most reproducible of these is the forced expiratory volume measured within the first second of flow, or FEV_1. Persons with normal lungs should be able to exhale 75%–80% of FVC within this time, whereas those with increased airway resistance will exhibit reduction in airflow rates and a decrease in both FEV_1, and the FEV_1/FVC ratio.

The early portion of the FVC curve is, however, more effort-dependent, and the terminal portion can be poorly recorded (as, for example, when the sensation of dyspnea causes the subject to terminate the FVC maneuver prematurely). The midportion of the curve—measured between the first and last quarters of the FVC spirogram—is considered the most effort-independent portion and is the most sensitive indicator of small airways obstruction. This parameter is known as the forced expiratory flow ($FEF_{25\%-75\%}$) or as the midmaximal expiratory flow rate (MMEFR).

Although other parameters can be generated, these are the most useful for detection of obstructive airway disease. Characteristics of obstructive pulmonary disease are normal or decreased vital capacity, decreased FEV_1, and decreased $FEF_{25\%-75\%}$. They are evaluated against a set of "normal" numbers based on the patient's age, sex, and height. If FEV_1, or $FEF_{25\%-75\%}$ is at least 75% of predicted, this is considered normal, mild obstructive disease is present at flow rates 65%–75% of predicted, moderate disease at 50%–65%, and severe obstructive disease if flow rates are less than 50% of predicted.[19]

Once the presence of airflow limitation has been established, the next step is to determine whether it is reversible. This is accomplished by repeating spirometry approximately 20min after administration of an inhaled bronchodilator. A significant bronchodilator response requires an improvement of at least 15% in at least two of the three parameters FVC, FEV_1, and $FEF_{25\%-75\%}$.

Many factors can affect the response to inhaled bronchodilators: the chronicity and severity of airflow limitation, dosage and technique of bronchodilator administration, complete withdrawal of preexisting therapy, smoking history, age of patient, presence of respiratory infection, even time of day. Therefore, more accurate results are obtained if testing is done in a pulmonary function laboratory, where more formalized and extensive testing of patients with respiratory disorders can be obtained.

Although seldom required, other pulmonary tests include static lung volume measurements, either by body plethysmography or by gas dilution, and serial lung volume measurements, which may indicate a bronchodilator response that is not evident by spirometry. Although flow rates cannot be directly assessed by the latter technique, relief of airway obstruction is inferred from decreases in residual volume and FRC.

Flow-volume curves also can be generated as a subject performs the FVC maneuver. Flow is recorded as a function of volume; it provides the same information as the standard spirogram but represents it in a different form. It is useful in differentiating between diffuse airways disease and central obstruction. Distinction also can be made between fixed and variable obstruction, and the location—either intra- or extrathoracic—can be pinpointed.

Additional tests include diffusing capacity, direct measurement of airways resistance, measurement of static and dynamic compliance of the lung, and measurements of closing volume. These are useful in screening for subclinical pulmonary disease and may have significant predictive value for the later development of overt disease, but they will rarely be specifically helpful to the anesthesiologist.

Evaluation

In evaluating an asthmatic patient for a proposed procedure, the anesthesiologist must be mindful of many factors.

Consideration must be given to (1) baseline pulmonary function, (2) current pulmonary function, (3) therapeutic regimen, (4) proposed surgical procedure and (5) anesthetic options.

A thorough history should be elicited. Information should be obtained regarding the duration and severity of the disease. At what age did the patient first become asthmatic? Has the severity changed over time? (Childhood asthma that has since disappeared is of little importance in the adult.) What factors precipitate attacks (eg, exercise, cold weather, aeroallergens)? How frequent are the attacks? How often has the patient required hospitalization or treatment in the emergency room? Have intubation, catecholamines, or steroids been required? Does the patient smoke, or is he exposed to others' smoking? (This question refers not only to tobacco but to any recreational drug that can be consumed by smoking). Is the patient currently experiencing any symptoms of respiratory infection? What is the patient's best peak flow? What current medications are being ingested?

On physical examination the anesthesiologist should pay particular attention to auscultation of the chest, remaining mindful of the fact that the physical findings in asthma may be both variable and misleading. Surgery should be postponed for a patient who is actively wheezing, if possible, so that pulmonary function can be optimized. The absence of wheezing, on the other hand, may indicate the quiescent period between attacks or, conversely, marked impediment to airflow (the "silent chest" sometimes noted in status asthmaticus). Other physical signs indicative of severe bronchospasm include tachypnea, use of accessory muscles of inspiration, pulsus paradoxus, cyanosis, or disturbances of consciousness.

The chest x-ray is more useful in excluding other causes of pulmonary disease than in evaluating asthma qua asthma. Emphysematous changes, bronchiectasis, tracheal obstruction, or pneumonia may be diagnosed. The chest x-ray and arterial blood gas analysis are unlikely to be helpful in the asymptomatic patient.

The electrocardiogram should be reviewed. Evidence of right axis deviation or right heart strain (p pulmonale, right ventricular hypertrophy) suggests persistently elevated pulmonary vascular resistance, hence more severe pulmonary disease. The anesthesiologist should also look for tachydysrhythmias including premature ventricular contractions or ischemic changes such as ST depression or T wave changes which may suggest either an acute attack or bronchodilator toxicity.

If a patient is taking theophylline, serum drug levels should be measured. The therapeutic range for theophylline is generally cited as 10–20μg/ml. Some patients respond outside this range;[16] there are some who require levels of 20–25μg/ml to achieve an adequate effect, as there are some who will respond at levels as low as 2–5μg/ml.

Simple spirometry should be a routine part of the preanesthetic assessment in asthmatlcs. A peak flow meter also may be used; the correlation with FEV_1 is about 90%. If measurements suggest moderate or severe disease, consultation with a pulmonologist or internist may be indicated and a therapeutic regimen designed to maximize the patient's pulmonary reserve prior to elective surgery.

Premedication

Patients on oral theophylline or beta agonists should continue their regular regimen on the day of surgery, except for those patients whose serum theophylline levels are high enough to raise concern about drug toxicity. Patients receiving inhaled drugs (beta agonists, cromolyn, ipratroprium, or steroids) also should continue their regular regimen. Those who use metered-dose inhalers only intermittently should be premedicated, if only prophylactically, with two or three puffs given several hours before the anticipated time of surgery.

$Beta_2$-selective agents are preferred over the nonselective agents such as isoproterenol. The exact timing depends on the individual agent used. Metaproterenol, for example, has a duration of action estimated at 2–4h, so it should be given closer to the time of surgery than is necessary for terbutaline, fenoterol, or albuterol, whose duration of action is 4–6h. Although any of these drugs can be given "on call" in the operating room, effects such as nervousness, tachycardia, or palpitations may exacerbate the patient's anxiety about surgery.

Patients on intravenous theophylline should have their dosage decreased by approximately one third on the morning of surgery and continued intraoperatively at this lower infusion rate. The rationale for this recommendation is decreased clearance of aminophylline secondary to diminished liver blood flow under general anesthesia, with the risk of toxic levels of aminophylline developing should the usual doses be continued.[20]

It is generally recommended that patients on steroids receive stress doses of steroids perioperatively. Those who take only inhaled steroids are at much lower risk of adrenal suppression and probably do not require supplementation.

It is popularly believed that anxiety can precipitate bronchospasm in susceptible individuals, although controlled studies are lacking. The preoperative visit by an anesthesiologist who can explain and reassure is often sufficient to allay anxiety. Those patients for whom additional preanesthetic sedation is required and who present no sign of impending respiratory failure can be given almost any of the common premedicants.[21–22]

Diphenhydramine is appealing; it provides both sedation and blockade of histamine-induced bronchoconstriction. Both hydroxyzine and droperidol decrease airway resistance. The barbiturates have been shown to increase bronchial tone in excised lung tissue, but there is no convincing evidence that this is true in vivo. For asthmatic patients who are sensitive to sulfur, avoidance of the thiobarbiturates is generally recommended. The histamine-releasing opioids, such as morphine and meperidine, also should theoretically be avoided in patients with bronchospastic disease. Should opioid premedication be required, fentanyl is suggested.

Finally, the use of H_2-receptor blockers (cimetidine, ranitidine) poses the risk of provoking bronchospasm. Histamine mediates bronchoconstriction via H_1 receptors, whereas bronchodilation is mediated by H_2 receptors; hence, selective blockade of H_2 receptors could unmask unopposed bronchoconstriction.

The routine use of antisialogog agents as premedication has all but disappeared now with the use of modern inhalation agents that are less sialorrhic than their predecessors, but these agents may be useful for the asthmatic patient. They inhibit release of cyclic GMP, which mediates bronchospasm; additionally, they dry up airway secretions, which can provoke bronchospasm in sensitive individuals. As many find the anticholinergic effects of flushing, dry mouth, and tachycardia distressing, it is probably better to avoid premedication with the anticholinergics but to include them as part of the induction of general anesthesia.

Anesthetic Plan

Conventional wisdom holds that regional or local anesthesia is preferred to general anesthesia for asthmatic patients. This caveat must be tempered with common sense. A particularly painful or difficult block, such as one involving repeated injections, or a block that is patchy or inadequate for the operation being performed offers little advantage over a well-conducted general anesthetic. If general anesthesia is elected, mask anesthesia is preferred where possible because it avoids the necessity of instrumenting the airway. The management and avoidance of bronchospasm has been recently reviewed.[23]

Any of the commonly used induction agents may be employed. Methohexital and ketamine are often used; ketamine has the unique advantage of providing significant bronchodilation. Among its disadvantages, however, are its tendency to stimulate excessive airway secretions, which themselves are bronchial irritants, and the occasional report of seizures following ketamine induction in patients receiving aminophylline. The concomitant administration of benzodiazepines is therefore advised in this setting. The thiobarbiturates also have been used successfully, although, as mentioned earlier, there are theoretical reasons to avoid them.

It is well appreciated that the presence of an endotracheal tube is an exceptionally potent stimulus to bronchospasm. This is further worsened by the application of surgical stimulus in an inadequately anesthetized patient. Therefore, the principle that the airway should not be instrumented until an adequate depth of anesthesia has been achieved is more important than the choice of anesthetic agent. Following an intravenous induction, it is good practice to continue with inhalation anesthesia with halothane or isoflurane in oxygen until a deep plane of anesthesia has been achieved.

Mask anesthesia is preferable, but if intubation should be required, it is important to assess depth before the tube is inserted. Muscle relaxants can be used prior to intubation, but they will make this assessment more difficult, and certainly their use often tempts the anesthesiologist to intubate before adequate conditions have been realized. Although coughing and bucking will be prevented, muscle relaxants have no effect on bronchial smooth muscle and bronchospasm often results.

A number of the commonly used muscle relaxants have been implicated in histamine release, specifically, d-tubocurarine, metocurine, atracurium, and succinylcholine. Pancuronium, which does not cause histamine release, promotes tachycardia, an effect that can be exaggerated in the presence of drugs commonly employed to treat asthma. This would seem to leave vecuronium as the optimal muscle relaxant in asthmatic patients. Anticholinesterase reversal agents have sometimes been reported to precipitate bronchospasm in asthmatics, but the simultaneous administration of anticholinergics is thought to be protective.

All of the potent inhalation agents are bronchodilators, and all have in fact been used to treat status asthmaticus. Mucociliary clearance is decreased, which may predispose the asthmatic to postoperative pulmonary complications. Inhalation anesthesia is associated with an increased risk of ventricular dysrhythmias in myocardium sensitized with epinephrine or other beta agonists, as well as with aminophylline. This risk is highest with halothane and lowest with enflurane; isoflurane occupies a middle ground. To avoid drying the airways, carrier gases should be warmed and humidified.

Nitrous oxide appears to be safe for clinical use in the asthmatic patients, although it is prudent to maintain concentrations below 60%. Patients with significant air-trapping or ventilation-perfusion mismatch are theoretically at more risk for diffusion of nitrous oxide into these air-filled spaces, with resultant dilution of oxygen concentration, or even pneumothorax. These concerns have been neither proved nor disproved.

Emergency surgery in the asthmatic patient poses particular problems.[22] The anesthesiologist must weigh the relative risks of aspiration of gastric contents against those of severe bronchospasm. If general anesthesia must be used, recommendations include preoxygenation, rapid-sequence induction with cricoid pressure, and intubation. The drugs most commonly employed in these circumstances are ketamine (alone or in combination with barbiturates) and succinylcholine. Because of the permutations possible—patient, situation, coexisting medical disease, and surgical procedure—blanket recommendations are impossible.

Management of Intraoperative Bronchospasm

Bronchospasm is easier to prevent than to treat. Once it occurs, precipitating factors must be identified and treated. For example, airway secretions and mucus plugging, improper endotracheal tube placement (against the carina), and too-light anesthesia are all easily addressed. Maintenance of adequate oxygenation is of cardinal importance. If bronchospasm persists, specific measures include discontinuing nitrous oxide and increasing the concentration of volatile agent. Specific bronchodilator drugs can be given.

The beta agonists can be delivered by aerosol or by nebulizer placed directly in the breathing circuit, and in cases where this is impossible, a dilute solution can be delivered directly into the trachea. This therapy should be pursued aggressively.

Glycopyrrolate (5μg/kg by inhalation or 0.2–0.4mg intravenously) or atropine (25–75μg/kg by inhalation) can be administered.[16] Steroids may be given. Although an effect is not expected until about 12 hours later, immediate action may occur. Aminophylline should be reserved for those cases in which all other therapies have failed. The pharmacokinetics of aminophylline in anesthetized patients have not been studied, but possible drug interactions with the volatile anesthetics are of concern, as is the risk of increased toxicity. Thus the loading dose should be decreased to approximately 3mg/kg, followed by a maintenance infusion of 0.4mg/kg/h.

If bronchospasm cannot be reversed, the operation should be terminated as expeditiously as possible. No attempt should be made to awaken the patient as this may only worsen the spasm. Blood gases must be checked; in all but the most severe cases oxygenation can usually be maintained. Ventilation and CO_2 exchange are impaired.

The anesthesiologist must evaluate airway pressures before deciding to increase tidal volume. If airway pressures are already high (ie, more than 50–60 mmHg) it is probably safer to accept hypercarbia than to risk tension pneumothorax. Arrangements should be made to continue postoperative mechanical ventilation and the patient should be transferred to an intensive care unit as soon as possible.

References

1. Bonner J.R.: The epidemiology and natural history of asthma. *Clin Chest Med* 1984, 5:557–65.
2. Evans R., Mullaly D.I., Wilson R.W., et al: National trends in the morbidity and mortality of asthma in the U.S.: Prevalence, hospitalization and death from asthma over two decades: 1965–1984. *Chest* 1987, 91:65S–74S.
3. Samet J.: Epidemiological approaches for the identification of asthma. *Chest* 1987, 91:74S–78S.
4. Drazen J.M., Boushey H.A., Hogate S.T., et al: The pathology of severe asthma: A consensus report from the workshop on pathogenesis. *J Allergy Clin Immunol* 1987, 80:429–36.
5. Hogg J.C.: The pathology of asthma. *Clin Chest Med* 1984, 5:567–72.
6. Reed C.E.: Immunologic aspects of asthma. *Clin Chest Med* 1984, 5: 599–606.
7. Lavietes M.H.: Ventilatory control in asthma. *Clin Chest Med* 1984, 5: 607–17.
8. Jackson L.K.: Functional aspects of asthma. *Clin Chest Med* 1984, 5: 573–87.
9. Corrao W.M., Braman S.S., Irwin P.S.: Chronic cough as the sole presenting manifestation of bronchial asthma. *N Engl J Med* 1979, 300:633–7.
10. Braman S.S., Corrao W.M.: Bronchoprovocation testing. *Clin Chest Med* 1989, 10:165–76.
11. Lillington L.A.: Differential diagnosis of asthma in adults: Bronchial asthma, occult asthma and pseudoasthma, in Gershwin M.E. (ed): *Bronchial Asthma: Principles of Diagnosis and Therapy,* 2nd ed., Orlando, Fla: Grune & Stratton, 1986, 139–151.
12. Shim C.: Adrenergic agonists and bronchodilator aerosol therpay in asthma. *Clin Chest Med* 1984, 5:659–68.

13. Dunlap N.E., Fulmer J.D.: Corticosteroid therapy in asthma. *Clin Chest Med* 1984, 5:669–83.
14. Jenne J.W.: Theophylline use in asthma: Some current issues. *Clin Chest Med* 1984, 5:645–58.
15. George R.B., Payne D.K.: Anticholinergics, cromolyn and other occasionally useful drugs. *Clin Chest Med* 1984, 5:685–93.
16. Seligman M.: Bronchodilators, in Chernow B. (ed): *The Pharmacologic Approach to the Critically Ill Patient,* 2nd ed. Baltimore: Williams & Wilkins, 1988, 436–50.
17. Welch M.H.: Ventilatory function of the lungs, in Guenter C.A., Welch M.H. (eds): *Pulmonary Medicine*, 2nd ed., Philadelphia: JB Lippincott, 1982, 171–95.
18. Gardner R.M., Crapo R.O., Nelson S.B.: Spirometry and flow-volume curves. *Clin Chest Med* 1989, 10:145–54.
19. Snider G.L., Kory R.C., Lyons H.A.: Grading of pulmonary function impairment by means of pulmonary function testing. Recommendations of the Committee on Pulmonary Physiology, American College of Chest Physicians. *Dis Chest* 1967, 52:270–5.
20. LoSasso A.M., Gibbs P.S., Moorthy S.S.: Obstructive pulmonary disease, in Stoelting R.K., Dierdort S.C. (eds): *Anesthesia and Coexisting Disease.* New York: Churchill Livingstone, 1983, 171–200.
21. Kingston H.G.G., Hirshman C.A.: Perioperative management of the patient with asthma. *Anesth Analg* 1984, 63:844–55.
22. Fung D.L.: Emergency anesthesia for asthma patients. *Clin Rev Allergy* 1985, 3:127–41.
23. Bishop M.J.: Bronchospasm: Managing and avoiding a potential anesthetic disaster in: Annual Refresher Course Lecures, *Amer Soc of Anes.,* Las Vegas, 1990, no. 272.

CHAPTER 12

The Patient Undergoing Neuroradiologic Procedures

Daran W. Haber, M.D.

Case History. *A 46-year-old woman presented with a history of focal seizures for 7 years, slowly progressive left-side weakness of the extremities and face, and gait disturbance beginning approximately 3 years prior to admission. She also had numbness of the left side of her body and face. Past medical history was otherwise remarkable only for occasional asthmatic attacks, relieved by nonprescription inhalers. She took no other medications. An allergy to shellfish was noted. On admission to the hospital, phenytoin (Dilantin®), dexamethasone (Decadron®), and ranitidine (Zantac®) were prescribed.*

CT was positive for a large contrast-enhancing mass in the right posterior parietal area (the scan was performed uneventfully). Angiography was subsequently scheduled and completed successfully. During angiography, however, the patient developed sudden shortness of breath and severe hypotension upon injection of contrast. Resuscitative measures required oxygen, subcutaneous epinephrine, I.V. crystalloid infusion, diphenhydramine, and steroids. She recovered without further complication.

The angiogram demonstrated a large, highly vascular meningioma in the right parasaggital posterior parietal area. Vascular supply to the tumor was noted to arise from posterior branches of the right middle meningeal artery. The patient was scheduled for endovascular embolization of feeding arterial branches to the tumor. Anesthetic consultation was requested. The patient was treated prophylactically with diphenhydramine and hydrocortisone sodium succinate (Solu-Cortef®).

Reviewed by Dr. Franklin Moser, Attending Physician, Department of Radiology, Lenox Hill Hospital, and Associate Professor, New York Hospital/Cornell Medical College, New York City.

Introduction

Neuroradiology has evolved to include sophisticated diagnostic and therapeutic procedures for central nervous system disease (Table 1.).

TABLE 1. Neuroradiologic Procedures

TABLE 1. Neuroradiologic Procedures
Computerized (axial) tomography (CT)
Magnetic resonance imaging (MRI)
Pneumoencephalography
Positron emission tomography (PET) and single photon emission computed tomography (SPECT)
Diagnostic angiography
Interventional neutroangiography

Computerized imaging and angiographic procedures provide detailed anatomy of intracranial and spinal lesions that previously could have been diagnosed only by exploratory surgery.

Certain interventional angiographic procedures may serve as an adjunct to surgery. Preoperative embolization of vascular lesions may help reduce intraoperative blood loss and make resection easier for the surgeon. More important, it may help to prevent catastrophic intraoperative rupture of the lesion. In some cases endovascular surgery under x-ray guidance may even take the place of more invasive surgical procedures.

The development of these procedures over the past two decades is the direct result of sophisticated computerized imaging systems that were previously unavailable. These procedures may be lifesaving and may dramatically reduce surgical morbidity. They are not, however, completely noninvasive and not without their own inherent risks (and problems).

Some of these procedures may be painful or discomforting. They may require immobility for extended periods of time, often in patients unwilling or unable to cooperate. They may be required for patients who are hemodynamically unstable or patients at risk for allergic reactions to contrast dye. Finally, the radiologic procedure itself, if invasive, may entail a risk of life-threatening complications such as intracranial hemorrhage.

For these reasons, the anesthesiologist is often called on to provide sedation or general anesthesia or simply to monitor the patient in the imaging or angiography suite (Table 2.).

TABLE 2. Indications for Anesthetic Monitoring During Neuroradiologic Procedures

Infants or children requiring immobilization or sedation
Patients with movement disorders requiring immobilization
History of allergic reaction to contrast dye
Patients with severe concurrent disease or hemodynamic instability
Interventional neuroangiography

Unlike the operating room, the radiology suite was often designed without accommodation for the anesthesiologist's array of monitoring, anesthetic, and resuscitative equipment. Space is often cramped, and accessibility to suction and oxygen lines may be limited.

Additionally, the x-ray equipment itself often poses logistic problems that may interfere with anesthetic management (for example, the magnetic field induced by magnetic resonance imaging makes the use of ferrous materials impossible). Once anesthetic equipment is added, the radiology suite may become a hazardously cramped maze. The anesthesiologist may find a unique set of interesting challenges that must be met for safe practice.

General Preanesthetic Considerations

In the imaging or angiography suite, equipment to ventilate, monitor, and resuscitate equal to that of the OR must be present. It must be situated in a way that provides maneuverability for the anesthesiologist and radiologist and easy access to the patient: a situation more easily imagined than realized.

Prior to the procedure the patient should be evaluated by an anesthesiologist, premedicated, and kept fasting, just as if he or she were undergoing any other surgical procedure. During the preanesthetic interview any signs or symptoms of intracranial hypertension should be sought (such as headache, nausea or vomiting, mental status changes, or papilledema). History of other concurrent diseases, use of medications, and any allergies should be elicited. Specifically, history of allergic reaction to iodinated contrast should always be sought. Considerations prior to and during anesthesia include stabilization of intracranial blood flow and metabolic rate, seizure prophylaxis, and protection against cerebral ischemia.[1]

The group of patients presenting for neuroradiologic procedures is diverse, and extremes of age are common. The patient may be very healthy and free of neurologic deficit or intracranial hypertension. On the other hand, the patient may be very sick, such as an elderly patient with multisystem disease or the victim of severe trauma. Children commonly require anesthesia for these procedures because they cannot cooperate by holding their heads still for the minutes to hours that are often required. This also holds true for psychotic or demented patients, or patients with disorders of uncontrolled movement, such as Parkinson's disease or seizure disorders.

Benzodiazepines are good choices as premedicants and as sedative adjuncts during surgery and during induction of anesthesia. Unlike narcotics, the benzodiazepines are usually not associated with respiratory depression (when given in low premedicant doses). Hypercapnia occurring with respiratory depression can increase cerebral blood flow and exacerbate intracranial hypertension. Short-acting barbiturates such as thiopental are particularly protective to the brain and are an excellent choice for induction of general anesthesia.

Both barbiturates and benzodiazepines decrease cerebral metabolic rate and blood flow, thus decreasing intracranial hypertension. Both classes of drugs also elevate the seizure threshold in a group of patients particularly prone to seizures, making their use beneficial. As can be imagined, the flood of metabolic activity associated with a generalized seizure may cause catastrophic brain damage in addition to the risks inherent in a tonic-clonic convulsion, including aspiration.

A recently introduced intravenous anesthetic agent, propofol, decreases cerebral blood flow and cerebral metabolic requirements for oxygen. Direct depression of the electroencephalographic pattern are observed with increasing dosage and prompt return to the awake state occurs with discontinuation of anesthesia.[2] Although the myocardial depressant effects of propofol may theoretically result in significant decrease in cerebral perfusion pressure, recent clinical studies suggest that this agent may prove to be a satisfactory alternative for neurosurgical anesthesia.[3] We have certainly used propofol (1–2 mg/kg) successfully to induce short periods of amnesia prior to painful injections. Prompt reversal of cerebral depression, ensures that patients are responsive within a few minutes and adequate neurologic assessment can be made.

Inhalation agents in high concentrations may cause a "cerebral steal" phenomenon by vasodilating vessels supplying healthy brain at the expense of poorly perfused brain tissue. Isoflurane seems to be less of an offender in this phenomenon than either enflurane or halothane, and in concentrations of 1 MAC or less the effect is probably clinically insignificant. Prior or concurrent hyperventilation further decreases this effect. Enflurane may lower the seizure threshold (especially in the presence of hypocapnia) and may also cause increased production of cerebrospinal fluid (CSF), and is therefore a poor choice of anesthetic agent.

The use of nitrous oxide in neurosurgical patients is still a fairly hotly debated issue because of its questionable effect of cerebral vasodilation as well as its poor solubility and tendency to fill and increase the pressure of air-filled cavities. Despite a long history of safe use, we prefer to avoid nitrous oxide in neurosurgical procedures, substituting air and low-dose fentanyl infusion.

Imaging

Computerized Tomography

Computerized (axial) tomography (CT) is at present the most commonly ordered neuroradiologic test.[1] CT scanning was introduced in the United States in the early 1970s. The procedure involves placement of the patient's head in the center of a large, rotating, doughnut-shaped gantry. The gantry projects an x-ray beam onto a set of detectors. Numerous tomographic films of horizontal sections of the head (or body) are taken and reconstructed by computer to form cross-sectional images.

The first commercial scanner, the EMI scanner, required the patient's head to be placed in a rubber bag surrounded by water. With newer machines this is no longer necessary. Early CT scanners required approximately 4 minutes for the full rotation of the gantry necessary for each "cut"; newer scanners can do this in less than 2 seconds.[1] The CT scanner emits less radiation than certain other x-ray machines, but anesthesia personnel may receive a radiation exposure of 1–2mrad/h.[1] Lead aprons should be worn by all personnel in the room and the use of either thyroid shields and lead-lined glasses or a lead glass screen is recommended.

Although CT is not as sensitive a test as magnetic resonance imaging (MRI) for many types of lesion, it is still an excellent test for cerebrovascular disease, neoplasms, cerebral degenerative disease, and spinal column disease. It is the study of choice for acute trauma and bone disease, and many clinicians consider it the test of choice for disease of the head and neck, reserving MRI for equivocal problems.[1] CT is now commonly used as a guiding tool for intracranial stereotactic biopsies.

Contrast (usually Hypaque®, a hypertonic iodinated solution) is often given intravenously during the study to identify areas of disruption of the blood-brain barrier as in infarcts, neoplasms, and abscesses.[4] It may, however, be associated with anaphylactic reactions. Additionally, because of its hyperosmolar nature, transient intravascular hypervolemia may occur upon injection, resulting in fluid overload and possibly progressing to pulmonary edema in patients with cardiomyopathy.

A vigorous diuresis may follow the initial intravascular hypervolemia, causing excessive bladder distention and hypovolemia in patients who were initially fasting and dehydrated. This hypovolemia combined with a direct glomerulotoxic effect of the dye may very rarely precipitate acute renal failure.[5]

Except for the injection of I.V. contrast the CT scan is a completely noninvasive test. If the patient has a known history of allergy to contrast (or to iodine), the anesthesiologist is often called to monitor the procedure. The patient should be prophylactically treated with steroids and antihistamines before the procedure. Many radiologists prescribe 150mg/day of prednisone or an equivalent steroid 18 hours prior to the procedure and continuing for 12 hours afterward.[4] Antihistamines such as diphenhydramine (50mg) may be given I.V. immediately prior to the procedure or orally about 1 hour before.

Emergency resuscitative equipment and medication should be available (as always) in the event of a full-blown anaphylactic reaction (this may occur in spite of prophylactic premedication). Use of epinephrine 0.25mg subcutaneously and steroids (such as hydrocortisone 100mg I.V.) is indicated, and full resuscitative measures are essential.

Because the patient must remain motionless during the scan, those who cannot voluntarily cooperate, such as young children, demented patients, and patients with movement disorders, may require sedation or general anesthesia. The fact that the new scanners can produce a cut in less than 2 seconds is helpful for patients who can voluntarily cooperate but sometimes choose not to. General anesthesia may be required in one third to one half of pediatric patients and in about 17 of patients with acute head injury.[1,6]

Careful attention must be focused on avoiding hypothermia, especially in infants and small children. The CT scanning area is usually kept cool to protect the integrity of electronic circuitry, and small children are particularly susceptible to conductive and evaporative heat loss. The incidence of hypothermia in children under 12 months of age (rectal temperature $< 35°C$) has been reported to be greater than 40%.[4]

CT scanning for lesions of the posterior fossa may present a unique set of problems. Visualization of the posterior fossa along a plane accessible to the CT scanner may require extreme head flexion (less common with the new scanners), which may be very discomforting to the conscious patient. In an anesthetized patient it may cause kinking of the endotracheal tube, resulting in ventilatory difficulties.

Extreme head flexion also may cause brain compression if a large mass is present in the infratentorial compartment.[1,4] In children with severe hydrocephalus visualization of the posterior fossa may be difficult, and in infants stabilization of the large head may be problematic.[4,6]

Magnetic Resonance Imaging

Magnetic resonance imaging (MRI) is a noninvasive diagnostic imaging procedure that uses a strong magnetic field and radiofrequency (RF) pulses to produce a high-resolution image in any plane. MRI is used for posterior fossa lesions, spinal column and cord lesions, cranial nerve abnormalities, and skull base tumors. It is highly sensitive in the diagnosis of white matter diseases and has been proposed as a screening test for patients with epilepsy.[1,7]

The theory behind MRI is based on the magnetic properties of hydrogen nuclei. Atomic nuclei with an odd number of total protons and neutrons, when placed in a magnetic field, will tend to align and rotate to lie parallel with that field.[8] A pulse of RF energy will disrupt the alignment of the nuclei. By varying the RF pulse duration, the degree of rotation varies. When the pulse is removed, the nuclei rotate back to their original position, emitting an RF signal.

Two acquisition sequences, T1 and T2, are used in imaging. T1, the longitudinal relaxation time constant pulse sequence, emphasizes anatomic detail and gives high spatial resolution.[8] Gadolinium, a contrast agent, is effective in using the T1 pulse sequence. T2, the spin-spin relaxation time constant pulse sequence, emphasizes free water. It is very sensitive for detecting pathology but results in loss of spatial resolution.[1]

During the MRI scan the patient is encased in a long, cylindrical magnetic gantry, similar in appearance to the CT gantry but longer and even more confining for the patient. The scan itself usually takes 30 minutes to 1 hour. Indications for anesthesia or for the presence of an anesthesiologist during an MRI study are the same as those for a CT scan. However, because of its strong magnetic field, the MRI scanner has inherent idiosyncrasies that make routine anesthetic monitoring impossible.

Any ferromagnetic substance will degrade the quality of the MRI image. It also may cause the object to be propelled through space into the center of the magnet. Affected items include key stethoscopes, pens, and many components of the standard anesthesia machine and monitors.[1]

Patients with implanted ferromagnetic devices are also in danger. For example it has been reported that vascular clips have been dislodged and pacemaker casings have been dislaced or rotated within the chest wall. Additionally, certain cardiac pacemakers are switched from demand to fixed-rate pacing when influenced by the MRI magnets.[8,9] For this reason, patients with pacemakers, cochlear implants, and operated cerebral aneurysms using ferromagnetic vascular clips should not be studied by MRI.[1]

The anesthesiologist may remain close to the patient during the MRI study because no x-ray radiation is involved. The patient in the MRI imager will be hidden from view, and it is necessary to monitor vital signs. Nonferromagnetic ventilators, monitoring devices, and anesthesia equipment are available.[10,11] A plastic precordial stethoscope may be used, but because the MRI machine emits a loud, rhythmic drumming noise, it is extremely difficult or impossible to detect and differentiate the sounds. Fiberoptic or telemetric ECG monitoring may be used to avoid ferromagnetic leads and cables. Although the laryngoscope is not ferromagnetic, the batteries contain iron plastic or paper-coated batteries must be used.

Recommended sedation techniques for children during MRI include one of the following regimes:

a. 50–60μg/Kg oral chloral hydrate.
b. 2μg/Kg meperidine and 4μg/Kg secobarbital intravenously.
c. 2μg/Kg meperidine, 1μg/Kg promethazine, 1μg/Kg chlorpromezine intravenously.[12]

Resuscitation may be affected by the MRI scanner. The magnetic field produced by superconduction magnets will cause resuscitative equipment to malfunction, and these magnets may not be turned off. The magnet may weigh as much as 100 tons and thus cannot be moved. Before resuscitation can be carried out, the patient may have to be physically moved from the area. Resistive magnets, on the other hand, may be quickly turned off if resuscitation is necessary, but several hours may be required to reestablish the magnetic field for subsequent studies.[1]

Myelography

Myelography is still used as a diagnostic test for degenerative disease of the spine, disk herniations, spinal column tumors, and metastatic disease. It is the most sensitive test for intradural extramedullary disease.[1] MRI, however, is rapidly taking the place of myelography in most cases.

Myelography is an invasive test requiring the injection of contrast material into the subarachnoid space, usually through a lumbar puncture. Occasionally, a lateral C1–C2 puncture may be used (for a high cervical lesion) or ventriculography, the injection of contrast material directly into a lateral ventricle through a burrhole craniotomy. After lumbar puncture and injection of the contrast medium the patient is placed in various positions in order to layer the contrast along the intrathecal space. To concentrate the contrast agent in the cervical area steep Trendelenburg's position may be required.

Contrast agents currently used include metrizamide, Pantopaque®, iohexol (Omnipaque®), and iopamidol (Isovue®). Metrizamide is a water-soluble contrast agent. When injected intrathecally, it has a very low rate of longterm complications but may be associated rarely with anaphylactic reactions manifest as severe headaches, seizures, and occasionally, occipital cortical blindness.[13] Phenothiazines may lower the seizure threshold predisposing patients to metrizamide-induced seizures.[1]

If general anesthesia is given to these patients, enflurane and ketamine are poor choices, as they too may induce seizures at higher concentrations.

Pantopaque is an ester-based, highly viscous contrast agent with a low rate of immediate complications, but it may have a very high rate of longterm complications such as arachnoiditis. Iohexol and iopamidol are nonionic contrast materials with low rates of short- and longterm complications.[1]

The discomfort from myelography occurs mostly during lumbar puncture and with prolonged or extreme positioning of the patient. In most cases, myelography is accomplished without assistance of monitoring by an anesthesiologist.

Pneumoencephalography

Pneumoencephalography is very rarely used since CT and MRI have become available. O_2 or air is injected intrathecally through a lumbar puncture, and the gas is allowed to pass up into the cranium. It is an extremely painful procedure, and complications such as air embolism or brain stem herniation have been reported. Because CT and MRI may be equally efficacious, without the high complication rate, the use of pneumoencephalography will continue to decline. Occasionally, the technique of CT-air cisternography may be used in the diagnosis of small acoustic neuromas.[1] If general anesthesia is performed for this procedure, nitrous oxide should be avoided to reduce the likelihood of tension pneumocephalus. Occasionally, nitrous oxide is included in the air mixture injected intrathecally to promote rapid diffusion of the gas bubble out of the subarachnoid space.

PET and SPECT

Positron emission tomography (PET) and single photon emission computed tomography (SPECT) are used for the measurement of regional cerebral blood volume. Both involve I.V. injection of radionuclides before placing the patient in a circumferential array of detectors. An image of a slice of tissue is reconstructed by the computer, based on the relative amount of radioactive tracer emitted by different points in the body.

Areas of decreased uptake are delineated, which may point to areas of infarction or decreased metabolism; 2-dioxyglucose is used for this type of metabolic imaging. PET scanning uses positron emitters of extremely short half-lives (such as carbon-11, oxygen-15, and fluorine-18). For PET scanning, an on-site cyclotron is necessary to manufacture these short-lived isotopes. SPECT scanning uses conventional radionuclides, which decay by single photon emission, such as technetium-99m and iodine-123. SPECT images are inferior to those of PET scanning, however, and its use is still predominantly for research purposes.

Diagnostic Angiography

Angiography involves the intraarterial or I.V. injection of contrast agent during x-ray. Before the 1960s direct carotid and vertebral arteriograms were performed. At present, cerebral angiography most commonly involves the passage of a catheter from the femoral artery up the aorta to the carotid or vertebral arteries. Angiography is used to diagnose intracranial vascular lesions such as aneurysms and arteriovenous malformations.

Digital subtraction angiography (DSA) is a process by which an image taken just prior to contrast injection is subtracted by computer from the postinjection image. Only the vessels containing contrast material are visualized on the reconstructed image. DSA may be arterial or venous. Venous DSA has the advantage of not requiring arterial catheterization, but it requires relatively large amounts of contrast dye and a much longer circulation time as the dye traverses the venous pulmonary and then the arterial system. Also, multiple arterial branches may be superimposed on the image because specific arteries are not selectively cannulated.[1]

Interventional Neuroangiography

Interventional or surgical neuroangiography involves the endovascular embolization of vascular lesions of the central nervous system. This procedure is particularly advantageous for arteriovenous fistulas (AVFs), arteriovenous malformations (AVMs), and cerebral aneurysms. It involves the selective catheterization of small branches of the cerebral or spinal vasculature and the placement of embolic agents to occlude aneurysms, fistulas, or feeding vessels to vascular lesions.[14,15]

The embolic agents include detachable balloons, silicon spheres, polyvinyl alcohol, and tissue adhesives such as isobutyl-2-cyanoacrylate (IBCA) and N-butyl-cyanoacrylate (NBCA). Chemotherapeutic or tissue-destructive substances may be infused directly into an unresectable lesion by catheterizing the vessels that feed it.

Localization of vessels is accomplished under angiographic guidance. Differentiation of diseased brain from adjacent healthy brain may be made by the direct injection of sodium amytal into the vessel, a maneuver that is extremely important in lesions involving the more "eloquent" portions of the brain.

Before operating on areas in the vicinity of the speech center a similar test called the Wada test, may be used.[17] It involves cannulation of an internal carotid artery and the subsequent injection of 150–200mg of sodium amytal. The patient is asked to hold up both hands and begin counting. Only the contralateral arm should fall. If the patient also becomes aphasic, the speech center has been localized to that hemisphere.[1] Sodium amytal also may be directly injected into a more distal branch of the cerebral circulation to detect the perfusion of healthy and clinically important areas of the brain

Once thus identified, feeding vessels of AVMs or vascular tumors may be embolized. This may reduce lesion size, decrease steal from vessels supplying surrounding healthy tissue, and aid subsequent surgery by minimizing vascularity.

Endovascular embolization of cerebral aneurysms as a definitive treatment is still held as a last resort for otherwise "inoperable" lesions. Detachable balloons are currently the treatment of choice for these lesions. In the future it is likely that endovascular embolization of aneurysms will provide a far less invasive alternative to open surgical techniques.

Interventional neuroangiography is endovascular neurosurgery. It therefore carries with it a much higher complication rate than do less invasive diagnostic procedures. A major complication rate of 3% was reported by Hieshima in the treatment of intracranial aneurysms.[18] Because of this higher complication rate, anesthetic monitoring, including pulse oximetry, continuous ECG, and noninvasive blood pressure monitoring, should be routinely employed during interventional neuroangiography.

Complications include infarction by dislodgment of thrombi or atheromatous plaque by the catheter, dislodgment and migration of the embolic agent, air injection, dissection through vessel walls with vascular compression, and arterial spasm.[1] Intracranial hemorrhage during these procedures may be accompanied by sudden rise in intracranial pressure, cardiac dysrhythmias, and sudden death by brain stem compression or herniation.[19] Contrast agents may exert a direct necrotizing effect on the cerebral vasculature. Vasoconstrictors may exaggerate this effect and should be used cautiously if at all during these procedures.[1]

Mental status changes indicative of developing complications may be assessed by maintaining continuous verbal contact with the patient.

Oxygen is given by nasal cannula, and the capnograph is connected to a plastic 14-gauge angiocath inserted through the cannula at one naris. This may not give a continuously accurate capnography reading because of the open system, changes in flow, and intermittent mouth breathing, but it does allow the anesthesiologist to follow a trend in end tidal CO_2 measurements.

When available, a Walkman®-type cassette player is brought into the suite and the patient is given a choice of music to listen to through headphones.

Anesthetic Plan

In our institution a portable anesthesia machine equipped with a ventilator and all of the monitoring devices is taken to the neuroangiography suite, where suction and oxygen are available. Adult patients are sedated by I.V. medication unless a specific contraindication exists or there is an indication for general anesthesia. Under general anesthesia, hyperventilation may slow the transit time of contrast media, thus improving the quality of angiographic study. With vasoconstriction of normal cerebral blood vessels, tumor vessels may be more easily identified.

Many anesthetic agents are suitable for sedation during neuroradiologic procedures. One technique involves the continuous infusion of alfentanil 0.5–2.0 μg/kg/min, supplemented with alfentanil boluses of 5–10μg/kg during painful or stressful periods. Small boluses of methohexital occasionally may be necessary. Also, appropriate doses of benzodiazepines, droperidol, or diphenhydramine may be added as sedatives or if anxiolytic, antinausea, or allergy prophylaxis is desired. The patient remains awake but sedated and ideally is pain-free. An alternative technique, as already mentioned, may be use of propofol (1–2mg/kg) prior to painful injections or manipulations. Blood pressure must be carefully monitored and ephedrine (5–10 mg bolus) readily available to treat hypotension.

Because of the relatively high rate of serious complications, especially following intervential neuroradiology, it may be prudent to have an operating room avilable and prepared should emergent neurosurgical decompression become necessary.

References

1. Moser F.G., Frost E.A.M.: Current neuroradiologic techniques and their anesthetic implications, in *Current Anaesthesia and Critical Care.* Edinburgh: Churchill Livingstone, 1990, pp 90–98.

2. Sebel P.S., Lowdon J.D.: Propofol: A New Intravenous Agent. *Anesthesiology* 1989, 71:260–77.
3. Van Hemelrijck J., Tempelhoff R., Jellish W.S., et al: Comparison of Thiopental-Isoflurane-N20, Profol-N2), and Propofol alone for Neuroanesthesia. *J Neurosurg Anesthesiol* 1990, 2:3–232.
4. Andrews I.C.: Anesthetic management for neuroradiologic diagnostic procedures, in Frost E.A.M. (ed): *Clinical Anesthesia in Neurosurgery,* 1st ed. Boston: Butterworth, 1984, pp 95–110.
5. Samra S.K.: Anesthesia for diagnostic procedures and nonsurgical treatment in children with brain tumors. *J Neurosurg Anesthesiol* 1989, 1:145–52.
6. Wolfson B., Hetrick W.D.: Anesthesia for neuroradiologic procedures, in Cottrell J.E., Turndorf H. (eds): *Anesthesia and Neurosurgery,* 2nd ed. St. Louis: CV Mosby, 1986, pp 104–13.
7. Buonanno F.: Magnetic resonance. *Current Opinion in Neurology and Neurosurgery* 1988, 1:977–84.
8. Nixon C., Hirsch M.P., Ormerod I.E.C. et al: Nuclear magnetic resonance: Its implications for the anesthetist. *Anesthesia* 1986, 41:131–37.
9. Pavlicek W., Geisinger M., Castle L. et al: The effects of nuclear magnetic resonance on patients with cardiac pacemakers. *Radiology* 1983, 147:149–53.
10. Roth J.L., Nugent M., Gray J.E. et al: Patient monitoring during MRI. *Anesthesiology* 1985, 62:80–3.
11. Chalapathi C.R., McNeice W.L., Emhardt J.: Modification of an anesthesia machine for use during MRI. *Anesthesiology* 1988, 68:640–1.
12. Davis P.J., Gillen C., Kretchman E., et al.: Experience with anesthesia for children requiring nuclear magnetic resonance imaging. *Anesthesiol Review* 1990, 27:6 pp 35–40.
13. Hanus P.M.: Metrizamide: A review with emphasis on drug interactions. *Am J Hosp Pharm* 1980, 37:510–3.
14. Choi I.S., Berenstein A.: Surgical neuroangiography of the spine and spinal cord. *Radiol Clin North Am* 1988, 26:1131–41.
15. Theron J.: Interventional neuroangiography. *Curr Opin Neurol Neurosurg* 1988, 1:996–1001.
16. Berenstein A., Choi I.S.: Surgical neuroangiography of intracranial lesions. *Radiol Clin North Am* 1988, 26:1143–51.
17. Wada J., Rasmussen T.: Intracarotid injection of sodium amytal for the lateralization of cerebral speech dominance. *J Neurosurg* 1960, 17:266–82.
18. Hieshima G.: Keystone interventional neuroradiology meeting. *Abstract.* Keystone, Colorado, Aug. 1988.
19. Brain J.E., Eleff S., McPherson R.W.: Immediate hemodynamic management following subarachnoid hemorrhage during embolization of cerebral vascular abnormalities. *J Neurosurg Anesthesiol* 1989, 1:63–7.

CHAPTER 13

The Pregnant Patient For Nonobstetric Surgery

Jill Fong

Case History. *At 32 weeks of gestation, a 33-year-old woman visited her obstetrician for a routine physical examination. A left breast mass was discovered. Her past medical and surgical history were unremarkable. There was no family history of breast carcinoma.*

Her physical examination revealed a healthy-looking pregnant patient. Her blood pressure was 120/60mmHg; pulse, 82/min; respiratory rate, 16/min. She was 5'6" tall and weighed 150lb.

Laboratory data revealed a hematocrit of 35% and normal SMA-12 and coagulation studies. The day after the breast lump was discovered, biopsy under local anesthesia was performed. The pathologic finding was adenocarcinoma. She was scheduled for a left modified radical mastectomy on a semiemergent basis.

Introduction

The incidence of nonobstetric surgery in pregnant patients is approximately 1.6%–2.2% of all pregnancies, or up to 50,000 cases per year in the United States.[1–3] Most commonly, nonobstetric surgery is performed in the pregnant patient for acute appendicitis, ovarian pathology, incompetent cervix, and cancer. Intracranial procedures and cardiac procedures requiring extracorporeal circulation also have been performed (see Table 1).

Acute appendicitis is the diagnosis in two thirds of the exploratory laparotomies performed during pregnancy; the incidence is 0.38–1.41 per 1000 deliveries.[4,5] The incidence of cancer detected during pregnancy

Reviewed by Dr. Steven Schwalbe, Director, Obstetrical Anesthesia, Albert Einstein College of Medicine/Montefiore Medical Center.

TABLE 1. Nonobstetric Surgery in Pregnancy
Appendicitis
Ovarian or myoma torsion
Incompetent cervix
Cancer
Ruptured intracranial aneurysm
Cardiac lesion

Several general surgical conditions may require emergent operation during pregnancy

is 1 per 1000 pregnancies, and that of breast cancer is approximately 1 per 3000 pregnancies.[6] Breast cancer during pregnancy was thought to have a grave prognosis. However, if age and stage of disease are taken into account, the prognosis for and behavior of breast cancer in the pregnant patient are similar to those of breast cancer in nonpregnant patients. The poor outcome of the disease in pregnant patients results from the advanced stage of the disease at the time of diagnosis, and perhaps from delay in treatment. Seventy-five percent to 85% of pregnant patients have nodal metastases at the time of diagnosis.[7]

During the first half of pregnancy, mammary tissue proliferates rapidly because of increased levels of progesterone, estrogen, and prolactin. Patients should perform frequent self-examinations to detect breast masses early. In both pregnant and nonpregnant women, masses and spontaneous nipple discharges are the most frequent signs of mammary neoplasm. If a lesion is found it should promptly be biopsied. In some centers mammography is not performed in pregnant women because the radiographic density of the pregnant woman's breast results in a poor-quality study and because of the risk of fetal radiation. However, some radiologists believe that mammography can be a useful tool if adequate uterine shielding is used.

Depending on the lesion, the primary treatment of breast cancer in pregnancy is often mastectomy combined with axillary dissection for staging and the determination of hormonal—estrogen and progesterone—receptors. If nodal metastases exist, chemotherapy may be given. If chemotherapy is required in the first trimester, the pregnancy should be aborted. In the second trimester, alkylating agents (with the exception of methotrexate) can be used with reasonable fetal safety, and the pregnancy may be continued. During the third trimester of pregnancy, chemotherapy is usually withheld until after delivery. Nodal metastases should not be irradiated unless it is possible to do so without exposing the fetus. Radiation as primary therapy, combined with lumpectomy, has no place in the pregnant patient; nor does therapeutic abortion or prophylactic castration, because these measures do not enhance cure rates.[6–8]

Regardless of the reason for nonobstetric surgery in the pregnant patient, the goal is, of course, safety of maternal and fetal outcome. To plan a rational anesthetic management, the changes in maternal physiology should be taken into account, and the potential hazards to the fetus from anesthesia, drug administration, maternal disease, and the operation must be minimized. Unlike anesthesia for obstetric surgery, anesthesia for nonobstetric surgery in the pregnant patient aims to prevent increased uterine tone and relies only on the integrity of the uteroplacental unit to provide nutrients, such as oxygen, to the fetus, without the concerns of depressed spontaneous newborn breathing.

Physiologic Changes of Pregnancy

Pregnancy alters the function of every major organ system. Those of most concern to the anesthesiologist include the cardiovascular, respiratory, hematologic, gastrointestinal, central nervous, and renal systems (see Table 2).

Cardiovascular System

During the first trimester of pregnancy, cardiac output increases until it reaches a plateau of 30%–40% above normal at 28 weeks of gestation, where it remains until the additional stress imposed by labor increases it further. The higher level results from a 30% increase in stroke volume and a 15% increase in heart rate.[9,10] Blood pressure, however, is not elevated in normal pregnancy, because of a decrease in peripheral vascular resistance of approximately 15% that is partially mediated by hormonal and prostacyclin changes, which increase uterine, renal, and extremity blood flow.[11,12]

Aortocaval compression syndrome can occur as early as the 20th week, although it is usually not pronounced until after the 24th week of gestation. It is caused by impingement of the gravid uterus on the inferior vena cava (IVC) and aorta in the supine position. Compression of the IVC decreases venous return and can cause severe maternal hypotension. When the IVC is compressed, blood from below the obstruction is returned to the heart via the paravertebral (epidural) veins into the azygos venous system.

Compression of the IVC and aorta may affect the well-being of the fetus by decreasing uteroplacental perfusion: compression of the former by diminishing cardiac output, of the latter by directly occluding flow to the uterus. Therefore, after the 20th week of gestation, pregnant patients should not be nursed in the supine position but should be turned partially

TABLE 2. Effects of Pregnancy on Other Systems

Pregnancy exerts many effects, both subtle and profound, on all other systems. Some are of particular importance to the anesthesiologist.

SYSTEM:
Cardiovascular
Cardiac output ↑
Peripheral vascular resistance ↓
Hematologic
Blood volume ↑ Relative anemia
Hypercoagulability
Respiratory
Capillary engorgement
Functional residual capacity ↓
Tidal volume ↑
Inspiratory reserve volume ↑
Minute ventilation ↑
$paCO_2$ ↓
Oxygen consumption ↑
Gastrointestinal
Gastric emptying ↓
Gastric acidity ↑
Central nervous
Sensitivity to anesthetic agents ↑
Renal
Plasma flow ↑
Glomerular filtration rate ↑

(15°) to the left side. Left uterine displacement can be accomplished by inserting a foam rubber wedge or sheets under the right hip or tilting the operating table.[13,14]

It should be noted that as rheumatic heart disease has decreased, congenital heart disease is now the most common type of heart disease in pregnancy. Patient risks include cardiac decompensation because of an inability to meet the increased cardiac demands, and thromboembolism. Heart disease that causes pulmonary hypertension and patients with tetralogy of Fallot are at greatest risk.

Hematologic System

Beginning in the 6th to 12th week of gestation, the blood volume expands. The increase is greatest in the second trimester but continues until the 28th–32nd week of gestation and reaches a plateau of approximately 35% above nonpregnancy values. The plasma volume increases 45%, whereas the red cell mass increases only about 20%, which accounts for the "relative" anemia of pregnancy.[15] Although there is a lower hematocrit, oxygen transport during pregnancy is enhanced by an increased maternal arterial oxygen tension, increased cardiac output, and a right shift of the hemoglobin oxygen dissociation curve.[16]

During gestation there is an increase in Factors I, VII, VIII, X, and XII, leading to hypercoagulability. Therefore, pregnant patients have a propensity for thromboembolism.[17] Antiembolism stockings should always be applied before anesthesia if it is anticipated that the legs will be immobile for several hours.

Pre-eclampsia associated with pregnancy may exert further changes in normal coagulation. A study of patients with pre-eclampsia compared to those with normal pregnancy showed that 34% of pre-eclamptic patients had prolonged bleeding times compared to 13% of controls. Also, 7.5% of pre-eclamptic patients had platelet counts lower than 100,000.[18]

Respiratory System

Anatomically, there is marked capillary engorgement of the respiratory system. Thus, pregnant patients often require smaller endotracheal tubes (7.0–6.0mm ID) than their nonpregnant counterparts, and any manipulation of the nasal passages can lead to profuse bleeding.

Lung volumes undergo progressive changes beginning in the 5th month of gestation. By term there is a 20% reduction in functional residual capacity (FRC) as a result of decreased expiratory reserve volume and tidal volume increase; total lung capacity and closing volume (CV) remain unchanged.[19]

The FRC is important for two reasons. First, during the expiratory phase, blood oxygenation is mainly dependent on the gas left in the lung at the end of a normal tidal volume (ie, the FRC). Therefore, when the FRC is low, hypoxemia occurs with apnea more quickly. Second, when the FRC is decreased below the closing capacity (CV+RV), airway closure occurs in the dependent regions of the lung during normal tidal volume ventilation. Early and increased airway closure results in shunting and decreased arterial oxygenation.[20] Also, in the supine position, FRC is reduced by general anesthesia in more than 50% of pregnant patients.

When these changes are combined with the increased oxygen consumption caused by pregnancy, it is apparent that pregnant patients are more prone to hypoxemia. Indeed, the pregnant patient may be extremely susceptible to hypoxemia during short periods of apnea, such as induction and intubation. Therefore, it is important to preoxygenate (ie, denitrogenate) these patients with 100% oxygen before beginning general anesthesia.

Ventilation increases early in pregnancy. By term, minute ventilation increases about 50%, mainly because of a rise in tidal volume (40%), with some increase in respiratory rate (15%). As a result of this greater minute ventilation and unchanged dead space, alveolar ventilation rises approximately 70% above nonpregnancy levels. Mean maternal arterial carbon dioxide tension is usually reduced to about 32mmHg. Maternal pH remains unchanged because of the metabolic compensatory decrease in serum bicarbonate of about 4mEq/L.[21]

The increase in alveolar ventilation and the decrease in FRC enhance the speed of induction with inhalation anesthetics.[22] Although an inhalation induction is usually not indicated in pregnancy, this increase in speed, combined with a decrease in minimum alveolar concentration (MAC), can result in a general anesthetic state when "mask analgesia" was anticipated.

Gastrointestinal System

The following changes occur in the gastrointestinal system during pregnancy: contents increase, gastric motility decreases, gastric acidity increases, lower esophageal sphincter tone decreases, and the angle of the gastroesophageal junction changes because of the shift in stomach position produced by the gravid uterus.[9]

By the 12th–14th week of gestation, gastric emptying is prolonged.[23] Because of these changes pregnant patients are at increased risk of aspiration pneumonitis and should always be treated as if they had "full stomachs." Patients are considered to be at risk for aspiration pneumonitis if the gastric volume exceeds 0.4ml/kg and the pH is less than 2.5.[24]

Elevated levels of serum glutamic oxaloacetic transaminase (SGOT), alkaline phosphatase, lactic dehydrogenase, and cholesterol are associated with pregnancy. Fifty percent of pregnant patients have abnormal bromsulphalein excretion tests.[9]

Plasma cholinesterase levels are decreased by 21%–24% throughout pregnancy. Although this is not associated with any structural abnormality and recovery from succinylcholine is rarely prolonged,[25] the availability of short-acting nondepolarizing muscle relaxants may make the use of these drugs preferable.

Also, as with coagulation, pre-eclampsia has been shown to decrease cholinesterase activity by 60% over normal levels, ie, 40% more than in healthy, pregnant controls.[26]

Central Nervous System

During pregnancy, anesthetic requirement is decreased. MAC is reduced 25%–40%,[27] probably as a result of the actions of progesterone and endorphins.[28] Similarly, the regional anesthetic requirements (ie, epidural and spinal anesthesia) are reduced, perhaps because of epidural venous sensitivity to drug effect.[29,30] These changes can occur as early as the first trimester. Therefore, drugs should be carefully titrated to effect.

Renal System

Renal plasma flow and glomerular filtration rate increase in pregnancy. Creatinine clearance is increased and there is a decrease in the normal upper limits for blood urea nitrogen (BUN) and serum creatinine levels of approximately 40%. Drugs may undergo rapid urinary excretion, and thus dosage increases of some agents may be needed.[31]

The renal calices, pelves, and ureters dilate, partly because of higher progesterone levels and compression by the uterus. The result is urinary stasis, which contributes to the increased incidence of urinary tract infections during pregnancy.[32]

Fetal Considerations

Whenever a pregnant patient undergoes surgery, two patients are at risk. To minimize the maternal risk, a thorough understanding of the maternal physiologic changes is needed. Similarly, if fetal hazards are to be minimized, it is important to maintain fetal oxygenation and to understand the fetal effects of maternal drug administration and of the operation itself.

Maintenance of Fetal Oxygenation

The most important acute fetal risk consequent to maternal surgery and anesthesia is intrauterine asphyxia. Therefore, maternal oxygen-carrying capacity, oxygen affinity, arterial oxygen tension and uteroplacental blood flow must be carefully maintained perioperatively. Maintaining uterine blood flow is crucial to providing adequate oxygen to the fetus, because survival of the fetus depends on an intact uteroplacental perfusion. Similar to all flow, pressure, and resistance relationships in the body, uterine blood flow is derived from Ohm's law. Simply stated: potential or pressure equals flow times resistance. Ohm's law applied to uterine blood flow is;

$$\text{uterine blood flow} = \frac{(\text{uterine arterial pressure} - \text{venous pressure})}{\text{uterine artery resistance}}$$

Normally, 10% of the pregnant patient's cardiac output goes to the uterus. The placenta receives 80% of the uterine artery blood flow, with the remainder going to the myometrium. There is no autoregulation of the uterine vessels, and therefore flow is directly proportional to pressure. According to the above equation, anything that increases uterine venous pressure or uterine artery resistance decreases uterine blood flow. A decrease in uterine blood flow can lead to placental hypoperfusion and fetal asphyxia.

Factors decreasing uterine arterial pressure include hypotension, alpha-adrenergic vasopressors, and endogenous catecholamines. Factors increasing uterine venous pressure include IVC compression and increased uterine tone, which may result from contractions, hypercarbia, oxytocin, ergot alkaloids, and small doses of ketamine (<1mg/kg). Uterine artery resistance is increased by catecholamines.[33]

The effect of hyperventilation on fetal well-being is controversial. However, mechanical hyperventilation, independent of changes in $PaCO_2$ or pH, reduces uterine blood flow and should be avoided.[34] Maternal alkalosis may compromise fetal oxygenation because of umbilical artery vasoconstriction and a leftward shift of the maternal hemoglobin oxygen dissociation curve.

Perhaps the adverse complications of anesthesia, such as maternal hypoxemia, hypotension, hyper- or hypocarbia, or electrolyte abnormalities, are stronger determinants of teratogenicity than the anesthetic agents themselves. Hypoxemia and hypercarbia have been shown to be teratogenic in animal models.[35,36]

Effects of Surgery and Anesthesia on the Embryo and Fetus

Several major studies in humans have attempted to delineate the relationship between surgery and anesthesia during pregnancy and fetal outcome, with respect to congenital anomalies, premature labor, and/or abortion.[1,2,37,38] None has correlated congenital anomalies with anesthesia and surgery. However, all have demonstrated an increased fetal mortality, especially when the operation occurred during the first trimester.

The most relevant factor affecting fetal mortality was the condition necessitating surgery; the highest mortality occurred with pelvic surgery. For instance, perinatal mortality in patients with peritonitis is 200–350 per 1000, and that for nonperforated appendicitis is approximately 50 per 1000.[4,5] In patients undergoing breast biopsy, with or without mastectomy, there is a 1.2% incidence of spontaneous abortion.[7]

To date, no anesthetic agent has been proved to be teratogenic in humans. Studies on the teratogenic effects of anesthetic agents present difficulties in interpretation for many reasons. First, studies are performed in vitro and in experimental animals, not all of which are mammals. Second, the teratogenic effects of substances vary markedly between species and also within species. Third, the concentration of the anesthetic and the duration of exposure are often far in excess of clinical relevance. Finally, the stage of development at the time of exposure is crucial, because although exposure at one time may have catastrophic effects, at other times no effect may be seen. Therefore, the extrapolation of animal data to humans is difficult if not impossible.

In humans, organogenesis occurs between the 15th and 56th days. Therefore, this is the period when the fetus is most susceptible to the teratogenic effects of drugs. However, animal data indicate that the central nervous system may be susceptible to external factors during the period of myelination, which lasts through the first few months after birth.[39]

Of the commonly used anesthetics, the inhalation agents have created the most questions about teratogenicity. Nitrous oxide (N_2O) has been extensively studied in animal models and has been variably associated with congenital abnormalities.

The mechanism by which nitrous oxide causes teratogenicity in animals is unclear but may be related to its inhibition of the enzyme methionine synthetase. Nitrous oxide irreversibly oxidizes vitamin B_{12}, a methionine synthetase cofactor.[40] Inhibition of methionine synthetase (failure to convert homocysteine to methionine) causes abnormalities of nerve fiber myelination and decreased DNA synthesis as a result of lower thymidine production.

These adverse effects notwithstanding, two recent studies have shown that methionine synthetase is not significantly affected after exposure to 60%–70% nitrous oxide for up to 4 hours or 50% nitrous oxide for up to 6 hours.[41,42] Two other reviews of exposure to nitrous oxide during cervical cerclage procedures showed that the agent had no effect on fetal outcome.[43,44] Thus, although questions remain, many feel that nitrous oxide may still be administered safely to the pregnant patient.

The halogenated inhalation agents—halothane, enflurane, and isoflurane—have produced variable results in animal models. Rats exposed to 0.75 MAC of halothane, enflurane, or isoflurane or to 0.55 MAC of nitrous oxide had no major structural abnormalities.[45] Single exposure to one of these agents is unlikely to produce fetal abnormalities. The halogenated agents can decrease uterine activity and therefore may be helpful in preventing premature contractions.

Similarly, the intravenous agents have not been proved to cause teratogenicity in humans. Several studies have implicated prenatal diazepam exposure in the development of cleft lip or palate, but this is still controversial.[46,47] The induction agents thiopental sodium and ketamine can be safely used, with the dose of ketamine limited to <1mg/kg.

Doses above that level increase uterine tone and can decrease uteroplacental perfusion. Morphine and meperidine may be safely used, as may the muscle relaxants succinylcholine and d-tubocurarine.[39]

The effects on pregnancy of chronic exposure to subanesthetic concentrations of inhalation agents are yet another issue. In 1974 a survey conducted by the American Society of Anesthesiologists found that there were higher incidences of cancer, spontaneous abortions, and congenital anomalies in the infants of female anesthesiologists.[48]

Other studies also have noted increased incidences of spontaneous abortion and congenital anomalies in health care professionals subjected to chronic subanesthetic doses of inhalation agents. All of these studies have been criticized for statistical inaccuracies or faulty methods.

More recently, two studies that avoided some of the previous design problems found that there was no difference in the expected rate of threatened abortion, low birth weight, perinatal mortality, or congenital abnormalities in women exposed to subanesthetic concentrations of inhalation agents versus unexposed women. However, the authors do point out that their results do not eliminate the possibility of reproductive hazards, because early abortions, late manifestations of congenital abnormalities, and infertility may have been missed.[49,50]

Because an anesthetic agent with proved teratogenic effects in humans has not been found, any may be used. However, it is common practice to use agents with a long history of safety in the pregnant population. These include thiopental sodium, ketamine, morphine, meperidine, d-tubocurarine, and low doses of nitrous oxide (50%), halothane, and enflurane.

Anesthetic Plan

In general, only operations that cannot be postponed because of the risks to the mother should be undertaken during pregnancy. The following anesthetic approach is based on the already noted maternal and fetal considerations.

Preanesthetic Visit

The patient should be visited preoperatively. She should be informed in a thorough yet reassuring manner about the minimal hazards of anesthesia for both herself and the unborn baby. The low incidence of fetal loss and the animal data on teratogenicity should be put into perspective. The patient should be informed that none of the commonly used anesthetic drugs has been shown to be teratogenic in humans and that, as far as is known, the mother's disease and the nature of the surgery affect the fetal outcome more than any particular anesthetic technique.

She should know that, if possible, the baby will be monitored before, during and after the operation. The well-being of all fetuses is assessed before surgery and postoperatively. Intraoperative fetal monitoring is possible in all cases of nonobstetric surgery, as in this case.

Intraoperative Anesthetic Management

A general anesthetic should be planned for mastectomy because of the site of surgery. (Regional anesthesia is perfectly acceptable if the site of surgery permits.) Because pregnant patients are at risk for aspiration pneumonitis, a nonparticulate antacid, such as sodium citrate should be given within 1/2 hour of induction of anesthesia. Pregnant patients should be transported to and placed on the operating table in the lateral decubitus or the supine position, with left uterine displacement to avoid aortocaval compression.

The patient should be monitored with the routine intraoperative monitors, including electrocardiogram, blood pressure monitoring, temperature probe, pulse oximeter, and capnogram. In addition, an external fetal heart rate Doppler monitor and tokodynamometer should be used. The directional Doppler apparatus makes it possible to monitor the fetal heart as early as the 16th week of gestation.

Early indicators of fetal distress include bradycardia, tachycardia, and decreased beat-to-beat variability. However, these changes may be related not to hypoxemia but to the drugs used, fetal immaturity, maternal fever, or fetal sleep cycles. For instance, ephedrine produces a dose-related increase in fetal heart rate and variability; the narcotics and inhalational agents decrease fetal heart rate and beat-to-beat variability. If changes do occur in the fetal heart beat pattern, a cause should be sought. Hypoxemia, hypotension, and abnormalities of arterial carbon dioxide and electrolytes should be ruled out before these changes are attributed to anesthetic use.[51,52]

Maternal uterine activity also should be monitored with a tokodynamometer, which detects premature contractions. Postoperatively, it is common to monitor patients for 12 to 24 hours.

The induction of general anesthesia requires a rapid sequence to minimize the likelihood of aspiration pneumonitis. Denitrogenation with 100% oxygen is indicated to prevent maternal and fetal hypoxemia. Thiopental sodium, 4mg/kg, and succinylcholine, 1–1.5mg/kg, are appropriate agents. Cricoid pressure (Selleck's maneuver) should be applied during induction and maintained until the endotracheal tube is properly placed and its cuff inflated. When properly applied, cricoid pressure withstands pressures of up to 94cmH_2O.[53]

A reasonable choice for maintenance anesthesia includes nitrous oxide (50%), morphine, or meperidine and one of the halogenated agents, such as isoflurane. Finally, on emergence, because these patients are considered to have "full stomachs," extubation should be delayed until airway reflexes are intact.

References

1. Shnider S.M., Webster G.M.: Maternal and fetal hazards of surgery during pregnancy. *Am J Obstet Gynecol* 1965, 92:891–900.
2. Brodsky J.B., Cohen E.N., Brown BW., Wu M.L., Whitcher C.: Surgery during pregnancy and fetal outcome. *Am J Obstet Gynecol* 1980, 138: 1165–7.
3. Cohen E.N., Bellville J.W., Brown B.W., Jr.: Anesthesia, pregnancy and miscarriage: A study of operating room nurses and anesthetists. *Anesthesiology* 1971, 35:343–7.
4. Babaknia A., Parsa H., Woodruff J.D.: Appendicitis during pregnancy. *Obstet Gynecol* 1977, 50:40–4.
5. Weingold A.B.: Appendicitis during pregnancy. *Clin Obstet Gynecol* 1983, 26:801.
6. Baden J.M., Brodsky J.B.: The Pregnant Surgical Patient. New York: Futura, 1985, pp 183–7.
7. Donegan W.L.: Breast cancer and pregnancy. *Obstet Gynecol* 1977, 50: 244–52.
8. Rubin P.: *Clinical Oncology: A Multidisciplinary Approach*, 6th ed. New York: American Cancer Society, 1983, pp 120–40.
9. Cheek T.G., Gutsche B.B.: Maternal physiologic alterations during pregnancy, in Shnider S.M., Levinson G. (eds): *Anesthesia for Obstetrics.* Baltimore: Williams and Wilkins, 1987, pp 3–13.
10. Walters W.A.W., MacGregor W.G., Hill M.: Cardiac output at rest during pregnancy and the puerperium. *Clin Sci* 1966, 30:1–11.
11. Goodman R.P., Killam A.P., Brash A.R., Branch R.A.: Prostacyclin production during pregnancy: Comparison of production during normal pregnancy and pregnancy complicated by hypertension. *Am J Obstet Gynecol* 1982, 142:817–22.
12. Ylikorkala O., Jouppila P., Kirkinen P., Viinikka L.: Maternal prostacyclin, thromboxane, and placental blood flaw. *Am J Obstet Gynecol* 1983, 145: 730–2.
13. Marx G.F.: Aortocaval compression: Incidence and prevention. *Bull NY Acad Med* 1974, 50:443–6.
14. Pitkin R.M., Perloff J.K.: Pregnancy and congenital heart. *Ann Inern Med* 1990, 112:445–54.
15. Lund C.J., Donovan J.C.: Blood volume during pregnancy. *Am J Obstet Gynecol* 1967, 98:393–403.
16. Kambam J.R., Handte R.E., Brown W.R., Smith B.E.: Effect of pregnancy on oxygen dissociation. *Anesthesiology* 1983, 59:A395.

17. Fletcher A.P., Alkjaetsig N.K., Burstein R.: The influence of pregnancy upon blood coagulation and plasma fibrinolytic function. *Am J Obstet Gynecol* 1979, 134:743–51.
18. Ramanathan J., Sibai B.M.: Correlation between bleeding times and platelet counts in women with pre-eclampsia undergoing cesarean section. *Anesthesiology* 1989, 71:188–91.
19. Russell I.F., Chambers W.A.: Closing volume in normal pregnancy. *Br J Anaesth* 1981, 53:1043–6.
20. Nunn J.F.: *Applied Respiratory Physiology,* 3rd ed. Boston: Butterworth, 1987, pp 36–41.
21. Anderson G.J., James G.B., Mathers N.P., Smith E.L., Walker J.: The maternal oxygen tension and acid-base status during pregnancy. *Obstet Gynaecol Br Commonw* 1969, 76:16–9.
22. Moya F., Smith B.E.: Uptake, distribution and placental transport of drugs and anesthetics. *Anesthesiology* 1965, 26:465–76.
23. Simpson K.H., Stakes A.F., Miller M.: Pregnancy delays paracetamol absorption and gastric emptying in patients undergoing surgery. *Br J Anaesth* 1988, 60:24–7.
24. Roberts R.B., Shirley M.A.: Reducing the risk of acid aspiration during cesarean section. *Anesth Analg* 1974, 53:859–68.
25. Shnider S.M.: Serum cholinesterase activity during pregnancy, labor and puerperium. *Anesthesiology* 1965, 26:335–9.
26. Kambam J.R, Mouton S.: Effect of pre-eclampsia on plasma cholinesterase activity. *Can J Anaesth.* 1987, 34(5): 509–11.
27. Palahniuk R.J., Shnider S.M., Eger E.I. II: Pregnancy decreases the requirements for inhaled anesthetic agents. *Anesthesiology* 1974, 41:82–3.
28. Csontos K., Rust M., Hollt V., Mahr W., Kromer W., Teschemacher H.J.: Elevated plasma endorphin levels in pregnant women and their neonates. *Life Sci* 1979, 25:835–44.
29. Fagraeus L., Urban B.J., Bromage P.R.: Spread of epidural analgesia in early pregnancy. *Anesthesiology* 1983, 58:184–7.
30. Dana S., Lambert D.H., Gregus J., Gissen A.J., Covino B.G.: Differential sensitivities of mammalian nerve fibers during pregnancy. *Anesth Analg* 1983, 62:1070–2.
31. Dignam W.J., Titus P., Assali N.S.: Renal function in human pregnancy: I. Changes in glomerular filtration rate and renal plasma flow. *Proc Soc Exp Biol Med* 1958, 97:512–4.
32. Van Wagenen G., Jenkins R.H.: An experimental examination of factors causing ureteral dilatation of pregnancy. *J Urol* 1939, 42:1010–20.
33. Parer J.T.: Uteroplacental circulation and respiratory gas exchange, in Shnider S.M., Levinson G. (eds): *Anesthesia for Obstetrics.* Baltimore: Williams and Wilkins, 1987, pp 14–40.
34. Levinson G., Shnider S.M., deLorimier A.A., Steffenson J.L.: Effects of maternal hyperventilation on uterine blood flow and fetal oxygenation and acid-base status. *Anesthesiology* 1974, 40:340–7.

35. Grabowski C.T., Paar J.A.: The teratogenic effects of graded doses of hypoxia on the chick embryo. *Am J Anat* 1958, 103:313–47.
36. Haring O.M.: Cardiac malformations in rats induced by exposure of the mother to carbon dioxide during pregnancy. *Circ Res* 1960, 8:1218–27.
37. Smith B.E.: Fetal prognosis after anesthesia during gestation. *Anesth Analg* 1963, 42:521–6.
38. Duncan P.G., Pope W.D.B., Cohen M.M., Greer N.: Fetal risk of anesthesia and surgery during pregnancy. *Anesthesiology* 1986, 64:790–4.
39. Pedersen H., Finster M.: Anesthetic risk in the pregnant surgical patient. *Anesthesiology* 1979, 51:439–51.
40. Chanarin I.: Cobalamins and nitrous oxide: A review. *J Clin Pathol* 1980, 33:909–16.
41. Nunn J.F., Sharer N.M., Bottiglieri T., Rossiter J.: Effect of short term administration of nitrous oxide on plasma concentration of methionine, tryptophan, phenylalanine and s-adenosyl methionine in man. *Br J Anaesth* 1986, 58:1–10.
42. O'Sullivan H., Jennings F., Ward K., McCann S., Scott J.M., Weir D.G.: Human bone marrow biochemical function and megaloblastic hematopoiesis after nitrous oxide anesthesia. *Anestheisology* 1981, 55:645–9.
43. Crawford J.S., Lewis M.: Nitrous oxide in early human pregnancy. *Anaesthesia* 1986, 41:900–5.
44. Aldridge L.M., Tunstall M.E.: Nitrous oxide and the fetus: A review and the results of a retrospective study of 175 cases of anaesthesia for insertion of Shirodkar suture. *Br J Anaesth* 1986, 58:1348–1356.
45. Mazze R.I., Fujinaga M., Rice S.A., Harris S.B., Baden J.M.: Reproductive and teratogenic effects of nitrous oxide, halothane, isoflurane, and enflurane in Sprague-Dawley rats. *Anesthesiology* 1985, 63:A439.
46. Saxen I., Saxen L.: Association between maternal intake of diazepam and oral clefts. *Lancet* 1975, 2:498.
47. Safna M.J., Oakley G.P.: Association between cleft lip with or without cleft palate and prenatal exposure to diazepam. *Lancet* 1975, 2:478–80.
48. American Society of Anesthesiologists, Ad Hoc Committee on the Effect of Trace Anesthetics on the Health of Operating Room Personnel: Occupational disease among operating room personnel: A national study. *Anesthesiology* 1974, 41:321–40.
49. Axelsson G., Rylander R.: Exposure to anaesthetic gases and spontaneous abortion: Response bias in postal questionnaire study. *Int J Epidemiol* 1982, 11:250–6.
50. Ericson H.A., Kallen A.J.B.: Hospitalization for miscarriage and delivery outcome among Swedish nurses working in the operating rooms 1973–1978. *Anesth Analg* 1985, 64:981–8.
51. Liu P.L., Warren T.M., Ostheimer G.W., Weiss J.B., Liu L.M.P.: Fetal monitoring in parturients undergoing surgery unrelated to pregnancy. *Can Anaesth Soc J* 1985, 32:525–32.

52. Wright R.G., Shnider S.M., Levinson G., Rolbin S.H., Parer J.T.: The effect of maternal administration of ephedrine on the fetal heart rate and variability. *Obstet Gynecol* 1981, 57:734–8.
53. Fanning G.L.: The efficacy of cricoid pressure in regurgitation of gastric contents. *Anesthesiology* 1970, 32:553–5.

CHAPTER 14

The Patient with Systemic Lupus Erythematosus

Michael Peck

Case History. *A 12-year-old black girl with no significant past medical history presented with the sudden onset of headache and lethargy. Her laboratory work showed thrombocytopenia (platelet count 8000/mm^3). The hematologic differential diagnosis included idiopathic thrombocytopenic purpura and systemic lupus erythematosus. During evaluation the platelet count decreased to 5000/mm^3. A CT scan showed a frontal intraparenchymal hemorrhage. The patient developed hemiparesis, aphasia, and a dilated pupil. Dexamethasone 20mg was given and she was scheduled for emergency exploratory craniotomy. The presumptive diagnosis was hematoma caused by disruption of an arteriovenous malformation, most probably related to systemic lupus erythematosus.*

Introduction

Systemic lupus erythematosus (SLE) is a chronic, multisystemic inflammatory disease that changes the morphology of the blood vessels and connective tissues of many organs. The disease occurs in up to 1 per 1000 of the population,[1] varying from 1 per 17,000 in low-risk populations to 1 per 250 in high-risk groups. It is more prevalent in women (9:1) and dark-skinned races.[2] The majority of cases are first diagnosed in the third to fourth decades of life; however, 18%–20% of cases occur in children, usually up to 8 years of age.[3] In younger children it is associated with familial lupus or complement defects.[4]

SLE is a member of a group of syndromes known as collagen vascular diseases. Nowadays the term "immune complex diseases" is used instead of "collagen diseases" because of the characteristic pathologic

Reviewed by Dr. Steven Schwalbe, Director of Obstetric Anesthesia and Assistant Professor, Department of Anesthesiology, Albert Einstein College of Medicine/ Montefiore Medical Center.

deposition of immune complexes in specific organs and tissues. Connective tissue is made up of cells and fibrils that contain elastin, collagen, and reticulum embedded in ground substance. Pathologic similarities in immune complex diseases include alterations in interfibrillary ground substance, proliferation of fibroblasts, predominantly mononuclear inflammatory reaction, and fibrinoid necrosis.

Patients with connective tissue diseases have multisystem involvement and are chronically ill. SLE affects the skin, joints, kidneys, nervous system, serous membranes, heart, and lungs. The course is one of remissions and relapses. Distinct immunologic abnormalities develop. Included under the umbrella of immune-complex diseases are SLE, scleroderma, scleredema, periarteritis nodosa, thrombotic thrombocytopenic purpura, and other disorders characterized by vasculitides and dermatomyositis.

Historical Background

The term lupus was used during the 13th century to describe skin conditions with malar erythema in a pattern resembling the facial markings of a wolf.[3] Kaposi first described systemic lupus in 1872, and in 1875 he noted that the rash resembled a butterfly.[5] The concept of collagen diseases was advanced in 1941 by Klemper.[6] In 1948 Hargraves described the LE cell test and Fricus developed the immunofluorescent antinuclear factor, that increased the accuracy of diagnosis.[3]

Pathophysiology

The cause of SLE is unknown although many theories as to the etiology of immune complex diseases have been postulated. Trauma to mast cells may release edema-producing substances that infiltrate cells and lead to fibrin formation.[3]

Another theory is the autoimmune hypothesis, that followed the discovery of rheumatoid and LE factors.[7] Impaired suppressor T cells may lead to the formation of excessive amounts of antibodies. Antigen-antibody complexes formed in the circulatory system are then deposited in the tissues of the target organs. They may activate potent mediators of inflammatory activity such as complement proteins that attract monocytes and polymorphonuclear leukocytes. Toxic products of oxygen metabolism as well as proteases and liposomal enzymes are released by these cells, causing tissue damage.

Histologically, fibrinoid substance is found along tissue fibers including synovium, serous membranes, pericardium, pleura, and blood vessels. Biopsy of a skin lesion shows acute fibrinoid necrosis of the dermis.

Immune (antigen-antibody) complexes cause a fall in serum complement levels. In patients with SLE, decreased complement and antigen levels have been found in the basement membrane of the kidney, skin, and choroid plexus.[8] The skin rash characteristic of SLE can appear promptly after exposure to sunlight or ultraviolet light. Other factors that can cause exacerbations are infections, surgery, pregnancy, and certain drugs (see Table 1).

Endocrine factors are probably influential in women. The disease can go into remission during the last trimester of pregnancy and relapse postpartum.[9] However, if the onset of systemic disease occurs during gestation, there is a high incidence of thrombocytopenia and nephrotic syndrome, with a very aggressive course.[10]

Genetics also may play a role. The increased incidence of SLE in women may be related to an X chromosome factor. Autoimmune disease may be seen in relatives and twins. A deficiency of IgA and C_2 has been recorded in SLE patients.[7] The immune response may also be genetically determined. The origin of the autoantibodies in SLE is not known but is believed to be a loss of tolerance to self antigens, perhaps caused by a virus; viral nucleocapsids or structures that look like nucleocapsids have been found in the cells of patients with SLE.[7] Also, viral antibody titers are elevated in SLE. Although mortality is highest in the first year after diagnosis, the prognosis has recently improved significantly.

TABLE 1. Drugs that can induce exacerbations of SLE

Hydralazine	Sulfonamides
Isoniazid	Oral contraceptives
Phenytoin	Penicillamine
Procainamide	Carbamazepine
Penicillin	Alpha-methyldopa

Diagnosis

Clinically, patients with SLE are young women who present with fever, malaise, anorexia, arthralgia, acute polyarthritis, weight loss, pleural effusion, nephritis, and a butterfly rash. Most patients are diagnosed before the classical clinical picture occurs. In children the disease may be episodic and insidious or acute and fatal. The skin lesions, which occur in most children, may extend over the face, nose, shoulders, and trunk; thay can become bullous and infected. Oral and nasal ulcerations occur in 50% of patients. Ischemic necrosis of the femoral head also has been described in children.[11]

The majority of patients have joint pain, which commonly affects the fingers, hands, wrists, knees, ankles, and elbows; but major joint deformities are not common. Arthritis is usually milder than that seen in rheumatoid arthritis. Muscle pain, weakness, and proximal muscle wasting often occur.

The erythrocyte sedimentation rate is classically elevated. Albumin levels are low, gamma globulin levels elevated. Other laboratory findings include a false-positive test for syphilis, which occurs in up to 15% of patients and can be an early sign of SLE. Rheumatoid factors have been found in patients who did not have clinical signs of rheumatoid arthritis.

The most characteristic findings are autoantibodies. The LE cell is formed in the lab when white cells are shaken, releasing DNA histone (nucleoprotein) that reacts with an IgG antibody. The remaining white cells phagocytize the nucleoprotein-IgG complex. Therefore, the LE cell consists of a leukocyte with a large purple-red homogeneous inclusion body. These LE cells are not present in all cases, and more sensitive tests have been developed.

An immunofluorescent technique detects antinuclear antibodies in 90% of patients. Antinuclear antibodies also have been found in Sjögren's syndrome, scleroderma, and rheumatoid arthritis.[7] However, the highest titers are in SLE. Antibodies to DNA can be found by using diffusion techniques, complement fixation, agglutination, or radioimmunoassays. Also, complement levels are decreased at some point, usually when the disease is most active. The activation and fixation of complement by circulating immune complexes lead to low levels of complement. Assessment of serial levels can be helpful in management. A cogenital deficiency of C_2 in some patients can cause SLE.[3]

The American Rheumatism Association's preliminary criteria for classification of SLE consist of 11 major criteria (Table 2). At least four of the following must be present to substantiate the diagnosis: (1) malar rash, (2) discoid rash, (3) photosensitivity, (4) oral or nasopharyngeal ulcerations, (5) arthritis, (6) serositis (pleuritis or pericarditis), (7) renal involvement (proteinuria or cellular casts), (8) CNS involvement (seizures or psychosis), (9) hematologic disorder (anemia, leukopenia, lymphopenia, or thrombocytopenia), (10) immunologic disorder (LE cells, anti-DNA antibodies, or chronic false-positive serologic test for syphilis), (11) high titer of antinuclear antibodies.[8]

Preanesthetic Assessment

As noted, SLE is a multisystem disease, requiring careful evaluation of each organ system. Assessment should include a careful history and physical exam. The history should include age at onset of the disease, original symptomatology, any subsequent manifestations, and medications used in the past and present, including immunosuppressive therapy, antimalarial drugs, steroids, and salicylates.

TABLE 2. Criteria for classification of SLE (at least 4 symptoms must be present to substantiate the diagnosis.)

Malar rash	Renal disorders (proteinuria, casts)
Discoid rash	
Photosensitivity	Neurologic disorders (seizures, psychosis)
Oral ulcers	
Arthritis	Hematologic disorders (anemias)
Serositis (pleuritis, pericarditis)	Immunologic disorders Antinuclear antibodies

Cardiovascular System

Almost any part of the heart may be involved. Over half of patients have involvement of the pericardium, endocardium, or myocardium. When fever and pericardial friction rub are present, the disease resembles rheumatic fever.

Verrucous nonbacterial endocarditis (Libman-Sacks endocarditis) consists of nonbacterial vegetations on the heart valves or chordae tendineae that are usually asymptomatic. Vasculitis of the myocardium may occur. Auscultation of the heart may reveal a friction rub or murmur, and chest x-ray may confirm pericardial effusion. If a significant effusion is present, pericardiocentesis is beneficial.

Myocarditis is associated with tachycardia out of proportion to the fever and anemia commonly seen in these patients. Precordial chest pain with an associated friction rub may be the result of pleurisy or pericarditis. Tamponade or constrictive pericarditis is rare. There may be mitral valve involvement. The ECG often shows nonspecific changes in ST and T waves. Congestive heart failure is rare. Recurrent thrombophlebitis may be the first sign of SLE.

Respiratory System

The lungs and pleura are involved in 50%-70% of patients.[12] Pleuritic pain may be the first system of the disease. There is often a friction rub. Accompanying pleural effusions are usually painless. Pulmonary infiltrates not caused by infection can shift between lobes, which may become atelectatic, resulting in cyanosis and pulmonary insufficiency. Other lesions are interstitial pneumonitis, acute vasculitis, focal alveolar hemorrhage, and pneumonia. Symptoms include dyspnea, cough, pleural pain, and, rarely, hemoptysis.

The x-ray findings are often minimal. Pulmonary function tests show a restrictive defect, with a loss of lung volume, out of proportion to the clinical picture. Flow rates are normal, with a reduced diffusing capacity. Arterial blood gas analyses on room air show a low pO_2 with a normal pCO_2. Lung compliance also is reduced. Most pulmonary complications respond well to steroid therapy. Intraoperative requirements include a high inspired O_2 concentration, low tidal volume, and a rapid respiratory rate.

If there is evidence of pulmonary hypertension, preanesthetic placement of a pulmonary artery catheter may be helpful. Cor pulmonale may exist and preoperative digitalization should be considered.

Gastrointestinal System

Symptoms include nausea, vomiting, and abdominal pain that may be due to pancreatitis, ileus, peritonitis, or enteritis. Vasculitis may cause intestinal perforation. A pseudo-obstruction caused by dilated loops of edematous bowel occasionally develops. The liver is often enlarged as a result of passive congestion, and "onion skin" lesions have been described in the spleen. A "lupoid hepatitis," characterized by jaundice, hepatomegaly, hyperglobulinemia, and abnormal liver function tests may be fatal.

Renal System

Renal involvement ranges from mild to severe. The mild form consists of immunoglobulin and complement deposits in the mesangium or basement membrane. Glomerulitis is the most common lesion, with increased cellularity, basement membrane thickening, and fibrinoid changes. Glomerulonephritis is severe and consists of hypercellularity of endothelial, mesangial, and epithelial cells as well as inflammatory cells, leading to crescent formations. Approximately 50% of patients have renal involvement within 2 years of the onset of SLE.[13] Acute nephritis or the nephrotic syndrome may be present. A mild glomerulitis is most common; it causes proteinuria and/or hematuria that usually responds well to therapy. Acute lupus glomerulonephritis can cause proteinuria, pyuria, hematuria, edema, hypoalbuminemia, hypertension, azotemia, and fluid retention. The occurrence of pyuria and fever may indicate a urinary tract infection rather than lupus nephritis. Most patients recover from renal damage; in others renal damage may progress to a chronic state with hypertension and proteinuria. However, acute lupus glomerulonephritis can be fatal.

Central Nervous System

Neurologic and psychologic changes occur in up to 50% of patients with SLE. The fears and anxieties produced by chronic illness and organic disturbances can lead to hyperirritability, confusion, hallucinations, paranoid reactions, or psychoses.[14] The psychosis of SLE may be difficult to distinguish from a steroid-induced psychosis. The EEG is often abnormal, but brain scans are usually normal. Seizures occur in up to 15% of patients.[14] Other less common symptoms include aphasia, ptosis, diplopia, hemiparesis, nystagmus, and peripheral neuropathy. Retinal findings include "cotton-wool spots" that are present during acute phases of the disease but usually dissappear with treatment.

Hematologic System

Painless enlargement of lymph nodes and splenomegaly are seen in 15% of patients, and mild anemia can occur as a result of infection, renal disease, or bleeding. About half of patients have leukopenia. Otherwise the cell count is usually normal. (Steroids cause leukocytosis.) Thrombocytopenia may occur years before other signs of the disease. A lupus anticoagulant that prevents the complex of phospholipid, calcium, and factors X and V from activating prothrombin has been found in up to 25% of patients with SLE.[15] Although this anticoagulant was first described in patients with SLE, it has since been associated with malignancy and inflammatory disorders.[16,17]

Menometrorrhagia may be problematic. A coagulation profile should be performed preoperatively and appropriate therapy instituted. Tests such as partial thromboplastin time (PTT) may be falsely elevated as the SLE antibodies react with phospholipids used to determine PTT. Blood products should be available. However, there is no correlation between coagulation tests and clinical hemorrhage.[18] Indeed, despite the prolongation of bleeding, induced hemmorrhagic complications are infrequent and there is a tendency to thromboembolism,[19,20] especially involving the central nervous system.[21–24] Hypothrombinemia also may be seen and can be a factor in cases of clinical bleeding.

Integument

Skin involvement may vary from the classic butterfly rash to only a swelling or maculopapular rash on the cheeks and bridge of the nose.

Any area of skin exposed to the sun can be affected. Discoid lesions can become scaly and may atrophy on healing, causing scarring and increased or decreased pigmentation. Areas of the scalp may be involved, resulting in hair loss. Periungual erythema and telangiectasia sometimes occur, and vasculitis can cause ulcerations at the fingertips that may progress to gangrene. There may be ulcers on the nose, mouth, gums, or buccal mucosa, making mask fit or intubation difficult.

Other skin manifestations include purpura or ecchymosis, which may result from a coagulation defect, corticosteroid therapy, renal insufficiency, or vasculitis. Joint mobility, particularly of the temporomandibular joint, should be assessed. The cartilaginous portion of the larynx may be narrowed, and small endotracheal tubes should be readily available. Fiberoptic intubation may be indicated. Involvement of the arytenoid cartilage can make vocal cord abduction minimal.[25]

There is an increased incidence of aseptic necrosis of the femoral head due to steroid therapy, but the bones are rarely affected.

Therapy

Ten-year survival is now listed at 90%, compared with 50% survival at 4 years some 30 years ago.[10] Treatment is limited to therapy during relapses and prevention of exacerbations. Avoidance of sunlight, ultraviolet light, blood transfusions, and provoking drugs is helpful. The development of signs such as anemia, leukopenia, thrombocytopenia, and decreased complement heralds relapses. They should be recognized, as they signal the need for therapy.

Active disease should be treated aggressively.[26] Therapy includes rest, sunscreens, physical therapy, salt restriction, and diuretics for those on steroids. Death from CNS involvement decreased from 26% in the years 1950–1955 to 8% in 1962–1972. Also, death from uremia decreased from 26% to 14% in the same periods.[10]

Other treatments include topical steroids for skin lesions, antimalarial drugs, salicylates for fever or arthralgia, azathioprine, cyclosporine, and plasmapheresis. Pericarditis and pleurisy respond to rest and steroids. Pericardial fluid, if more than minimal, should be removed. Steroids, fluid restriction, and digitalis are helpful in patients with myocarditis.

Anemia and thrombocytopenia need not be treated unless symptomatic. Renal involvement usually responds to steroids, but renal failure may require hemodialysis. Immunosuppressive therapy is reserved for patients with disease intractable to steroids and in life-threatening situations. A complete blood count and urinalysis should be done, and electrolyte and creatinine levels and complement levels should be reported at frequent intervals. Antiplatelet therapy (aspirin) may be useful for patients with lupus anticoagulant and basal ganglion lacunar infarctions.[27] Aspirin and non-steroidal anti-inflammatory drugs also have been useful in the management of arthritis, serositis, and fever.

Anesthetic Plan

The anesthetic plan of management for the patient with SLE must take into consideration the organs affected by the disease and the medications taken to combat its progress.

Patients receiving longterm corticosteroids should have steroid coverage. Although some patients on maintenance levels of steroids can still activate their pituitary-adrenal axis, the identification of such patients is not easy.[28] Signs and symptoms of acute adrenal insufficiency include hypotension, restlessness, weakness, anorexia, and abdominal pain. One recommendation for perioperative steroid supplementation is hydrocortisone, 100mg I.V. q8h, starting 3 hours before anesthesia and continued for 1 to 3 days.[29]

In patients with an established intravenous route, hydrocortisone sodium succinate (Solu-Cortel®) 100mg is more useful, and following intravenous injection, demonstrable effects are evident within 1 hour and persist for 4–6 hours, a response similar to that following intramuscular injection.

Antibiotic prophylaxis is indicated if valvular disease exists.

Although no specific anesthetic agents have been judged superior, drugs that are metabolized or cleared by the kidney should be avoided if there is renal involvement. Such agents include gallamine, curare, methoxyflurane, and enflurane.[30] Regional anesthesia may be preferable if the surgical site permits and if the coagulation profile is normal and there is no evidence of bleeding.

References

1. Dubois E.L.: *Lupus Erythematosus* (monograph). Los Angeles, University of Southern California Press, 1978.
2. Eisele, J.H.: Connective tissue diseases, in Katz J., Benumof J., Kadis L.B. (eds): *Anesthesia and Uncommon Diseases.* Philadelphia: WB Saunders, 1990, p 654.
3. Schur P.H.: Systemic lupus erythematosus, in Wyngaarden J., Smith L.H. Jr (eds): *Cecil Textbook of Medicine.* Philadelphia, WB Saunders, 1982, pp 1852–7 .
4. Schaller J.: Lupus in childhood. *Clin Rheum Dis* 1982, 8:219–25.

5. Hebra F., Kaposi M.: *On Diseases of the Skin, Including the Exanthemata.* Vol 4. Fagge H., Smith P, Tay W. (transl., eds): London: New Sydenham Society, 1874.
6. Klemperer P., Pollack A.D., Baehr G.: Pathology of disseminated lupus erythematosus. *Arch Pathol* 1941, 32:569–631.
7. Tan E.M.: Autoantibodies to nuclear antigens: Their immunobiology and medicine. *Adv Immunol* 1982, 33:167–72.
8. Muller S.B.: Systemic lupus erythematosus, in Hurst J.W. (ed): *Medicine for the Practicing Physician,* 2nd edition. Boston: Butterworths, 1988, pp 228–32.
9. Abouleish E.: Obstetric anesthesia and systemic lupus erythematosus. *Middle East J Anaesthesiol* 1988, 9:435–46.
10. Fine L.G., Barnett E.V., Danovitch G.M., et al.: Systemic lupus erythematosus in pregnancy. *Ann Intern Med* 1981, 94:667–76.
11. Smith M.F.: Skin and connective tissue disorders, in Katz J., Steward D. (eds): *Anesthesia and Uncommon Pediatric Diseases.* Philadelphia: WB Saunders, 1987, pp 406–7.
12. Divertie M.B.: Lung involvement in the connective tissue disorders. *Med Clin N Amer* 1964, 48:1015–30.
13. Balow J.E., Austin H.A. III, Muenzl R., et al.: Effective treatment of the evolution of renal abnormalities in lupus nephritis. *N Engl J Med* 1984, 311:491–5.
14. Feinglass E.J., Arnett F.C., Dersch C.A., et al.: Neuropsychiatric manifestations of systemic lupus erythematosus: Diagnosis, clinical spectrum and relationship to other features of the disease. *Medicine* 1976, 55:323–39.
15. Feinstein D.I., Rappaport S.I.: Acquired inhibitors of blood coagulation. *Prog Hemost Thromb* 1972, 1:75–95.
16. Conley C.L., Hartmann R.C.: A hemorrhagic disorder caused by circulating anticoagulants with disseminated lupus erythematosus. *J Clin Invest* 1952, 321:621–22.
17. Schleider M.A., Nachman R.L., Jaffe E.A., et al.: A clinical study of lupus anticoagulant. *Blood* 1976, 48:499–509.
18. Boxer M., Ellman L., Carvalo A.: The lupus anticoagulant. *Arthritis Rheum* 1976, 19:1244–8.
19. Mueh J.R., Herbst K.D., Rappaport S.I.: Thrombosis in patients with the lupus anticoagulant. *Arch Intern Med* 1980, 92:156–9.
20. Elias M., Eldor A.: Thromboembolism in patients with the "lupus"-type circulating anticoagulant. *Arch Intern Med* 1984, 144:510–15.
21. Landi G., Calloni M.V., Sabadini M.G., et al.: Recurrent ischemic attacks in two young adults with lupus anticoagulant. *Stroke* 1983, 14:377-9.
22. Englert HJ, Hawkes CH, Boey ML, et al.: Degos' disease: association with anticardiolipin antibodies and the lupus anticoagulant. *Br Med J* 1984, 289:576.
23. Ke R.E., Gillman P.B., Kovacs A.D.: Cerebral ischemia in the presence of lupus anticoagulant. *Arch Neurol* 1984, 41:521–3.

24. Fisher M., McGehee W.: Cerebral infarct, TIA, and lupus inhibitor. *Neurology* 1986, 36:1234–7.
25. Sourander L.B., Pulkkinen K.: Simultaneous occurrence of ankylosis of the cricoarytenoid joints with dyspnea and LE syndrome in rheumatoid arthritis. *Acta Rheum Scand* 1962, 8:2557.

CHAPTER 15

The Patient With A History of Anaphylaxis

Kenneth J. Abrams, M.D.

Case History. *A 50-year-old woman with a medical history significant for essential hypertension was admitted for total abdominal hysterectomy for uterine leiomyomata. Her surgical history revealed cholecystectomy 5 years previously under general anesthesia without complications. Medications included atenolol 50mg b.i.d. and hydrochlorothiazide 50mg o.d. with potassium supplementation. She had a history of allergy to shellfish and seasonal rhinitis, but there were no known drug allergies. Prior exposures to penicillins were without incident.*

On examination vital signs were blood pressure, 150/90mmHg; pulse, 60/min; respiratory rate, 20/min; temperature, 37.0°C; weight, 70kg; height, 5'4"; airway assessment, Molampati grade I; lungs, clear to auscultation; heart, normal tones without murmurs or gallops.

Laboratory results were as follows: hemoglobin, 11gm/dl; hematocrit, 34%; SMA 6/12, within normal limits. Chest x-ray was normal. ECG showed left ventricular hypertrophy without acute ischemic changes.

Preoperative medication included atenolol 50mg, diazepam 10mg p.o. The patient opted for epidural anesthesia with planned postoperative epidural narcotics for pain control.

On arrival in the operating room routine monitoring, including automatic blood pressure cuff, 5-lead ECG, pulse oximeter, and skin temperature, was initiated. A vein was cannulated and prehydration commenced. A lumbar epidural catheter was placed under local anesthesia with 1% lidocaine at the L3–4 level. Anesthesia to the T6 level was achieved with 75mg of 0.5% bupivacaine in incremental doses.

At the surgeon's request, cephazolin 2gm was administered, divided into 1gm by slow I.V. bolus and 1gm by I.V. infusion. Approximately 5 minutes after antibiotic administration the patient began to complain of

Reviewed by Dr. Erlina Farcon, Department of Anesthesiology, Albert Einstein College of Medicine.

difficulty breathing, lightheadedness, and diffuse pruritus. Evaluation revealed appropriate anesthetic level, maculopapular rash across the chest, bilateral wheezes, and blood pressure of 70/40mmHg. After successful resuscitation, the planned surgery was postponed for 3 days.

Introduction

The majority of medications administered during the course of anesthesia are nontherapeutic and potentially lethal, with a narrow range of efficacy.[1] In addition, many, if not all, of these agents represent serious allergic potential. Levy suggests that approximately 1 per 2700 hospitalized patients will experience severe medication-related allergy.[1] It has been estimated that 4% of all hospital admissions are the result of adverse drug responses.[2] Associated mortality is approximately 0.005%. Moreover, the incidence of adverse reactions to anesthetic agents appears to be increasing.[3] This can be attributed in part to the greater numbers of drugs being administered during anesthesia and the potential for cross-reactivity.

The administration of multiple drug regimens over a short time course is unique to anesthetic practice. Induction agents, muscle relaxants, and antibiotics represent just a few of these medications.

A major problem facing anesthesiologists is the identification of patients likely to manifest severe allergic reaction. Clinically, these reactions vary from trivial to catastrophic, and may even be fatal. The anesthesiologist should be aware of the pathophysiology, clinical presentations, management strategies, and diagnostic modalities involved in severe allergic reactions. An understanding of these entities will enhance preanesthetic preparation for the allergic patient.

Types of Reactions

Generally, there are two types of reaction to medications, the predictable and the unpredictable. Patients often refer to adverse reactions as "allergic." For example, seizures from unintentional overdose of intravenous local anesthetics may be termed a bad reaction or an "allergy" to the anesthetic by the patient. The incidence of adverse drug reactions has been estimated at 30% for hospitalized patients.[4]

Predictable reactions are best described as being dose-dependent and related to the pharmacologic properties of the drug. These represent approximately 80% of adverse reactions.[5] In contrast, unpredictable responses occur independently of dosage and often involve the immunologic response of the patient. These reactions pose the greatest risk for the patient and the greatest challenge for the physician.

Certain groups of patients are at increased risk for severe allergic reactions, and their history should provide warning to the anesthesiologist. Patients with a history of chronic atopy (asthma, hay fever, previous drug allergy, certain food sensitivities) have greater likelihood of developing allergic reactions to intravenous anesthetics.[6] IgE antibodies are often chronically elevated in patients with a history of allergy, compared to normal persons. It is this elevation that predisposes the person to severe reactions.

Another vital factor in predicting the risk of an allergic reaction is the number of exposures to a particular substance or chemically similar agent. This factor suggests that highly susceptible people should not receive the same drug more than once in their lifetime. Fox et al have demonstrated that a 2-week incubation period is optimal for development of drug allergy.[7]

It is important to bear in mind that prior exposure to a drug without deleterious effect does not eliminate the possibility of a severe allergic reaction on subsequent exposure.[8] Prior sensitization is required for true anaphylaxis to occur via an IgE-mediated mechanism; however, severe reactions (ie, anaphylactoid reactions) can also occur without prior exposure to the responsible agent by one of two mechanisms, complement activation[9] and nonimmunologic histamine release.[10] Although these causal differentiations are important for the diagnosis of anaphylactoid versus anaphylactic reactions after the event and in preventing future reactions in individual patients, they are of little importance clinically because the reactions are indistinguishable.

Certain other factors have been found to increase the risk of specific drug allergies. For example, in protamine-insulin dependent diabetics the presence of anti-protamine antibodies is a significant factor for life threatening reactions when protamine is used intravenously.[11] Also, significant increase in anti-protamine IgG antibody in the sera of vasectomized males increases their risk for the development of a severe allergic response to protamine.[12]

Pathophysiology

Severe, life-threatening allergic reaction can occur in several different ways. Mechanisms demonstrated to date include antibody-mediated anaphylaxis (type I hypersensitivity), complement activation, and nonimmunologic release of histamine.

Type I Hypersensitivity

As noted, true anaphylaxis requires prior exposure to the allergenic substance or a chemically similar substance. The initial exposure stimulates IgE antibody production from lymphocytes, a process termed sensitization. To produce IgE antibodies, an antigenic molecule must be capable of reacting with lymphocytes. Drugs are generally of low molecular weight, and by themselves they are incapable of initiating antibody formation.[6] A substance that cannot stimulate an antibody response by itself is called a hapten. Thus, a drug or drug product must bind with a carrier, usually a protein, to become a complete antigen. It is the hapten-carrier (drug-protein) complex that then stimulates lymphocytes to produce IgE antibodies.

Once the antibodies are produced, they attach to receptor sites on the cell membranes of plasma basophils and tissue mast cells.[13] After being sensitized, the basophils and mast cells are capable of participating in an anaphylactic reaction on reexposure to the antigen. Reexposure to the antigen or a structurally similar substance bridges two cell-bound immunospecific IgE antibodies. Bridging causes cross-linking, with resultant cell membrane alterations and ultimately degranulation of these cells (Figure 1).

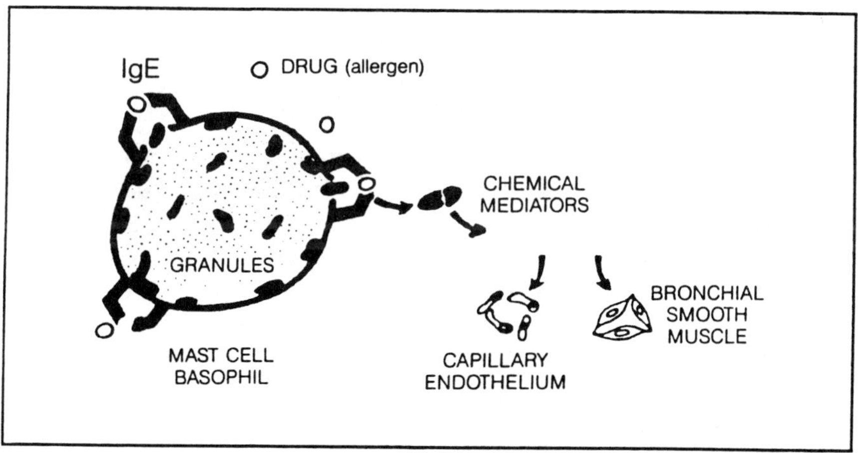

FIGURE 1. Schematic depiction of an allergic reaction due to anaphylaxis. Adapted from Stoelting R.K.: Allergic reactions during anesthesia. Anesth Analg 1983, 62:341.

Degranulation, an energy-dependent process, releases a multitude of chemical mediators. The extent of degranulation is affected by several factors, namely, the dose of the drug, affinity of the antigen for antibody, number of antibodies present, and concentration of intracellular nucleotides.[8]

The clinical manifestations are the result of pharmacologic action of the chemical mediators, which are either preformed or synthesized after degranulation. Preformed mediators include histamine, eosinophilic chemotactic factor of anaphylaxis (ECF-A), and neutrophil chemotactic factor of anaphylaxis (NCF-A). Leukotrienes, prostaglandins, and kinins are examples of newly synthesized mediators.[13]

Chemical mediator release can be modified by the autonomic nervous system and by alterations in cyclic nucleotide concentrations, particularly cyclic adenosine monophosphate (cAMP).[14] Inhibition may be achieved by increasing cAMP levels via beta-adrenergic stimulation. Conversely, decreasing cAMP concentrations by alpha-adrenergic stimulation or increasing cyclic guanosine monophosphate (cGMP) concentration enhances release of histamine and of slow releaseing substance of anaphylaxis (SRA-A).[14]

SRS-A is a leukotriene, thus a product of arachidonic acid. It is rapidly synthesized after degranulation and serves to potentiate bronchoconstriction and capillary leak.

Histamine is the only mediator that is essential for anaphylaxis. By itself, histamine increases capillary permeability, dilates arterioles, and causes bronchial smooth muscle constriction. These pharmacologic effects of histamine are responsible for the syndrome of vasodilation, bronchospasm, and urticaria.[1]

Complement Activation

The complement cascade is composed of nine globulin protein compounds (Figure 2). This pathway can be activated immunologically (classical pathway) or nonimmunologically (alternative pathway). Stimulation of this cascade produces the activated forms of these protein components, particularly C3a and C5a. C3a and C5a are anaphylatoxins capable of inducing mast cell and basophil degranulation. The specific effects of C3a and C5a are outlined in Table 1.

Classical pathway activation requires antigen-antibody reaction with IgM or IgG as the antibody. The antigen-antibody interaction initiates the cascade by activating circulating C1. Alternative pathway activation is independent of antibody reaction. The drug or offending agent can directly stimulate C3 to the activated form, C3a, thereby propagating the complement pathway.

In distinct contrast to IgE-mediated anaphylaxis, complement-induced reactions do not require prior sensitization. The initial exposure of the immune system can cause catastrophe.

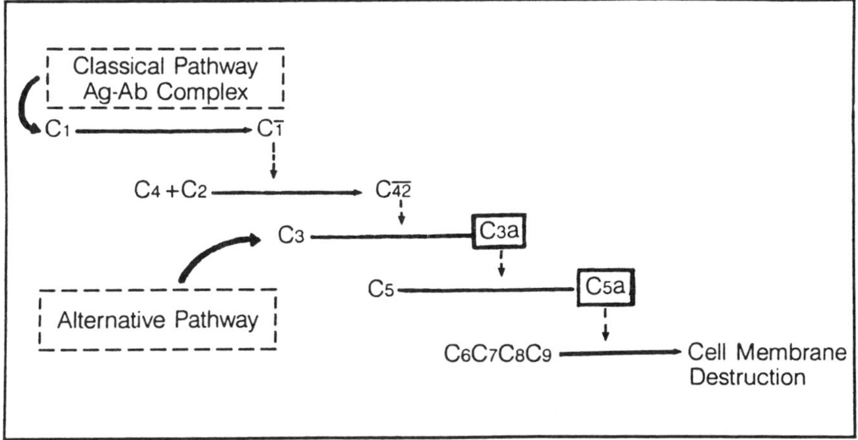

FIGURE 2. Schematic diagram of complement activation.

TABLE 1. Biologic Effects of Anaphylatoxins

Effect	C3a	C5a
Histamine release	+	+
smooth muscle contraction	+	+
Increased vascular permeability	+	+
Chemotaxis		+
Leukocyte and platelet aggregation		+

From: Barash PG, Cullen BF, Stoelting RK: Clinical Anesthesia (p 1386). J.B. Lippincott Company, Philadelphia, 1989. Reprinted with permission of publisher.

Clinical Manifestations

Allergic responses vary greatly from individual to individual and are highly unpredictable. People with a history of chronic atopy tend to have more severe reactions than do normal patients.[6] Reactions can range from minor cutaneous manifestation to profound cardiovascular collapse.[1] Several factors influence the intensity of a reaction: dose of drug injected, reactivity of mast cells and basophils, smooth muscle sensitivity, and autonomic nervous system activity.[8] Emotional well-being may help maintain autonomic nervous system balance and attenuate allergic reactions. The dose of drug injected is the only controllable variable.

The clinical picture tends to occur suddenly, but it may be delayed. Anaphylactic and anaphylactoid reactions are clinically identical with regard to the end-organ responses to chemical mediators released during degranulation. Generally, cutaneous manifestations occur most frequently, followed by vasodilatory hypotension with tachycardia, and finally bronchospasm and hypoxemia (Table 2).

Skin changes exhibit the classic triple response of erythema, wheal, and flare. Most notably, the awake patient complains of itching, burning, and tingling sensations. Signs include urticaria, flushing, and periorbital and perioral edema. These changes are a direct result of tissue mast cell degranulation.

Cardiovascular manifestations are less subtle. They range from dizziness and malaise to vasodilatory shock. Increased capillary permeability with an acute drop in intravascular volume, as well as vascular smooth muscle relaxation, combine to produce profound hypotension. Resultant reflex tachycardia occurs in the patient with a responsive sympathetic nervous system capable of releasing catecholamines. Additional dysrhythmias may occur as a consequence of catecholamine or histamine release.

The most life-threatening characteristic of an allergic reaction is bronchospasm regardless of the presence or absence of laryngeal edema. Less severe signs of respiratory involvement can occur, including coughing and sneezing.

The prudent anesthesiologist recognizes the early signs of anaphylaxis, the diagnosis of which rests on a high level of suspicion. Anaphylaxis occurs unpredictably, with varying severity and frequently in persons without prior history.

Agents Associated With Severe Allergic Reactions

Virtually all of the anesthetic agents used today—including, but not limited to, intravenous induction agents, muscle relaxants, local anesthetics, and opioids—have been implicated in anaphylactic/anaphylactoid reactions. In addition, many of the adjunct agents administered during the course of anesthesia can cause life-threatening allergic reactionst[15] (Table 3). Even surgical gloves recently have been reported to cause an allergic

TABLE 2. Manifestations of Anaphylaxis

Cardiovascular
Hypotension (resulting from decreased SVR and increased vascular permeability)
Tachycardia, atrial and ventricular dysrhythmias
Altered cardiac output (usually decreased but change depends on treatment [eg, volume status], preexisting cardiac disease, and specific effects of released mediators on the heart)
± Pulmonary hypertension
Respiratory
Bronchoconstriction, (increased airway pressures, difficulty with oxygenation)
Upper airway obstruction caused by soft tissue edema
Cutaneous
Generalized rash or flushing
Urticaria
Edema (generalized, perioral, periorbital)
Hemostasis
Excessive bleeding (as a result of coagulation abnormalities)

reaction.[16] Other latex products such as catheters may also cause intraoperative anaphylaxis, especially in children with congenital urologic abnormalities and/or spina bifida.[17]

Reactions to local anesthetics are quite rare. More common is a predictable adverse response to the local anesthetic. Usually, unintentional intravascular injection can explain this response. Careful questioning can elicit the mechanism involved in most local anesthetic adverse responses. Allergic reactions will be characterized by difficulty in breathing, hives, or intense itching. Adverse effects such as confusion or seizures are the result of drug overdose or accidental intravascular injection.

Of the local anesthetics, esters are associated more frequently with allergic reactions. The metabolite para-aminobenzoic acid (PABA) is usually responsible. In addition, the preservative methylparaben may cause a true allergy because it is structurally similar to PABA.

TABLE 3. Agents Implicated in Anaphylactic/Anaphylactoid Reactions During Anesthesia

Anesthetic agents implicated in allergic reactions
Induction agents
cremophor solubilized drugs, barbiturates, etomidate
Local anesthetics
para-aminobenzoic ester agents
Muscle relaxants
succinylcholine, gallamine, pancuronium, d-tubocurarine, methocurine, atracurium, mevicurium, vecuronium
Opiods
meperidine, morphine, fentanyl
Other agents implicated in allergic reactions
Antibiotics
cephalosporins, penicillin, vancomycin
Atropine
Blood products
whole blood, packed cells, FFP, platelets, cryoprecipitate
Bone cement
Chymopapain
Cyclosporine
Drug additives
Furosemide
Mannitol
Methyl methacrylate
Protamine
Radiocontrast dye
Rubber products
Volume expanders (colloid)
dextrans, protein fractions, albumin, hydroxyethyl starch

From Levy JH: Anaphylactic Reactions in Anesthesia and Intensive Care. Butterworth, Boston, 1986. With permission of publisher.

Identification of local anesthetic allergy is difficult. It involves extensive intradermal testing, is associated with false positives and false negatives, and requires the use of preservative-free anesthetics—since the preservative may be the allergen. Therefore, it would seem more reasonable to select a preservative-free preparation of the alternative type of local anesthetic (ie, amides for those allergic to esters; esters for those allergic to amides).

Depolarizing and nondepolarizing muscle relaxants have been implicated in anaphylactic and anaphylactoid reactions.[18,19] Bronchospasm during intubation requires prompt evaluation and therapy while associated allergic manifestations are sought.

At present, protamine is the only available drug that reverses heparin. Insulin-dependent diabetes patients who are treated with neutral protamine Hagedorn (NPH) or protamine zinc insulin may be sensitized to protamine and have an increased risk of anaphylactic reactions.[20] Protamine is derived from salmon sperm. Patients allergic to fish may exhibit cross-reactivity to protamine. As already noted, vasectomized patients may have autoantibodies to sperm, increasing their risk of severe allergy. Sperm that cannot be ejaculated are absorbed into the circulation, causing sensitization.

Treatment

Severe allergic reactions require immediate therapy. The goals of therapy are the same for anaphylactic and anaphylactoid reactions; they include security and maintenance of the airway, restoration of circulating blood volume, and preservation of end-organ function. (see Table 4.).

All anesthetic agents should be discontinued. Ventilation should be maintained with 100% oxygen. Prompt evaluation of the respiratory system, with specific attention to airway edema and bronchospasm, will determine whether endotracheal intubation is required (if it has not already been performed).

Epinephrine is the specific drug of choice for the treatment of anaphylaxis (2–10μg I.V. bolus for hypotension, titrated to maintain blood pressure, and 0.1–0.5mg I.V. for cardiovascular collapse). Epinephrine leads to an increase in intracellular cAMP. The inverse relationship between cAMP and chemical mediator release probably explains the lifesaving effect of epinephrine. If hypotension is present, the initial dose should be administered as an I.V. bolus. Because perfusion to muscular and subcutaneous tissues will be variable, these parenteral routes are not recommended for treatment of severe reactions. It is important to titrate the dose of epinephrine to obtain cardiopulmonary stability.

Infusion of intravenous fluids (balanced salt solutions) is critical for hemodynamic stability. The increase in capillary permeability disperses intravascular fluid to the extracellular space.

Secondary therapy involves other drugs with either documented or theoretical value. Intravenous antihistamines (0.5–1mg/kg diphenhydra-

TABLE 4. Management of Anaphylaxis

Initial therapy

1. Stop administration of antigen.
2. Maintain airway with 100% oxygen.
3. Discontinue all anesthetic agents.
4. Start intravascular volume expansion (2–4L crystalloid for hypotension).
5. Give epinephrine (4–8μg I.V. bolus for hypotension, titrate as needed; 0.1–0.5mg I.V. infusion for cardiovascular collapse).

Secondary treatment

1. Antihistamines (0.5–1mg/kg diphenhydramine)
2. Catecholamine infusions (starting doses: epinephrine 2–4μg/min, norepinephrine 2–4μg/min, or isoproterenol 0.5–1μg/min as a drip, titrated to desired effects)
3. Aminophylline (5–6mg/kg over 20min for persistent bronchospasm
4. Coricosteroids (0.25–1gm hydrocortisone; alternatively, 1–2gm methylprednisolone)*
5. Sodium bicarbonate (0.5–1mEg/kg for persistent hypotension or acidosis
6. Airway evaluation (prior to extubation)

Methylprednisolone may be the drug of choice if the reaction is suspected of being mediated by complement.
From Levy J.H.: Anaphylactic Reactions in Anesthesia and Intensive Care. Boston, Butterworth, 1986. With permission of publisher.

mine) should be administered to increase competition for free histamine receptors. Aminophylline (5–6mg/kg as loading dose over 20min, followed by 0.9mg/kg) has been useful in the management of persistent bronchospasm. It inhibits phosphodiesterase activity, leading to an increase in cAMP. The use of sodium bicarbonate should be dictated by serial blood gas analyses.

Successful treatment of severe allergic reactions depends on an organized plan. However, even after resuscitation, patients should be observed in an intensive care setting for 24 hours. Judicious use of invasive hemodynamic monitoring should be employed. Patients with evidence of persistent hypotension and acidosis should have intraarterial monitoring. Monitoring via a pulmonary artery catheter may be of value in assessing the adequacy of intravascular volume resuscitation.

Diagnosis

History

To prevent subsequent severe allergic reactions in patients with a prior history of anaphylaxis, the responsible agent must be identified. A prior anesthetic reaction, manifested by cardiopulmonary compromise, should always be taken seriously. The previous anesthetic records should be obtained and carefully reviewed. In today's practice of polypharmacy, incriminating a particular drug is difficult and often temporal. It is, therefore, imperative to confirm allergy by laboratory testing.

Plasma Measurements

Serial measurements of plasma concentrations of IgE antibodies and complement C3 and C4 during the 72 hours after clinical manifestations will help attribute the reaction to an allergic response. Previously sensitized persons will demonstrate a decreased plasma level of IgE antibodies.[21] Similarly, consumption of complement C3 and C4 suggests classical pathway activation.

In contrast, anaphylactoid reactions do not demonstrate these immunologic alterations and the adverse response occurs with initial drug exposure. Although these measurements identify an anaphylactic reaction, they provide no indication as to the agent.

Intradermal Testing

Intradermal testing has been the most commonly used modality in identifying specific drugs.[19] The technique is relatively simple. It usually employs 0.1ml of 1:1000 dilution of specific agents injected intradermally into the anterior forearm. Because allergy may be related to a solvent or preservative, these agents must be evaluated independently. Positive intradermal testing produces the characteristic wheal-and-flare response. Strict protocols for intradermal testing have been developed by Fisher.[22] Although the risks of intradermal testing are low, they are not absent. Extremely sensitive persons may experience anaphylaxis from such in vivo testing. Therefore, full resuscitation equipment and trained personnel should be available should an untoward event occur.

Radioallergosorbent Test (RAST)

The RAST employs a plasma sample in order to generate information similar to that obtained by skin testing. This in vitro test detects specific

IgE antibodies to specific antigens. Basically, the process involves two steps. In the first step, antigens are attached to insoluble materials to make an immunoabsorbent and then incubated with the patient's serum, allowing antigen-antibody complex formation.[22] Next, radiolabeled anti-IgE antibodies are incubated with the complexes, and washed. The use of a gamma counter quantifies IgE levels to the agent under examination. Currently available RASTs include succinylcholine, meperidine, and thiopental. Limitations are related to the lack of commercially available tests and to false-positive results.[24]

Nevertheless, a recent report on the evaluation of 58 patients identified as having had a life-threatening anaphylactic reaction, indicates the value of isolation of allergic components and of skin testing.[25] Using the RAST test and human basophilic degranulation test, the specific allergens were identified (most frequently, succinylcholine). The patients were given a card and advised to avoid the offending drugs. In patients allergic to one muscle relaxant, if skin testing showed crossed reactivity, all muscle relaxants were avoided. If skin testing was negative, the alternative agent was used. Patient compliance was excellent and no adverse incidents occurred, confirming a predictive value of skin tests in the choice of anesthetic after a first anaphylactic reaction.

Pre-Anesthetic Evaluation

Evaluation of the allergic patient relies heavily on a detailed history. Prior allergic manifestations must be clarified because such reactions may be related to known pharmacologic principles, not allergy. Patients with a history of asthma, hay fever, food allergies, or previous drug reaction represent high risk. Old medical and anesthetic records should be carefully reviewed to confirm or discount an allergic reaction. For medicolegal reasons, it is best to avoid any agent thought to produce allergy. When the responsible agent cannot be avoided, as in the case of protamine for heparin reversal, it should be administered very slowly and in reduced amounts if at all possible. It should be remembered that cross-reactivity does exist and that prior exposure is not necessary to produce severe allergic reactions.

Prophylaxis has been recommended for patients with a history of contrast-medium allergy undergoing contrast studies. A typical regimen consists of corticosteroids, prednisone 50mg q6h for 24 hours, last dose 1 hour before study, and H_1 and H_2 antihistamines (diphenhydramine 50mg, cimetidine 300mg) 1 hour before study.

The evaluation is complicated when a multiple agent regimen has been used to induce anesthesia. In this situation, skin testing and

RAST may be necessary to ascertain which medications may be safely administered.[1]

Because of the unpredictability of the reactions, quick recognition with aggressive proper management can prevent a disastrous outcome.

References

1. Levy J.H.: Allergy and Anesthesia. 1990 Annual Refresher Course Lectures. No. 142, San Francisco.
2. Levy M., Lipshitz M., Eliakin M.: Hospital admissions due to adverse drug reactions. *Am J Med Sci* 1979, 277:49–56.
3. Beaven M.A.: Anaphylactoid reactions to anesthetic drugs. *Anesthesiology* 1981, 55:3–5.
4. Jich H.: Adverse drug reactions: The magnitude of the problem. *J Clin Immunol* 1984, 74:555–60.
5. Levy J.H.: Allergic reactions during anesthesia. *J Clin Anesth* 1988, 1: 39–46.
6. Van Arsdel P.P.: Diagnosing drug allergy. *JAMA* 1982, 247:2576–81.
7. Fox G.S., Wilkinson R.D., Rabow F.I.: Thiopental anaphylaxis: A case and a method for diagnosis. *Anesthesiology* 1971, 35:655–7.
8. Stoelting R.K.: Allergic reactions during anesthesia. *Anesth Analg* 1983, 62:341–56.
9. Watkins J., Clark A., Applegard T.N., et al: Immune mediated reactions to althesin (alphaxalone). *Br J Anaesth* 1976, 48:881–6.
10. Lorenz W., Dornicke A.: Histamine release in clinical conditions. *Mt Sinai J Med (NY)* 1978, 45:357–86.
11. Adourian U.A., Hirshman C.A., Adkinson N.F.: Immunoreactivity of protamine preparations used to reverse heparin anticoagulation. *Anesthesiology* 1990, 73: 328–31.
12. Adourian U., Fuchs E., Adkinson N.F., et al.: Incidence of anti-protamine antibody in vasectomized males. *Anesthesiology* 1990, 73 (3A):1257.
13. Altman L.C.: Basic immune mechanisms in immediate hypersensitivy. *Med Clin North Am* 1981, 65:941–57.
14. Norrow D.H., Luther R.R.: Anaphylaxis: Etiology and guideline for management. *Anesth Analg* 1976, 55:493-99.
15. Levy J.H.: Anaphylactic/anaphylactoid reactions during cardiac surgery. *J Clin Anesth* 1989, 1:426–30.
16. Gerber A.C., Jorg W., Abinden J., Seger R.A., Dangel P.H.: Severe intraoperative anaphylaxis to surgical gloves: Latex allergy, an unfamiliar condition. *Anesthesiology* 1989, 71:800–2.
17. Brande B.M., Gold M., Swartz J.S. et al.: Intraoperative anaphylaxis to latex: an identifiable population at risk. *Anesthesiology* 1990, 73 (3A):1106.
18. Ravidran R.S., Klemm J.E.: Anaphylaxis to succinylcholine in a patient allergic to penicillin. *Anesth Analg* 1980, 59:944–5.

19. Cohen S., Liu K.H., Marx G.F.: Upper airway edema—an anaphylactoid reaction to succinylcholine. *Anesthesiology* 1982, 56:467–8.
20. Levy J.H., Zaidan J.R., Farat B.: Prospective evaluation of risk of protamine reactions in patients with NPH insulin-dependent diabetes. *Anesth Analg* 1986, 65:739–42.
21. Etter M.S., Helrich M., Mackenzie C.F.: Immunoglobulin E fluctuation in thiopental anaphylaxis. *Anesthesiology* 1980, 52:181–3.
22. Fisher M.M.: Intradermal testing after anaphylactoid reaction to anaesthetic drugs: Practical aspects of performance and interpretation. *Anaesth Intensive Care* 1984, 12:115.
23. Nyhan D., Weiss M., Hirshmann C.A.: Immunologic aspects of anesthetic agents. *Anesth Rev* 1989, 16(Suppl):10–18.
24. Barash P.G., Cullen B.F., Stoelting R.K.: *Clinical Anesthesia*. Philadelphia: JB Lippincott, 1989, p 1386.
25. Ocelli G., Amedeo J., Raucoules M., et al.: Predictive value of skin tests in the choice of anesthetic drugs after a first anaphylactic reaction. *Anesthesiology* 1990, 73 (3A):1033.

Chapter 16

The Cocaine Abuser

Michael R. Seidel, M.D.

Case History. *A 28-year-old man was admitted to the emergency department with stab wounds in multiple sites, including his arms, hands, left lower part of the chest, and abdomen. There were also stab wounds on his neck through which secretions appeared to be bubbling. He was neurologically intact, with equal but dilated pupils, and 4+ patellar tendon reflexes.*

During the physical examination the patient appeared calm and was even joking with the orderlies. Blood pressure was 280/100mmHg, pulse 110/min, and respiratory rate 25/min. Temperature was 38.2°C. Track marks were visible on both arms, although both arms were considerably bloodstained and dirty.

Attempts to obtain an electrocardiograph reading were hampered by the patient's sudden belligerence and agitation: he simply would not sit still. He vehemently denied having used alcohol or other illegal drugs. Because of the bubbling neck wounds and increasing abdominal girth, emergency abdominal surgery was planned.

Introduction

The coca plant has been cultivated since the time of the Incas, when it was grown for religious ceremonies in the high Andes of South America.[1] After chewing coca leaves, the native Indian workers of the region supposedly could perform great feats of strength and endurance. The plant was brought to Europe in 1749, and in 1786 Lamarck described the shrub under the name *Erythroxylon* coca. Interest in the active ingredient began in 1860, when Albert Niemann of Austria isolated from the coca leaves an alkaloid that he named cocaine.[2]

Reviewed by Dr. Erlina Farcon, Assistant Professor, Department of Anesthesiology, Albert Einstein College of Medicine/Montefiore Medical Center.

By the mid-1880s the substance's physiologic effects, as well as its local anesthetic effects, were well known,[3] and cocaine was hailed as a miracle drug. Although the European production of cocaine faded because it was difficult to obtain fresh leaves, the availability of good-quality leaves in America, coupled with low-cost manufacturing, resulted in an epidemic of cocaine use just before the turn of the century.

It soon became apparent, from the many deaths and adverse psychological effects associated with casual and medical cocaine use, that it was an extremely dangerous drug. In 1906 the Food and Drug laws removed cocaine from the open market.

Cocaine has once again become a major public health threat, with social, economic, and political overtones.[2] Its high abuse potential and few recognized medical uses have led to its classification as a Schedule D drug. The recent surge in abuse began in the 1970s,[3] and the National Institute of Drug Abuse reported a fivefold increase in hospital admissions for cocaine-related problems during that decade.[4]

Lower street price, easier availability, and increased purity have contributed to its use by more than 30 million Americans.[5] Of the 5 million regular users in the United States, the highest prevalence is seen in young white men between the ages of 18 and 25. The impact of this recent surge in abuse is best demonstrated by a study of motor vehicle fatalities in New York City between 1984 and 1987,[5] which found that at least 1 of 4 drivers who were killed had documented use of cocaine within 48 hours of death.

Deaths are known to occur after administration of cocaine by any route. The severity of danger inherent in this drug is not yet fully recognized by the public.

Characterization of the Alkaloid

Cocaine is benzoylmethylecgonine ($C_{17}H_{21}NO_4$), with a molecular weight of 339.81 for the hydrochloride.[6] The amino alcohol base, ecgonine, is chemically related to tropine, the amino alcohol of atropine. The alkaloid is extracted from the leaves of *Erythroxylon coca* and is dissolved in hydrochloric acid to form a salt, achieving an extraction efficiency of between 0.65% and 1.2% by weight. The hydrochloride form exists as water-soluble crystals, granules, or white powder, which decomposes at 195°C.

This relatively pure cocaine is then adulterated ("cut" or "stepped on") with various impurities, including sugars—mannitol, lactulose, glucose, and inositol—or local anesthetics such as lidocaine, benzocaine, procaine, and tetracaine. Other drugs, including caffeine, amphetamines, heroin, phencyclidine, and quinine also are commonly found as impurities. The purity of street samples may vary from 15% to 60% cocaine.

The alkaloid form (known as "freebase," "crack," or "rock" exists as a colorless, odorless, transparent crystal that is soluble in alcohol, acetone, ether, and oils. The molecular weight of free-base is 303.36, and aqueous solutions are alkaline.[6] It is commonly reextracted from the hydrochloride salt with baking soda, then mixed with a solvent such as ether. Contaminants may persist even after such repurification. In this form, the crystals will melt at 98°C and vaporize at higher temperatures. Since it is not destroyed by heating, the free-base alkaloid may be added to cigarettes or marijuana and smoked.

Pharmacokinetics

Virtually every mode of drug administration has been tried with cocaine.[6] The first documented use of cocaine for subcutaneous infiltration to produce local nerve block was in 1880 by Von Anrep.[7] Other routes of administration include intramuscular, intravenous, intrathecal, and gastrointestinal (both intentional and by "body-packing" for smuggling[8]). Topical administration, with absorption via the mucous membranes, gives various rates of onset and durations of action.[9] The water-soluble hydrochloride, as a powder or in solution, can be applied to conjuctival membranes, nasal mucosa (as in snorting), oral mucosa (both sublingual and buccal), rectal mucosa, and vaginal mucosa. Sustained use is invariably associated with mucosal inflammation and sloughing.

Mucous membrane absorption is slow, with delayed onset and sustained duration of action. Snorting can produce euphoria that lasts as long as 1 to $1\frac{1}{2}$ hours. One "line" of street coke weighs between 5 and 10mg. Other routes give variously quicker onsets, with shorter durations. The inhalation route provides a dose equivalent to that of the intravenous route, with comparable onset and duration. A single inhalation, or "hit," of the free alkaloid (base) form delivers about 67mg of free-base, with euphoria lasting about 20 minutes.

However, this route requires deep, forced, prolonged inspiration, with Valsalva maneuver. The redistribution of physiologic forces by this mode of administration can lead to many different complications, including syncope, cerebral hemorrhage, and aortic rupture.

After administration, the drug may persist in the urine of an adult for 24 to 36 hours. Plasma and liver cholinesterases metabolize cocaine, most of the work being done in the liver within the first 2 hours of administration. The principal metabolites are rapidly excreted in the urine. In patients who have decreased plasma cholinesterase activity, such as those with severe liver disease or congenital cholinesterase deficiency, the geriatric population, parturients, and fetuses, there is the risk of sudden death from overdose.

Pharmacodynamics

Cocaine acts primarily as a local anesthetic by blocking conduction in nerve fibers. The rapid increase in sodium ion permeability normally seen during depolarization is prevented by cocaine. In addition, cocaine alters synaptic transmission. It blocks the reuptake of norepinephrine and dopamine by nerve terminals, causing excess catecholamines to accumulate at the postsynaptic receptor sites.

The sympathetic nervous system is stimulated, causing local vasoconstriction, acute rise in arterial pressure, tachycardia, ventricular dysrhythmias, seizures, mydriasis, hyperglycemia, and hyperthermia. The dopaminergic effects cause euphoria. There is also a direct action: vascular smooth muscle contraction, which is independent of the sympathetic nervous system and is not affected by phentolamine or prazosin. This activity is inhibited by diltiazem.

The physiologic derangements caused by cocaine can produce many life-threatening side effects (see Table 1), including cardiac dysrhythmias, acute myocardial infarction, cerebrovascular accidents, seizures, acute pulmonary edema, hyperthermia, gastrointestinal ischemia, spontaneous abortion, abruptio placentae, and acute rhabdomyolysis. The hyperthermia seen with cocaine abuse is attributed to generalized vasoconstriction coupled with increased muscle activity.[10]

TABLE 1. Life Threatening Side Effects Produced by Cocaine

• Cardiac dysrhythmias → arrest	• Acute pulmonary edema
• Acute myocardial infarction	• Hyperthermia
• Cerebrovascular accidents	• Gastrointestinal ischemia
Thrombosis	• Uterine hyperactivity
Hemorrhage	Abortion
Intracerebral	Abruptio placentae
Subarachnoid	Induction of labor
• Seizures	• Acute rhabdomyolysis

Not only is there a loss of heat dissipation, but there also appears to be a change in the thermoregulatory center, permitting acute rises in temperature to persist. This change is possibly due to local vasoconstriction at the thermoregulatory center itself. The rhabdomyolysis associated with free-base cocaine overdose and intravenous abuse has resulted in nonoliguric renal failure, and in one patient the renal failure progressed to hyperkalemia and death.[11,12]

Occasional deaths occur from accidental rupture of swallowed cocaine packages carried by smugglers.[8,13,14] Like oral dosages of cocaine, this can easily induce norepinephrine stimulation of intestinal vasculature, followed by reduced blood flow and bowel ischemia. This diagnosis must be considered in any cocaine abuser complaining of abdominal pain. A marked leukocytosis in such a patient may indicate gangrenous bowel.

In light of the enhancement of catecholamine reuptake, it is important to avoid using sympathomimetic drugs for cocaine abusers. This caveat must also extend to the use of adrenalin-containing solutions for injection prior to incision or for irrigation. Sensitivity to the effects of epinephrine is especially increased after acute administration of cocaine.

Chronic Cocaine Abuse and Associated Medical Problems

Many chronic minor medical problems plague cocaine abusers, and these should serve as clues to the presence of chronic cocaine use. Very often, the abuser complains of fatigue, increased hunger, and mental depression but appears nervous and agitated, exhibiting paranoid thinking. Chronic cocaine users frequently complain of sexual dysfunction. Although low doses of cocaine may delay ejaculation and orgasm, higher doses and chronic use may alter dopamine neurotransmission, resulting in inability to achieve orgasm in both sexes.[15]

Physical exam may reveal nasal septal atrophy, and possibly necrosis and septal perforation.[16] Nasal verrucae are another finding associated with the snorting habit.[17] Generalized noncarious tooth destruction may result from loss of both dentin and enamel as the mineral becomes dissolved by the hydrochloride solution applied to the gingiva. There may be bilateral loss of eyelashes and eyebrows secondary to burning in freebasing.[18]

There is a resting tachycardia, and the patient may show ECG changes consistent with findings of myocardial ischemia. Careful attention to the cardiac physical exam is very important because various heart diseases have been linked to cocaine abuse. Not only is the cocaine abuser at risk for ischemia and dysrhythmias, but myocarditis, cardiomyopathy, and valvular heart disease also have been associated with cocaine abuse.[19]

The chest x-ray may be abnormal with findings of atelectasis or localized parenchymal opacification. The syndrome of fever, transient pulmonary infiltrates, eosinophilia, pruritus, elevated IgE levels, and bronchospasm following inhalation of cocaine has been dubbed "crack lung."[20]

Use of inhaled cocaine in the freebase form results in systemic effect similar to intravenously administered cocaine. Reported pulmonary effects include edema, hemorrhage, barotrauma, hypersensitivity phenomena, and obliterative bronchiolitis.[21]

Neurologic examination may be normal apart from heightened reflexes and a loss of the sense of smell.

Chronic cocaine abuse has proved particularly refractory to treatment. The addiction has a major psychological and a minor physical component. Recently, increased success has been shown for treatment of patients with flupenthixol decanoate.[22]

Acute Emergencies Associated With Cocaine Use

The cardiovascular system, central nervous system, and pulmonary system are most often involved in acute emergencies following cocaine use. Most deaths result from cardiac dysrhythmias, acute myocardial infarction, acute pulmonary edema,[23] and seizures with respiratory complications. In addition, hepatonecrosis,[24] retinal artery occlusion,[25] and various obstetric emergencies are known to occur after cocaine use.

The cardiac effects of cocaine may be divided into dysrhythmic and ischemic consequences. The dysrhythmias result from both the direct effect cocaine has on conducting tissues and the enhancement of catecholamine levels. Hyperpyrexia and seizures also may be associated with the onset of dysrhythmias: sinus tachycardia, ventricular premature contractions, ventricular tachycardia, ventricular fibrillation, and asystole. The rhythm disturbances have been classically managed with propranolol, but recently amitriptyline has been used, and nitrendipine also has been found effective.

It is now well documented that cocaine induces coronary artery vasospasm and reduces coronary artery blood flow. Myocardial ischemic changes appear to be chronic; and during a study of withdrawal from chronic cocaine abuse,[26] 40% of the patients had ST-T wave elevations, which resolved after 6 weeks of withdrawal. Angina pectoris is associated with acute cocaine abuse.[27] Acute myocardial infarctions have been found temporally related to cocaine use.[28–31] Myocardial infarction has been documented after cocaine administration by intranasal,[32] intravenous, and inhalational routes.

The potential hazard of using cocaine in the face of fixed coronary artery disease is obvious when one considers the predictable rise in heart rate, systolic blood pressure, and myocardial oxygen demand created by using cocaine. The importance of cocaine-induced coronary artery vasospasm is not as obvious. In studies of cocaine-induced vasoconstriction,[33,34] the physiologic effects of 2mg/kg of intranasal cocaine were increased heart rate, mean arterial pressure, and myocardial oxygen demand.

Simultaneous cardiac catheterization data showed a mean decrease of 17% for coronary sinus blood flow, an increase of 33% for coronary vascular resistance, and a decrease in left coronary artery diameter of 8%–12%. The increase in resistance was attributed mostly to vasoconstriction within the intramural vasculature. These responses were the same for patients with left coronary artery disease, with and without left coronary artery disease.

The peripheral systemic vasculature is adversely affected by cocaine, as a result of both acute vasoconstriction and the sudden exaggerated rise in systemic pressure. Vasoconstrictive events, with local ischemia and decreased blood flow, are demonstrated by the findings of ischemic stroke pattern, anterior spinal artery syndrome, and gastrointestinal ischemia with gangrenous bowel. The acute onset of hypertension may show up as a hemorrhagic event, such as the acute rupture of the ascending aorta in a previously healthy man who has just smoked free-base cocaine. Other hemorrhagic events include intracranial hemorrhage, hemorrhagic stroke, and rupture of intracranial aneurysms as already noted.[35]

Crack abusers who present with acute chest pain must also have a thorough pulmonary exam.[36] Acute pneumothorax, hemopneumothorax, acute pulmonary edema, spontaneous pneumomediastinum, and pneumopericardium must be ruled out after examination of a chest x-ray. Arterial blood gas analysis also may prove helpful for the diagnosis of pulmonary complications.

The central nervous system complications of cocaine abuse range from serious to life-threatening. A strong temporal relationship between the use of alkaloidal cocaine and cerebrovascular events has been documented.[37] Cerebral infarction, subarachnoid hemorrhage, intraparenchymal and intraventricular bleeds have all been described, usually associated with headache and developing within one hour of drug ingestion.

Subarachnoid hemorrhage may follow intranasal cocaine use within minutes.[38] The sudden surge in blood pressure has been implicated in acute intracranial hemorrhage associated with the presence of aneurysms and arteriovenous malformations. Cocaine also may be involved in perinatal cerebral infarctions.

Seizures can be induced by a single dose of cocaine. Cocaine lowers the seizure threshold, or seizures may even be secondary to a myocardial event.[39] In 13 of 36 deaths attributed to cocaine abuse, seizures occurred before death. A factor contributing to the development of seizures is the combination of hyperpyrexia and neurotransmitter intensification.[40] Anticonvulsant medications are not useful in preventing seizures in the cocaine abuser, although intravenous diazepam may help manage acute seizures. Unexplained seizures that occur after age 12 should alert the physician and prompt the investigation of urine or blood for evidence of cocaine.[41]

An ischemic stroke pattern has been associated with crack cocaine abuse.[42] Although the etiology is uncertain, it may be related to disordered neurogenic control of cerebral circulation, coupled with the acute onset of hypertension.[43] Other neurologic diseases associated with cocaine abuse include anterior spinal artery syndrome, lateral medullary syndrome,[44] middle cerebral and vertebrobasilar artery transient ischemic attacks, and partial motor seizures. Like other types of intravenous drug abusers, cocaine abusers have been observed with fungal cerebritis (with no other predisposition to acquire this infection). Cerebral vasculitis has been associated with cocaine abuse.[45]

There has been a recent fad among full-term parturients of inhaling "crack" cocaine in the erroneous belief that it will make labor and delivery quicker and easier. However, abruption of the placenta has been shown to occur immediately after intravenous use and within hours of intranasal use. There is always possible compromise to the unborn baby because increased norepinephrine levels may lead to uterine contraction, uterine artery constriction, and placental vasoconstriction, all resulting in decreased uteroplacental blood flow.

Chronic use of crack during pregnancy is associated with poor birth outcome.[46] A significant number of patients have delivered at or earlier than 37 weeks, and crack-exposed infants are more likely to show intrauterine growth retardation. Newborn infants exposed to crack are at risk for higher rates of congenital malformations, perinatal mortality, and neurobehavioral impairments. Birth outcomes for infants of crack abusers were worse than for those infants exposed to other forms of cocaine.[47]

Children may be passively exposed to the smoke of free-base cocaine used by their adult caretakers.[48] They may present to the physician with drowsiness, unsteady gait, or seizures with no other known cause. The incidence of sudden infant death syndrome (SIDS) in infants exposed to cocaine prenatally was 15%, compared to 4% in preterm infants exposed to opiates. These infants show a higher incidence of cardiorespiratory pattern abnormalities when compared to infants exposed to methadone or to those who had no prenatal drug exposure.[49]

Anesthetic Considerations

As noted, catecholamine containing solutions should be avoided. Normo- or hypertension should not be considered indicative of adequate volume status. Although more reliable information may be obtained by placement of a pulmonary artery flow directed catheter, time may not be available to accomplish this maneuver. Radial artery cannulation and the placement of several large peripheral venous access routes are indicated. The mind altering effects of the drug may block pain sensation. Clinical assessment is extremely important. If blood loss appears to be extensive, rapid exploration is indicated. Adequate quantities of blood should be available. Sudden hypotension requires very rapid fluid replacement. Administration of sympathomimetics may prove counter-productive and resuscitation efforts also may not be successful when the chronic abuser's catecholamine stores are fully depleted.[50] Phenylephrine infusion is preferable to treat cardiac decomposition because of its postsynaptic alpha-receptor stimulant effects and its lack of action on cardiac beta receptors resulting in decreased heart rate and increased stroke output.

Conclusion

In light of the ongoing cocaine epidemic, and considering the newer, more potent forms of self-administration of cocaine, such as smoking free base, it is important for the anesthesiologist to assess the level of both chronic and acute cocaine abuse and intoxication. Many patients coming for elective procedures and those in ambulatory care settings are chronic abusers. The clinician must be alert to signs of cocaine use and tailor the choice of anesthetics accordingly to prevent complications. The history taking, if carefully approached, very often yields a positive picture. The patients usually have their own interests in mind and will provide information about their drug habits when they feel secure that the information is confidential.

References

1. Freud S.: Ueber Coca. *Centralbl F d Ges Therap* 1884, 2:289–314.
2. Gay G.R.: Clinical management of acute and chronic cocaine poisoning. *Ann Emerg Med* 1982, 11:562–72.
3. Smart R.G.: "Crack" cocaine use in Canada: A new epidemic? 1988, 127(6):1315–7.
4. Fishburn P.M.: National Survey on Drug Abuse: Main Findings: 1979. Rockville, Md: National Institute of Drug Abuse, 1980 (DHHS Publication No. ADM 80–976).
5. Marzuk P.M., Tardiff K., Leon A.C., Stajic M., Morgan E.B., Mann J.J.: Prevalence of recent cocaine use among motor vehicle fatalities in New York City. *JAMA* 1990, 263:250–6.
6. Cregler L., Mark H.: Medical complications of cocaine abuse. *N Engl J Med* 1986, 315:1495–9.

7. Gilman A.G., Goodman L.S., Gilman A. (eds): *The Pharmocological Basis of Therapeutics*, 6th ed. New York: Macmillan, 1980.
8. Caruana D.S., Weinbach B., Goerg D., Gardner L.B.: Cocaine-packet ingestion: Diagnosis, management, and natural history. *Ann Intern Med* 1984,100:73–4.
9. AHFS Drug Information 1989. Bethesda, Md: American Society of Hospital Pharmacists, 1989, pp 1508–9.
10. Roberts J.R., Quattrocchi E., Howland M.A.: Severe hyperthermia secondary to intravenous drug abuse. *Am J Emerg Med* 1984, 2:373.
11. Jandreski M.A., Bermes E.W., Leischner R., Kahn S.E.: Rhabdomyolysis in a case of free-base cocaine ("crack") overdose. *CLCHA Clinical Chemistry* 1989, 35:1547–9.
12. Parks J.M., Reed G., Knoche J.P.: Cocaine-associated rhabdomyolysis. *Am J Med Sci* 1989, 297:334–6.
13. McCarron M.M., Wood J.D.: The cocaine "body packer" syndrome: Diagnosis and treatment. *JAMA* 1983, 250:1417–20.
14. Sinner W.N.: The gastrointestinal tract as a vehicle for drug smuggling. *Gastrointest Radiol* 1981, 6:319–23.
15. Stoelting R.K., Dierdorf S.E., McCammon R.L. (eds): *Anesthesia and Co-Existing Disease*, 2nd ed. Edinburgh: Churchill Livingstone, 1988.
16. Smith D.E., Wesson D.R., Apter-Marsh M.: Cocaine-and-alcohol induced sexual dysfunction in patients with addictive diseases. *J Psychoactive Drugs* 1984, 16:359–61.
17. Schwartz R.H., Estroff T., Fairbanks D.N., Hoffman N.G.: Nasal symptoms associated with cocaine abuse during adolescence. *Arch Otolaryngol Head Neck Surg* 1989, 115:63–4.
18. Schuster D.S.: Snorters' warts [Letter]. *Arch Dermatol* 1987,123:571.
19. Tames S.M., Goldenring J.M.: Madarosis from cocaine use. *N Engl J Med* 1986, 314:1324.
20. Karch S.B., Billingham M.E.: The pathology and etiology of cocaine-induced heart disease. *Arch Pathol Lab Med* 1988, 112(3):225–30.
21. Kissner D.G., Lawrence W.D., Selis J.E., Flint A.: Crack lung: Pulmonary disease caused by cocaine abuse. *Am Rev Respir Dis* 1987, 136:1250–2.
22. Gawin F.H., Allen D., Humblestone B.: Outpatient treatment of "crack" cocaine smoking with flupenthixol decanoate: A preliminary report. *Arch Gen Psychiatry* 1989, 46:322–5.
23. Allred R.J., Ewer S.: Fatal pulmonary edema following intravenous "free-base" cocaine use. *Ann Emerg Med* 1981, 10:441–2.
24. Perino L.E., Warren G.H., Levine J.S.: Cocaine-induced hepatotoxicity in humans. *Gastroenterology* 1987, 93:176–80.
25. Devenyi P., Schneiderman J.F., Devenyi R.G., Lawby L.: Cocaine-induced central retinal artery occlusion. *Can Med Assoc J* 1988, 138:129–30.
26. Nadamanee K., Gorelick D.A., Josephson M.A. et al: Myocardial ischemia during cocaine withdrawal. *Ann Intern Med* 1989, 111:876–80.

27. Pasternack P.F., Colvin S.B., Baumann F.G.: Cocaine-induced angina pectoris and acute myocardial infarction in patients younger than 40 years. *Am J Cardiol* 1985, 55:847.
28. Coleman D.L., Ross T.F., Naughton J.L.: Myocardial ischemia and infarction related to recreational cocaine use. *West J Med* 1982, 136:444–6.
29. Wang T., Hadidi F., Triani F., Bargout M.: Myocardial infarction associated with the use of cocaine. *Am J Med Sci* 1988, 295:569–71.
30. Kossowsky W.A., Lyon A.E., Chou S.Y.: Acute non-Q wave cocaine-related myocardial infarction. *Chest* 1989, 96:617–21.
31. Smith H.W., Liberman H.A., Brody S.L., Donohue B.C., Morris D.C.: Acute myocardial infarction temporally related to cocaine use. *Ann Intern Med* 1987, 107:13–8.
32. Rod J.L., Zucker R.P.: Acute myocardial infarction shortly after cocaine inhalation. *Am J Cardiol* 1987, 59:161.
33. Lange R.A., et al: Cocaine-induced coronary-artery vasoconstriction. *N Engl J Med* 1989, 321:1557–62.
34. Schachne J.S., Roberts B.H., Thompson P.D.: Coronary-artery spasm and myocardial infarction associated with cocaine use. *N Engl J Med* 1984, 310:1665–6.
35. Henderson C.E., Torbey M.: Rupture of intracranial aneurysm associated with cocaine use during pregnancy. *Am J Perinatol* 1988, 5:142–3.
36. Eurman D.W., Potash H.I., Eyler W.R., Paganussi P.J., Beute G.H.: Chest pain and dyspnea related to "crack" cocaine smoking: Value of chest radiography. *RADLA Radiology* 1989, 172:459–62.
37. Mody C.K., Miller B.L., McIntyre H.B., Cobb S.K., Goldberg M.A.: Neurologic complications of cocaine abuse. *Neurology* 1988, 38:1189–93.
38. Nolte K.B., Gelman B.B.: Intracerebral hemorrhage associated with cocaine abuse. *Arch Pathol Lab Med* 1989,113:812–13.
39. Alldredge B.K., Lowenstein D.H., Simon R.P.: Seizures associated with recreational drug abuse. *Neurology* 1989, 39:1037–9.
40. Merriam A.E., Medalia A., Levine B.: Partial complex status epilepticus associated with cocaine abuse. *Biol Psychiatry* 1988, 23:515–8.
41. Schwartz R.H.: Seizures associated with smoking "crack": A survey of adolescent "crack" smokers. *West J Med* 1989, 150:213.
42. Levine S.R., Washington J.M., Jefferson M.E., et al: "Crack" cocaine-associated stroke. *Neurology* 1987, 37:1849–53.
43. Golbe L.I., Merkin M.D.: Cerebral infarction in a user of free-base cocaine ("crack"). *Neurology* 1986, 36:1602–4.
44. Rowley H.A., Lowenstein D.H., Rowbotham M.C., Simon R.P.: Thalamomesencephalic strokes after cocaine abuse. *Neurology* 1989, 39:996–7.
45. Kaye B.R., Fainstat M.: Cerebral vasculitis associated with cocaine abuse. *JAMA* 1987, 258:2104–6.
46. Cherukuri R., Minkoff H., Feldman J., Parekh A., Glass L.: A cohort study of alkaloidal cocaine "crack" in pregnancy. *Obstet Gynecol* 1988, 72(2): 147–51.

47. Kaye K., Elkind L., Goldberg D., Tytun A.: Birth outcomes for infants of drug abusing mothers. *NY State J Med* 1989, 89:256–61.
48. Bateman D.A., Heagarty M.C.: Passive freebase cocaine ("crack") inhalation by infants and toddlers. *Am J Dis Child* 1989, 143:25–27.
49. Chasnoff I.J., Hunt C.E., Kletter D.: Prenatal cocaine exposure is associated with respiratory pattern abnormalities. *Am J Dis Child* 1989,143:583–7.
50. Fleming J.A., Byck R., Barash P.G.: Pharmacology and therapeutic applications of cocaine. *Anesthesiolog* 1990, 73(3):518–31.

CHAPTER 17

The Patient With Adrenal Disease

Nirmala Balan, M.D.

Case History. *A 26-year-old man was admitted to the hospital with acute abdominal pain. Past history revealed that he had suffered from pulmonary tuberculosis at age 20. He had undergone a course of treatment at that time, but it was aborted after 4 weeks when he moved to another state. Since then, he had received no medical care because he felt reasonably well, despite some easy fatigability that had necessitated his changing from a construction job to an office position. He noted that he was generally thinner than he had been as a teenager. He had also noted development of some pigmented areas on his body.*

Examination showed acute right-lower-quadrant tenderness; temperature, 100°F, white cell count, 13,000/mm^3. *Blood pressure was 80/50mmHg; pulse rate, 120/min. SMA 6 revealed* K^+ *5.6mEq/L;* Na^+, *125mEq/L; glucose, 65. A diagnosis of acute appendicitis was made. The possibility of adrenal abnormality was also considered.*

Physiology

The adrenal cortex secretes three groups of hormones, the glucocorticoids, mineralocorticoids, and androgens. Cholesterol, which is derived from acetate, is the building block for the steroid nucleus necessary for the synthesis of these three hormones. It is converted to Δ^5-pregnenolone. This compound is changed either to progesterone or to 17-hydroxypregnenolone. Progesterone can be converted to aldosterone, the principal mineralocorticoid, only in the zona glomerulosa of the adrenal cortex. In the zona fasciculata, progesterone is converted into 11-deoxycortisol and finally to cortisol, the principal glucocorticoid. Androgens are produced in only trace amounts in the zona reticularis of the adrenal cortex[1] (see Figure 1).

Reviewed by Dr. Irene Osborn, Assistant Professor, Department of Anesthesiology, Montefiore Medical Center.

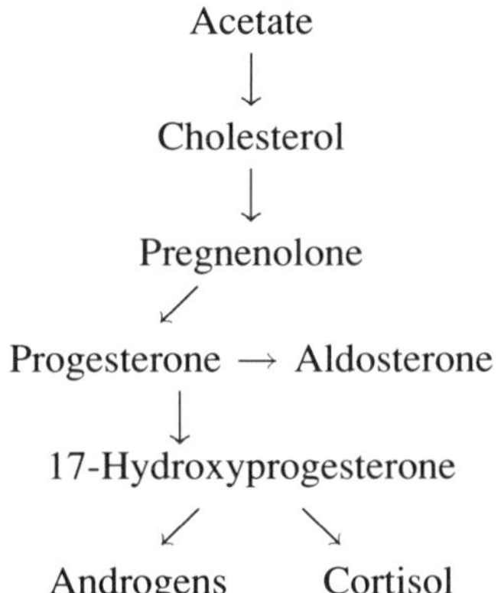

FIGURE 1. Schematic diagram of the synthesis of the primary hormones (androgens, cortisol, and aldosterone) secreted by the adrenal cortex.

Glucocorticoids

Cortisol is the principal endogenous glucocorticoid. Its synthesis and release is regulated by adrenocorticotropic hormone (ACTH), which is produced by the anterior pituitary. ACTH, in turn, is controlled by the corticotropin-releasing factor (CRF) present in the hypothalamus. Cortisol, through a negative feedback mechanism, inhibits both ACTH and CRF. In addition, hypoglycemia and stress can stimulate the release of ACTH and thereby increase plasma concentrations of cortisol. Most of the cortisol is inactivated in the liver and excreted as 17-hydroxycorticosteroids. Approximately 15–20mg of cortisol is secreted each day by the adrenal cortex.

The function of cortisol is complex but in general involves anti-inflammatory and immune responses and reactions to stress.[1] It is an important regulator of carbohydrate, protein, and lipid metabolism. Blood glucose levels are maintained by stimulation of hepatic gluconeogenesis and by inhibition of peripheral utilization of glucose.

Maintenance of blood pressure by cortisol reflects its importance in facilitating the conversion of norepinephrine to epinephrine in the adrenal medulla. In addition, it has some mineralocorticoid properties, thereby promoting retention of sodium and excretion of potassium.[2]

Endogenous and synthetic corticosteroids are compared with cortisol with respect to their glucocorticoid anti-inflammatory effect and their mineralocorticoid (salt-retaining) potency (see Table 1).

Five milligrams of prednisone will produce a glucocorticoid effect equivalent to 20mg of hydrocortisone. Methylation of prednisolone to

TABLE 1. Relative Potency and Equivalent Doses for Commonly Used Glucocorticoids

Steroids	**Relative glucocorticoid potency***	**Equivalent glucocorticoid dose(mg)**
Short-acting		
Cortisol (hydrocortisone)	1	20
Cortisone	0.8	25
Prednisone	4	5
Prednisolone	4	5
Methylprednisolone	5	4
Intermediate-acting		
Triamcinolone	5	4
Long-acting		
Betamethasone	25	0.60
Dexamethasone	30	0.75

* Based on data in Axelrod L.: Glucocorticoid therapy. Medicine (Baltimore) 1976; 55:39.

produce methylprednisolone or dexamethasone results in increased anti-inflammatory effects with reduced salt-retaining effects.

Mineralocorticoids

The principal mineralocorticoid is aldosterone, the secretion of which is regulated by the renin-angiotensin system rather than by ACTH. Aldosterone regulates the extracellular fluid volume by promoting the reabsorption of sodium from the distal renal tubule; in addition, it promotes the renal tubular excretion of potassium and hydrogen ions. A physiologic function of aldosterone is maintenance of blood pressure in the upright position.

Androgens

Production of androgens by the adrenal system increases markedly at puberty and declines with advancing age. Adrenal androgens are converted in the liver to more potent androgens such as testosterone. As with cortisol, androgen production is controlled by ACTH.

An important aspect of ACTH function is diurnal variation. Levels are higher in the morning. Thus, hormones dependent on this stimulant may be at levels earlier in the day that are twice as high as in the evening.

Adrenocortical Hormone Deficiency

Hypofunction of the adrenal gland (Addison's disease) may result from acute or chronic disease. Some of the causes are destruction of the adrenal cortex by tuberculosis, cancer, histoplasmosis, hemorrhage or an autoimmune mechanism; deficiency of ACTH; and prolonged administration of exogenous corticosteroids (see Table 2).

TABLE 2. Abnormal Adrenal Function

Adrenocortical hormone excess:
Glucocorticoid excess:
Cushing's syndrome
Mineralocorticoid excess:
Primary aldosteronism
Adrenocortical hormone deficiency:
Glucocorticoid deficiency:
Addison's disease
Mineralocorticoid deficiency:
Hypoaldosteronism

Perhaps the most common cause of Addison's disease is an autoimmune disorder that produces adrenal antibodies. Simultaneously, antibodies to thyroid tissue may develop, causing thyroiditis.[3] This combination of primary adrenalitis and thyroiditis is known as Schmidt's syndrome. Diabetes mellitus may be a complicating disorder.[4]

Adrenocortical insufficiency may also follow abrupt withdrawal of steroids, which suppresses the hypothalamic-pituitary-adrenal axis. Usually, higher doses of steroids used for longer intervals results in a longer period of suppression of the entire pituitary-adrenal axis following cessation of therapy. However, cases have occurred in which patients were treated for only 2 weeks and yet demonstrated significant hypothalamic-pituitary-adrenal axis suppression for up to 8 to 12 months following cessation of therapy. Alternate-day pharmacologic steroid therapy results in less suppression of the hypothalamic-pituitary-adrenal axis.

Primary adrenal axis deficiency usually develops slowly, producing hyperpigmentation, weight loss, hypotension, skeletal muscle weakness, hyponatremia, hyperkalemia, hemoconcentration secondary to hypovolemia, and, occasionally, hypoglycemia. (Patients with adrenal hypofunction secondary to pituitary hypofunction are not pigmented.) Any stress, such as surgical trauma, can result in an Addisonian crisis with circulatory collapse.[5]

Occasionally, acute adrenal insufficiency may be caused by sepsis and is associated with high fever. Anticoagulants such as heparin and coumadin cause acute bleeding into both adrenal glands. A critical-care problem results, characterized by severe hypotension.

Another syndrome involves the failure of the juxtaglomerular apparatus to secrete renin. The renin-angiotensin-aldosterone axis is interrupted, and a relative or secondary lack of aldosterone results. Patients present with hyperkalemia and postural hypotension. This syndrome is particularly common in patients with diabetes mellitus. Patients with aldosterone deficiency alone, who have adequate cortisol reserve, occasionally present only with hyperkalemia and a normal extracellular volume. Cortisol and cortisol precursors apparently compensate because of weak mineralocorticoid effects.

Diagnosis

Low plasma or urinary concentrations of corticosteroids with an elevated plasma concentration of ACTH suggest Addison's disease. However, a single plasma cortisol determination is not diagnostic, as both ACTH and cortisol are secreted in a pulselike fashion, and the variation in plasma cortisol can be 20μ/ml in the early morning and only 5μ/ml in the late evening. The most specific test is finding of low plasma cortisol or 24-hour urine 17-hydroxycorticoid excretion after parenteral administration of ACTH.

Another test involves intravenous administration of a synthetic derivative of ACTH (Cortrosyn®) 25U.[6] Plasma cortisol is measured 30 to 60 minutes later. Normally, there is an elevation of at least 18μg/100ml from baseline. In patients with prolonged suppression of the hypothalamic-pituitary-adrenal axis, sustained stimulation with ACTH gel (40U IM twice daily for 5 days) may be required before appropriate urine or blood samples can be obtained.

In case of severe electrolyte imbalance and volume depletion, patients should be given steroids such as dexamethasone (0.75mg orally daily). At this small dose, dexamethasone does not significantly contribute to 17-hydroxycorticoid urinary excretion. Therefore, urinary 17-hydroxycorticoid excretion in response to ACTH is an indicator of endogenous cortisol production. Urinary 17-hydroxycorticoid excretion should be 2 to 3 times higher than baseline as a normal response.

Treatment

In a patient who is clearly very ill, immediate testing is not necessary. Routine prompt therapy with volume replacement, intravenous glucose, and large doses of steroids is indicated. In case of circulatory collapse, 100mg of hydrocortisone is given intravenously, followed by 50mg every 4 to 6 hours during the first 48 hours after the crisis. Intravascular volume should be restored with the infusion of glucose in saline, colloid solutions, or whole blood, if necessary.[7] Patients tend to develop hyponatremia easily because of the inability of the kidneys to excrete free water. Therefore, it is prudent to avoid hypotonic fluids; give normal saline. Because of a propensity to develop hypoglycemia, glucose should be included in the intravenous fluid regimen.

Perioperative Corticosteroid Supplementation

Corticosteroid supplementation should be increased whenever patients with Addison's disease or panhypopituitarism undergo surgery because of the inability to release additional endogenous cortisol in response to the stress of surgery.[8] This also applies to patients who may manifest suppression of adrenal function resulting from current or previous administration of corticosteroids for the treatment of diseases such as asthma or rheumatoid arthritis.

Topical, aerosol, oral, or parenteral administration of steroids may also cause adrenocortical insufficiency. However, neither the dose nor the duration of therapy with a corticosteroid that will produce suppression of the pituitary-adrenal axis is known.

In addition, it is not possible to identify those patients requiring steroid supplementation without biochemical testing of adrenal function. The clinical approach with patients who are being treated with steroids or who have been treated for more than 1 month in the past 6 to 12 months is to administer supplemental corticosteroids empirically in the perioperative period.

Many studies[9–11] indicate the following regarding the adrenal response to the perioperative period in normal patients and in those taking steroids for other diseases: perioperative stress relates to the degree of trauma and the depth of general anesthesia. Deep general anesthesia postpones the usual intraoperative glucocorticoid surge to the postoperative period. Few patients who have suppressed adrenal function have perioperative cardiovascular problems if they do not receive supplemental steroids perioperatively. Although an occasional patient who takes steroids chronically becomes hypotensive perioperatively, only rarely can this event be attributed to glucocorticoid deficiency.

The precise amount of glucocorticoid required in the perioperative period has not been established. Many authors recommend giving the maximum amount of glucocorticoids that the body manufactures in response to a maximal stress (ie, 300mg/day of hydrocortisone per 70kg of body weight).[12] The possible adverse effects of perioperative steroid supplementation include delayed wound healing, increased susceptibility to infections, glucose intolerance, and gastrointestinal hemorrhage secondary to stress ulcers. In addition, the endogenous cortisol production during stress produced by major surgery or by extensive burns has been reported to range from only 75 to 150mg/day.[13] During the subsequent 2 to 3 days cortisol production declines, returning to normal in uneventful cases.

In view of these two facts, Symreng and colleagues have attempted to define an appropriate but minimal dosage schedule for corticosteroid in the perioperative period.[10] For major surgery, such as colectomy, they advocate the administration of 25mg of intravenous hydrocortisone during the induction of anesthesia, followed by a continuous infusion of 100mg/24h until resumption of gastrointestinal function, and only 25mg of intravenous hydrocortisone during the induction of anesthesia for minor surgery, such as herniorrhaphy. This low-dose cortisol substitution was found to maintain plasma cortisol concentrations closely paralleling the natural cortisol response to major surgery in patients receiving chronic treatment with steroids and manifesting subnormal responses to preoperative infusions of ACTH.

In addition, patients receiving daily maintenance doses of corticosteroids should also receive this dose with the preoperative medication on the day of surgery. In these instances when postoperative events could exaggerate the need for exogenous corticosteroid supplementation, the continuous infusion of hydrocortisone, 100mg every 12 to 24 hours, should be sufficient.[14,15] Some physicians have advocated the administration of a longacting steroid preparation such as methylprednisolone intramuscularly the night before surgery.

Other Adrenal Gland Abnormalities

Hypoaldosteronism

Hypoaldosteronism is a less common condition. Patients present with severe hyperkalemia, hyponatremia, and cardiac conduction defects; some patients are hypertensive. These symptoms can be treated successfully by the administration of mineralocorticoids (0.05–0.1mg of 9-fluorocortisol) preoperatively.

Cushing's Syndrome

Cushing's syndrome, due to excess of glucocorticoid hormones, can result from excess production of ACTH, excess production of cortisol, and exogenous administration of corticosteroid.[15] The most common cause is iatrogenic: administration of glucocorticoids for other conditions, such as asthma, arthritis, and allergies. Forty percent of endogenous cases of Cushing's syndrome are caused by bilateral adrenal hyperplasia resulting from an ACTH-secreting tumor of the anterior pituitary (Cushing's disease).[16] Another 15% stem from non-endocrine ectopic ACTH production, principally from tumors of the lung or the pancreas. Primary adrenal tumors (adenomas and carcinomas), with excess cortisol production and suppressed ACTH, account for only 10% to 20% of the cases.

Patients with Cushing's syndrome present with hypertension, which is caused by the mineralocorticoid effects of excess glucocorticoids, the increase in renin substrate, and the increased vascular reactivity induced by the glucocorticoids. Hypertension is usually benign, but in untreated patients it may be severe and be associated with congestive heart failure. Glucose intolerance is common, and some 20% to 30% of patients exhibit overt diabetes mellitus. This results from inhibition of peripheral glucose utilization, as well as an anti-insulin action and the concomitant stimulation of gluconeogenesis.

Patients often have osteoporosis as a result of decreased formation of bone matrix and impaired calcium absorption. Hypokalemia, hypernatremia, increased intravascular fluid volume, myopathy, and peptic ulcers also may be present in patients with Cushing's syndrome, as well as plethoric and moon facies, centripetal obesity, thin skin, abdominal striae, hirsutism, easy bruisability, poor wound healing, and psychiatric disturbances. Thromboembolism from hypercoagulability is a hazard. Incidence of bacterial and fungal infections is also increased.

The diagnosis is based on the demonstration of increased plasma and urinary concentrations of cortisol. However, plasma cortisol levels may also be elevated because of estrogen ingestion during acute stress and in alcoholism. Absence of diurnal variation indicates autonomous cortisol production. Elevated steroid levels are also found in obesity.

The single best screening test is the short dexamethasone suppression study that distinguishes obese patients from those with Cushing's syndrome in over 95% of cases. One milligram of dexamethasone is given by mouth at midnight. A normal response is suppression of the baseline plasma cortisol by at least 50%.

The treatment is surgical: adrenalectomy for tumors of the adrenal cortex and transphenoidal hypophysectomy for adenomas of the pituitary gland.[17] Medical therapy of small pituitary adenomas also has been successful. Bromocriptine mesylate is an ergot derivative with potent dopamine receptor agonist activity. The drug activates postsynaptic dopamine receptors. The dopaminergic neurons in the tubero-infundibular process modulate the secretion of prolactin from the anterior pituitary by secreting a prolactin inhibiting factor. Clinically, bromocriptine significantly reduces plasma levels of prolactin in patients with hyperprolactinemia. Effect is obtained at levels that do not significantly alter secretion of other anterior pituitary hormones.

Preanesthetic Preparation

Local specific considerations are required prior to anesthetizing the patient with Cushing's syndrome. Preoperative management may require treatment of hypertension and congestive heart failure, stabilization of diabetes mellitus with insulin, and reversal of hypokalemia.

The total blood volume and the blood pressure are increased in patients with Cushing's syndrome. These changes often lead to congestive heart failure, which mandates monitoring with a pulmonary artery catheter intra- and postoperatively. Significant elevations of the blood sugar are best controlled with small amounts of insulin given intravenously, with monitoring of the blood glucose concentration preoperatively.[18] Use of cimetidine and antacids is indicated for gastric hyperacidity.

The choice of preoperative sedation and the anesthetic agent is not influenced by the presence of Cushing's syndrome. However, the dose of muscle relaxants can be reduced, in view of the skeletal muscle weakness that occurs. Furthermore, the presence of hypokalemia may potentiate the response to muscle relaxants. Mechanical ventilation of the lungs during surgery is recommended because of the patient's inherent muscle weakness. Tracheal intubation may be difficult due to obesity.

Patients should be moved with care because they are prone to pathologic fractures secondary to osteoporosis. The atrophic skin is easily bruised, and hematomas develop frequently at the sites of venipuncture and after minimal trauma. Pneumothorax is always a possible complication during adrenal surgery, and the incidence may be as high as 20%.

Although adrenal tumors are almost always unilateral, supplemental steroids are required postoperatively as the contralateral gland is usually suppressed. Glucocorticoids should be given at the start of the tumor resection (100mg of hydrocortisone every 8 hours intravenously). This amount is reduced over 3 to 6 days until a maintenance dose is reached.

Chronic maintenance therapy with corticosteroids will be necessary in patients who have undergone a hypophysectomy or bilateral adrenalectomy. Postoperative pancreatitis from surgical retraction during a left adrenalectomy is a serious complication. The incidence of postoperative wound infections and delayed healing is also high.

Primary Aldosteronism

Hyperaldosteronism results from excess secretion of aldosterone from the adrenal cortex.[19] Primary aldosteronism or Conn's syndrome, is caused by a unilateral, benign adenoma of the adrenal gland in approximately 75% of cases.[20] The remaining 25% result from bilateral adrenocortical hyperplasia; rarely, an adrenocortical carcinoma is the cause. In secondary aldosteronism the increased renin secretion is responsible for the excess secretion of aldosterone, as in patients with renovascular hypertension.

Patients with Conn's syndrome present with hypertension due to sodium retention. Arterial hypertension is usually mild; it has been estimated that the frequency of this syndrome in the hypertensive population approximates 1%.[15] The renal excretion of potassium results in hypokalemic metabolic alkalosis. The hypokalemia also causes muscle weakness and a hypokalemic nephropathy, with polyuria and an inability to concentrate urine. An impairment of glucose tolerance occurs in 50% of patients.

The diagnosis is confirmed by the demonstration of hypokalemia, an increased plasma concentration of aldosterone, and elevated urinary potassium excretion. Initial treatment is with potassium supplementation and the administration of spironolactone, which is a competitive aldosterone antagonist. The hypertension may require treatment with antihypertensive drugs. Definitive treatment for aldosterone-secreting tumors is surgical excision.

Preanesthetic Preparation

Preoperative preparation includes correction of the hypokalemia and treatment of the hypertension. Serum electrolytes and the blood glucose concentration should be measured frequently in the perioperative period.

Inhaled or intravenous drugs are acceptable for the maintenance of anesthesia. The use of enflurane may be questioned if hypokalemic nephropathy and polyuria exist preoperatively. Other recommendations include the cautious use of muscle relaxants in patients with persistent hypokalemia and the avoidance of hyperventilation, which may aggravate metabolic alkalosis and hypokalemia. Despite the use of a wide variety of anesthetic agents, few anesthetic problems have been reported in patients with primary aldosteronism.

Surgical exposure may be difficult, and good muscle relaxation is therefore necessary. This necessitates control of the airway with endotracheal intubation and mechanical ventilation. Pneumothorax is always a possible complication during adrenal surgery.

References

1. Roizen M.F.: Diseases of the endocrine system, in Katz J., Benumof J., Kadis L.B. (eds): *Anesthesia and Uncommon Diseases.* Philadelphia: WB Saunders Co, 1990, pp 245–92.
2. Stoelting R.K., Dierdorf S.F., McCammon R.L.: *Endocrine Disease, Anesthesia and Co-Existing Disease.* New York, Churchill Livingstone, 1988, pp 491–508.
3. Graf G., Rosenbaum S.: Anesthesia and the endocrine system, in Barash P.G., Cullen B.F., Stoelting R.K. (eds): *Clinical Anesthesia.* Philadelphia: JB Lippincott Co, 1989, pp 1185–1210.
4. Stevens A., Roizen M.F.: Patients with diabetes mellitus and disorders of glucose metabolism, in Roizen M.F. (eds): *Anesthesiology Clinics of North America.* Philadelphia: WB Saunders Co, 1987, 5:339–55.
5. Roizen M.F.: Anesthetic implications of concurrent diseases, in Miller R.D. (ed): *Anesthesia.* New York: Churchill Livingstone, 1986, 1:263–71.
6. Spechart P.F., Nicoloff J.T., Bethune J.E.: Screening for adrenocortical insufficiency with cosyntropin. *Arch Intern Med* 1971, 128:761–3.
7. Goldmann D.R.: The surgical patient on steroids, in Goldmann D.R., Brown F.H., Levy W.K. (eds): *Medical Care of the Surgical Patient: A Problem-Oriented Approach to Management.* Philadelphia: JB Lippincott Co, 1982, pp 113–25.
8. Byyny R.L.: Preventing adrenal insufficiency during surgery. *Postgrad Med* 1980, 67:219–25.
9. Knudsen L., Christiansen L.A., Lorentzen J.E.: Hypotension during and after operation in glucocorticoid-treated patients. *Br J Anaesth* 1981, 53:295–301.
10. Symreng T., Karlberg B.E., Kagedal B., et al: Physiological cortisol substitution of long-term steroid-treated patients undergoing major surgery. *Br J Anaesth* 1981, 53:949–54.
11. Chernow B.: Perioperative management of endocrine problems in anesthesia, in Barasch P. (ed): *1988 Annual Refresher Course Lectures.* Philadelphia: JB Lippincott Co, pp 216, 1–4.
12. Roizen M.F.: Endocrine abnormalities and anesthesia, in *1985 Annual Refresher Course Lectures*, pp 253, 1–7.
13. Chernow B., Alexander H.R., Smallridge R.C., et al: Hormonal references to surgical stress. *Arch Intern Med* 1987, 147:1273–8.
14. Finch J.S.: Primary aldosteronism: Review of the anaesthetic experience in sixty patients. *Br J Anaesth* 1969, 41:880.
15. Gangat Y., Turner L., Baer L., et al.: Primary aldosteronism with uncommon complications. *Anesthesiology* 1976, 45:542.
16. Carpenter P.C.: Diagnostic evaluation of Cushing's syndrome, in Young W.F., Klee G.G. (eds): *Endocrinology and Metabolism Clinics of North America.* Philadelphia: WB Saunders Co, 1988, 17:445–65.
17. Tyrrell J.B., Brooks R.M., Fitzgerald P.A., et al: Cushing's disease. Selective transsphenoidal resection of pituitary microadenomas. *N Engl J Med* 1978, 298:753.

18. Chernow B., Cheung A.: Perioperative management of non-diabetic endocrine problems (#141), in *40th Annual Refresher Course Lectures and Clinical Update Program*, Chicago: American Society of Anesthesiologists, 1989.
19. Young W.F., Klee G.G.: Primary aldosteronism, in Young W.F., Klee G.G. (eds): *Endocrinology and Metabolism Clinics of North America.* Philadelphia: WB Saunders Co, 1988, 17:367–88.
20. Malhotre M.: Adrenocortical tumor, in Yao F., Artusio J.F. Jr. (eds): *Anesthesiology: Problem-Oriented Patient Management.* Philadelphia: JB Lippincott Co, 1983, pp 253–9.

CHAPTER 18

The Ex-Premature Infant

Sharda Dave, M.D.

Case History. *A 3-month-old male infant, 48 weeks postconception, was brought to the ambulatory care clinic for evaluation prior to bilateral inguinal hernia repair. The mother reported that she had suffered from toxemia of pregnancy and that the baby had been born at 36 weeks gestation by cesarean section; birth weight, 2000gm. The 1- and 5-min Apgar scores were 7–8. The baby required mask oxygenation and suctioning but intubation was not required. He was observed in the neonatal intensive care unit for 7 days. During this time, bilirubin level was measured at 12–15mg%; blood sugar, 45mg%; calcium, 7.5mg%; $PaCO_2$, 35mmHg. There was no evidence of acidosis or infection, and funduscopic examinations were normal.*

Transferred to the ambulatory care clinic, the baby weighed 4.3kg; hemoglobin was 14gm%; glucose, 50mg%; calcium, 9.5mg%. Examination of heart and lungs revealed no abnormalities. The pediatric surgeon requested an anesthetic consult for the possibility of operating on an ambulatory basis.

Assessment of Prematurity

A diagnosis of prematurity is made if the infant is born before 38 weeks of gestational age or weighs less than 2500gm at birth. In general, the earlier the birth date and lower the weight of the infant, the greater are the risks. Most body organs undergo continuous structural and functional development in the last 3 months of gestation. In the premature infant the ability to maintain normothermia, ventilate, and ingest food is impaired and is inversely related to the postconceptional age.

Preterm infants can be differentiated from term infants by the absence of creases on the posterior two thirds of the soles of the feet.[1] Several degrees of prematurity have been defined (Table 1).

Reviewed by Dr. Tatyana Katsnelson, Assistant Professor, Department of Anesthesiology, Albert Einstein College of Medicine.

TABLE 1. Classification of Prematurity

Degree	Weeks of Gestation	Weight	% of Live Births
Borderline	37–38	<2500gm	16
Moderate	31–36	1500–2500gm	6–7
Extreme	24–30	500–1500gm	1

Group 1: Borderline Prematurity

Infants of 37–38 weeks' gestation and birth weight<2500gm account for 16% of live births.[2] These babies should be carefully observed for 12 hours, because they have difficulty in maintaining their body temperature without external heat and may not have adequate sucking reflex; gavage feeding may be necessary for a few days. Birth weight may be regained only slowly. Eight percent of premature infants delivered by cesarean section have hyaline membrane disease,[3] compared to 1% of those born vaginally. Intercostal retraction, tachypnea, and cyanosis suggest respiratory distress syndrome, pneumonia, or pneumothorax in large preterm infants. Such infants requiring surgery in the first few days of life must be carefully evaluated preoperatively.

Group 2: Moderate Prematurity

Infants of 31–36 weeks' gestation and birth weight of 1500–2500gm account for 6%–7% of live births.[2] The neonatal mortality of those born at 31 weeks of gestation is less than 5%, whereas it is nearly zero at 36 weeks.[1] The major causes of death are intracranial hemorrhage, sepsis, and respiratory distress.

Group 3: Extremely Premature Infant

Infants of 24–30 weeks' gestation and a weight of 500–1500gm make up 1% of live births but account for more than 70% of neonatal mortality,[2] and constitute the major portion of infants with neurologic damage.[4] Causes of death include birth asphyxia, acidosis, respiratory failure, congestive heart failure, patent ductus arteriosus, hyaline membrane disease, infection, necrotizing enterocolitis, respiratory distress, and intracranial hemorrhage.[3]

Problems of Prematurity

Several key issues arise in the premature infant that must be considered and reevaluated during preanesthetic assessment (Table 2).

TABLE 2. Special Problems of Prematurity

Asphyxia
Respiratory manifestations:
- respiratory distress
- bronchopulmonary dysplasia
- apnea

Temperature instability
Patent ductus arteriosus
Infection
Necrotizing enterocolitis
Hematologic manifestations:
- anemia
- left shift of oxygen-dissociation curve

Metabolic disorders:
- metabolic acidosis
- hypoglycemia
- hypocalcemia
- hyperbilirubinemia

Retrolental fibroplasia

Asphyxia

Preterm infants are more prone to asphyxia because their hemoglobin concentration and oxygen-carrying capacity are reduced. Decreased oxygenation leads to anaerobic metabolism, causing metabolic and respiratory acidosis, which in turn reduces cardiac output and increases cerebral blood flow. Neonatal asphyxia occurs in 1 of 200 term infants, 1 of 20 moderately preterm infants and 1 of 2 extremely premature infants.[5] Causes of asphyxia include antepartum hemorrhage, intrauterine infection, and breech delivery.

The treatment of asphyxia in preterm infants requires airway establishment and maintenance, oxygenation, tracheal intubation, and assisted hyperventilation to correct respiratory acidosis. To correct metabolic acidosis, blood volume expansion is necessary. Slow, cautious infusion of sodium bicarbonate should be given to correct the pH to 7.30. Sodium bicarbonate should never be infused more rapidly than 1mEq/kg/min because it may cause intracranial hemorrhage by rapidly expanding the intraventricular volume and increasing $PaCO_2$. Fifty milliliters of $NaHCO_3$ produces 1250ml CO_2 when fully reacted with hydrogen ions. Supported ventilation is needed until infants can sustain their own ventilation and maintain a normal $PaCO_2$.

Respiratory Manifestations

Respiratory distress is common in preterm infants and three times more common in those born by cesarean section than in those born vaginally.[6]

Moderately premature babies with respiratory distress require more ventilatory support and have a lower survival rate than do larger preterm infants.

Bronchopulmonary dysplasia is a result of mechanical ventilation, oxygen toxicity, infection, or a combination of all three. It progresses through four stages, as shown in Table 3.

TABLE 3. Stages of Bronchopulmonary Dysplasia

Stage	Onset (days of age)	X-ray	Pathology
I	2–3	Patchy areas of collapse	Atelectasis, hyaline membranes, hyperemia, lymphatic dilatation, metaplasia, necrosis of bronchiolar mucosa
II	4–10	Obscure cardiac shadow, nearly complete opacification of lung fields	Necrosis and repair of epithelium, persistent hyaline membranes, emphysematous coalescence of alveoli, thickening of alveolar capillary membrane
III	10–20	Small rounded areas of radiolucency (spongelike)	Few hyaline membranes, regeneration of clear cells, bronchiolar metaplasia, mucous secretions, emphysematous alveoli, local thickening of basement membranes
IV	>30	Enlargement of radiolucent areas seen in III, alternating with areas of radiodensity	Emphysematous alveoli, marked hypertrophy of epithelium

The end result is maldistribution of ventilation and perfusion, hypercarbia, and hypoxemia. Occasionally, prolonged mechanical ventilation is required. In addition, positive end-expiratory pressure (PEEP) and large doses of furosemide, 5–10mg/kg q6h, may be necessary to decrease pulmonary edema and improve gas exchange. Furosemide therapy induces metabolic alkalosis, which is compensated for by retention of CO_2.

Apnea and periodic breathing are common in preterm infants after the first week of life.[7] Causes include hypo- and hyperthermia, hypo- and hyperglycemia, hypo- and hypercalcemia, hypo- and hypervolemia, anemia, decreased functional residual capacity, patent ductus arteriosus, hypothyroidism, poorly developed control of respiration, and excessive handling. Repeated apnea increases the likelihood of central nervous system damage, secondary to hypoxemia.

Moderately premature infants, especially those recovering from respiratory distress syndrome and those who require mechanical ventilation, may have chronic lung disease with elevated $PaCO_2$ and decreased PaO_2 at room air oxygen concentration. These children may require mechanical ventilation for months to years.

Temperature Instability

In preterm infants hypothermia increases the metabolic rate and oxygen consumption and leads to hypoxemia, acidosis, apnea, and respiratory distress. Body heat is dissipated by conduction, convection, radiation, and evaporation. The ratio of surface area to volume in preterm infants is high, and the flaccid and open posture of preterm infants increases heat loss. The lack of fat insulation reduces the surface-air insulation and the amount of tissue through which heat must be transferred readily from the core to the surface.

Preterm infants are less able to maintain their body temperature when exposed to a cold environment. Hypoglycemic infants and those with central nervous system damage also have difficulty in maintaining their body temperature, probably on a central basis. Preterm infants lose heat and water through their thin, transparent skin and easily become dehydrated.

Patent Ductus Arteriosus

A patent ductus arteriosus that becomes apparent on the third to fifth day after birth is common in preterm infants.[8] Pulmonary vascular resistance and pulmonary blood flow increase.[9] The murmur is usually heard at the left upper sternal border and is often continuous. It is loudest during apnea or during exhalation, and its intensity can be increased by hyperventilation. The pulses are bounding, with wide pulse pressure, and a gallop rhythm is often detectable.

As the shunt increases, so does pulmonary blood flow. The heart, therefore, is unable to keep pace with demand, which leads to congestive heart failure. Signs and symptoms include tachycardia, gallop, increasing respiratory failure with intercostal retraction, diminished breath sounds, poor air entry, and râles. The $PaCO_2$ increases as PaO_2 decreases.[10]

The initial treatment of patent ductus arteriosus is medical and includes fluid restriction, diuretics, and indomethacin (Indocin®).[11] Digitalis should not be given to preterm infants because it is ineffective in improving stroke volume and ventricular emptying. Moreover it decreases heart rate, and because tachycardia is the infant's major means of increasing cardiac output, such a change in heart rate is detrimental.[12]

If medical treatment fails, surgical ligation of the ductus is advised, to reduce morbidity and mortality. After the closure of the ductus, blood gases usually improve, and the patient can be weaned from assisted ventilation and fed within a few days.

Infection

Infections (pneumonia, sepsis, and meningitis) are common in preterm infants who are moderately or extremely premature, because cellular and tissue immunity are markedly reduced. Sepsis should be suspected if an infant becomes hypo- or hyperthermic in a neutral thermal environment, lethargic, mottled, or apneic. Preterm infants can develop sepsis in the absence of positive blood cultures and elevated white blood count with leftward shift. Blood counts that exceed 15% bands are abnormal and are good indicators of infection in preterm infants.[13]

Preterm infants respond appropriately to antibiotics, although the dosage and interval between doses may have to be altered. Aminoglycosides may cause muscle weakness or paralysis and act synergistically with nondepolarizing muscle relaxants to increase the effectiveness of the latter.[14]

Necrotizing Enterocolitis

Necrotizing enterocolitis is much more common in preterm infants than in term babies.[15] The infant suddenly develops abdominal distention, vomiting, bloody stools, and shock. Shock is the result of movement of large amounts of fluid into the third space, the peritoneal cavity, and the gut. X-rays of the abdomen show distended loops of bowel and air in the bowel wall. Free air in the peritoneal cavity indicates perforated gut.

Such infants are moribund. They require fluid resuscitation with blood and large volumes of Ringer's lactate solution prior to surgery. Large infusions usually increase respiratory failure and necessitate assisted ventilation and intravenous broad-spectrum antibiotics preoperatively.

Hematologic Manifestations

Anemia is common in preterm infants because of their reduced ability to produce red blood cells. Failure to introduce iron into the diet allows the condition to become more severe; hemoglobin of 8–10gm/dl is likely to cause apneic periods and congestive heart failure associated with patent ductus arteriosus. Stress tends to elevate the white counts, often to 40,000–50,000/mm^3. The following are seen in cases of vitamin E deficiency: increased red blood cell destruction, jaundice, anemia, lethargy, apnea, and occasional congestive heart failure. Such a deficit usually occurs in infants born before 30 weeks' gestational age.

Preterm infants have an increased percentage of fetal Hb, decresaed 2–3 DPG and an ability to significantly increase cardiac output. A recent study showed that in expremature infants, 44–60 weeks postconceptional age, with preoperative Hct$<$30%, the incidence of postoperative apnea was 80% compared to 20% in expremature infants with a Hct$>$ 30%.[16]

Metabolic Disorders

Because of a decreased ability to reabsorb bicarbonate and secrete ammonia, renal tubular acidosis is common in preterm infants. In moderately preterm infants, the respiratory rate and the work of breathing increase in response to acidosis. Lactic metabolic acidosis is suspected if weight gain is inadequate despite adequate caloric intake. Lethargy and/or apneic spells, pallor, and tachycardia occur. It may be necessary to correct acidosis with sodium bicarbonate. Moderate-size or small preterm infants may have relative respiratory acidosis with $PaCO_2$ $>$35mmHg despite a pH of $<$7.30 because of the limited ability of the chest wall to increase minute ventilation.

Preterm infants tolerate glucose loads poorly. Excessively high serum glucose levels ($>$500mg/dl) or very low levels ($<$10mg/dl) can be deleterious to the central nervous system. Most infants can tolerate 5–7mg/kg glucose per minute without developing hyperglycemia, glucosuria, polyuria, and dehydration. A blood glucose level of 40mg% is common in preterm infants.

Preterm infants normally have lower serum calcium levels. If the serum calcium level is 7mg/dl and the infant develops twitching, seizures, and hypotension, calcium gluconate 100–200mg/kg should be given immediately. Relative hypocalcemia and respiratory alkalosis with hyperventilation may also occur.

The serum sodium level is also quite labile. It rises quickly with dehydration and falls with overhydration. Both hypernatremia and hyponatremia may have adverse central nervous system effects. Hyponatremia of $<$120mEq/L is related to water intoxication. Fluid restriction is the treatment. Infusion of hypertonic saline to correct hypernatremia is rarely necessary.

Because of the reduced ability to conjugate substances, serum bilirubin levels remain high. The relative hyponatremia, an immature blood-brain barrier, and acidemia increase the susceptibility to kernicterus. Even low levels of bilirubin (10–15mg%) may be dangerous if the infant is acidotic. At this age, if the patient requires surgery, exchange transfusion is indicated as intraoperative hypoxemia and acidosis may prove disastrous.

Retrolental Fibroplasia

Retrolental fibroplasia develops in 10%–17% of preterm infants if PaO_2 is >100mmHg for a few minutes to hours, depending on gestational age. It is probably a result of multiple factors such as apnea requiring ventilation, prolonged parenteral nutrition, blood transfusion, vitamin E deficiency, hypoxemia, and hypercarbia rather than due solely to hyperoxia. It usually occurs in infants who weigh less than 2000gm and are of less than 35 weeks' gestational age.

Retrolental fibroplasia begins with some vascular obliteration and is followed by increased vascularity, hemorrhage, and, in the worst cases, retinal detachment (Table 4). The retinal vasculature spreads outward from the optic disk as the fetus grows. It reaches the nasal side of the retinal periphery by 36 weeks, and the temporal side by 40 weeks' gestational age. Hyperoxia constricts the retinal arterioles and causes swelling and degeneration of the endothelium of both the arterioles and the capillaries.[17]

TABLE 4. Stages of Retrolental Fibroplasia

STAGE I
Dilatation and tortuosity of retinal vessels; some obliteration
STAGE II
Neovascularization and some peripheral clouding; spontaneous regression may occur
STAGE III
Peripheral retinal detachment; regression unlikely
STAGE IV
Hemispheric or circumferential retinal detachment
STAGE V
Total retinal detachment

Vitamin E protects against retrolental fibroplasia by its membrane-stabilizing and antioxidant action.[18] Patients with retrolental fibroplasia come to the attention of the anesthesiologist because they require eye examination under anesthesia, photocoagulation, or scleral buckling procedure. It is wise to maintain PaO_2 within normal limits during anesthesia and surgery. As it is often necessary to inject air into the eye during surgery, nitrous oxide should be avoided. Air should be used as the carrier gas. Enough oxygen should be added to maintain the PaO_2 at normal levels.

Preanesthetic Evaluation and Preparation

Operations on post-intensive care babies are usually urgent. For example, the incidence of inguinal hernias is 6%–18% in the older child versus 31% for the postpremature baby.[19] Delay of surgery may result in bowel or gonadal infarction. Therefore, expedient preparation is essential.

An accurate history and physical examination are essential. Laboratory data include hematocrit, urinalysis, x-ray of the chest if there is a history suggestive of respiratory problems, and sickle cell prep as indicated. It is useful to have a second person available in the operating room to assist in ventilation, fluid resuscitation, and monitoring and charting of the events. The operating room temperature should be kept at 35–37°C. Infrared heaters should be placed over the operating table. A circulated-water heating blanket should be placed under the table sheet and its temperature kept at 37–38°C.

Intravenous solutions should be infused through an infusion pump or microdrip with buret that contains an adequate supply of fluids. Care must be taken to ensure a continuous infusion as embolization of even small amounts of air may prove catastrophic if a right-to-left shunt exists. (Patent foramen ovale, patent ductus arteriosus, chronic lung disease causing pulmonary hypertension and elevated right atrial pressure, and high pulmonary vascular resistance all increase the chance of air bubbles entering the systemic circulation through a right-to-left shunt.)

Lactated Ringer's solution is appropriate for the replacement of third space losses and for maintenance fluid. The basic rate is 4ml/kg/h. One half of the estimated fluid deficit can be replaced quickly with lactated Ringer's solution. Thereafter, D5 $\frac{1}{4}$NS may be used to ensure adequate glucose supplementation. If hypotension occurs, a bolus of lactated Ringer's solution (5–10ml/kg) may be given to maintain blood pressure.

Required monitoring includes precordial stethoscope, blood pressure cuff, electrocardiogram, pulse oximeter, and temperature.

Induction of Anesthesia

Although the requirements of preterm infants are lower than those of older patients, such infants need some anesthesia. Failure to provide adequate pain relief predisposes them to hypertension and intracranial hemorrhage if they lack the ability to autoregulate cerebral blood flow.[20] Adequate anesthesia prevents or attenuates these increases.

Deep anesthesia reduces the blood pressure of infants and children more than that of adults. The normal systolic blood pressure for a full-term neonate is 65mmHg; for a preterm infant, 50mmHg. A decrease of 30% may be caused by volatile anesthetic agents, resulting in pressures of 45mmHg and 35mmHg, respectively.

Intraoperative hypotension may be due to myocardial depression, a blunted neonatal baroreceptor response, the presence of coexisting congenital heart disease, or an anesthetic overdose resulting from overestimation of neonatal MAC.[21] Although MAC is greater in children than in adults, it is less in the preterm infant.

Anesthesia can be induced with inhaled anesthetics, usually halothane. Volatile anesthetics usually severely depress respiratory drive in preterm infants with a history of apnea. These babies often require ventilatory support postoperatively. Halothane, isoflurane, and enflurane are myocardial depressants. Halothane has a negative inotropic effect without negative chronotropic or vasodilator effect. Isoflurane reduces both after-load and contractility with a greater effect on afterload.

Intravenous anesthetics have a major advantage in that deleterious hemodynamic side effects are fewer. Fentanyl 30–50μg/kg[22] and sufentanil 10–15μg/kg[23] are preferred. Because an incomplete blood-brain barrier may allow passage to the central nervous system of circulatory opioid peptides that could synergistically act with narcotics, close postoperative observation and ventilatory support may be required.

Ketamine is an excellent analgesic and amnesic agent. The primary advantage is support of the sympathetic nervous system. It has a markedly prolonged half-life in neonates less than 3 months old, and it is poorly metabolized.[24] Clearance depends on renal excretion, so the duration of action is prolonged. Ketamine increases heart rate, blood pressure, cardiac index, and pulmonary vascular resistance. In intubated patients, ketamine should be used with an antisialagogue because of its potential for causing copious oral secretions.

In preterm infants in whom cranial sutures are not fused, mild increases in intracranial volume with ketamine, halothane, or isoflurane can be accommodated.

Regional Anesthesia

Regional anesthesia may be used in combination with general anesthesia in younger children or alone in older children. It can be used as an intraoperative anesthetic and as postoperative analgesia. A recent study showed that infants born after less than 37 weeks' gestation who underwent elective inguinal hernia repair at ≤ 51 weeks postconception had less apnea, bradycardia, and periodic breathing postoperatively under spinal anesthesia than did those who received general anesthesia or supplemental ketamine.[25]

For inguinal hernia operations the commonly performed blocks are caudal and local infiltration of ilioinguinal/iliohypogastric nerves.[26]

A recent study has reported over 90% success rate with caudal epidural anesthesia in premature and high risk infants for surgery below the umbilius.[27] Although the incidence of postoperative apnea is lower than following general anesthesia, it may still occur.

Caudal anesthesia is usually performed via a 23-gauge hypodermic needle inserted at a 60° angle until the sacrococcygeal membrane is pierced. A distinct "pop" should be felt as the needle enters the sacral canal. The needle is then advanced an additional 2mm in a plane parallel to the spinal axis. Following negative aspiration, the local anesthesia is introduced into the caudal epidural space.[28] Usually, a 0.25% solution in a dose of 3mg/kg bupivacaine is used, producing a plasma level below the accepted toxic levels and also providing postoperative analgesia without motor weakness.

If caudal anesthesia is used in combination with general anesthesia, a light plane of general anesthesia maintains adequate surgical anesthesia. The risk of reflex laryngeal spasm in response to surgical stimulation is reduced if caudal anesthesia is added, and there is often no need for tracheal intubation. Postintubation croup is eliminated.

Ilioinguinal/iliohypogastric nerve blocks provide both adjunct operative anesthesia and postoperative analgesia. The nerves to the groin can be blocked by infiltrating the abdominal wall muscles medial to the anterior superior iliac spine with 0.25% bupivacaine (4–6ml). A second technique involves a fan-shaped subcutaneous infiltration close to the anterior superior iliac spine.[29]

Outpatient versus Inpatient Care

During assessment of the premature infant, attention must focus on the ongoing development of the organ systems and their ability or lack thereof to adapt to extrauterine environmental demands. Premature infants are at risk for episodes of apnea and bradycardia,[30] resulting partially from lack of development and incomplete myelination in the central nervous system. Both central and obstructive apnea occur in about 30% of expremature infants postoperatively. At least one study has indicated that even brief apnea postoperatively may not be inconsequential as profound desaturation and hypoxemia may follow quickly.[31] Caffeine has been used successfully in the treatment of apnea of prematurity.[32] It is a potent central nervous system and respiratory stimulant and has fewer side effects than the previously used methylxanthines. There is, however, a significant reduction in the elimination rate of drugs in infants: neonates have an elimination half life of caffeine that varies from 37 to 231 hours.

Metabolic factors such as hypothermia, hypoglycemia, hypoxia, hypocalcemia, and acidosis may all contribute to apnea. Postoperatively, any residual anesthetic agents, whether administered by inhalation or by intravenous means, may increase the problem. Thus, narcotics are often avoided if postoperative ventilatory support is not anticipated.

Many surgical and diagnostic procedures on infants can be performed on an outpatient basis. The following are guidelines that must be modified as more experience is gained:

For premature nursery graduates 48 to 50 weeks or less of conceptual age, admission and apnea monitoring for 24 hours are advised. Otherwise healthy premature nursery graduates, more than 48 to 50 weeks postconception, with no residual lung disease, can be managed as outpatients. If they have bronchopulmonary dysplasia, they must be admitted until 1 year postnatal age because of a greater incidence of sudden infant death syndrome. Generally, full-term and healthy infants ASA I can be treated as outpatients by a month after birth. The major determinant of admission is the judgment of the surgeon and anesthesiologist.

References

1. Usher R.H., McLean F., Scott K.E.: Judgment of fetal age: 2. Clinical significance of gestational age and an objective method for its assessment. *Pediatr Clin North Am* 1966, 13:835–48.
2. Avery G.B. (ed): *Neonatology,* 3rd ed., Philadelphia: JB Lippincott, 1987, pp264–98.
3. Gregory G.A.: Anesthesia for premature infants, in Gregory G.A. (ed): *Pediatric Anesthesia*, 2nd ed., Edinburgh: Churchill Livingstone, 1989: pp803–31.
4. Usher R.H.: Clinical implications of perinatal mortality statistics. *Clin Obstet Gynecol* 1971, 14:885–925.
5. O'Brien J.P., Usher R.H., Maughan G.B.: Causes of birth asphyxia and trauma. *Can Med Assoc J* 1966, 94:1077–85.
6. Usher R.H., Allen A.C., McLean F.H.: Risk of respiratory distress syndrome related to gestational age, route of delivery and maternal diabetes. *Am J Obstet Gynecol* 1971, 111:826–32.
7. Greely W.: Anesthesia for neonates and expremature infants. *Annual Refresher Course Lectures,* Amer. Soc. Anes., Las Vegas, 1990, No. 174.

8. Kitterman J.A., Edmunds L.H. Jr.: Patent ductus arteriosus in premature infants: incidence, relation to pulmonary disease and management. *N Engl J Med* 1972, 287:473–77.
9. Siassi B., Blanco C., Cabal L.A. et al.: Incidence and clinical features of patent ductus arteriosus in low birth weight infants: A prospective analysis of 150 consecutively born infants. *Pediatrics* 1976, 57:347–51.
10. Thibeault D.W., Emmanouilides G.C., Nelson R.J., et al.: Patent ductus arteriosus complicating the respiratory distress syndrome in preterm infants. *J Pediatr* 1975, 86:120–6.
11. Heymann M.A., Rudolph A.M., Silverman N.H.: Closure of the ductus arteriosus in premature infants by inhibition of prostaglandin synthesis. *N Engl J Med* 1976, 295:530–3.
12. Berman W. Jr., Dubynsky O., Whitman V., et al: Digoxin therapy in low birth weight infants with patient ductus arteriosus. *J Pediatr* 1978, 93:652–5.
13. Spector S.A., Ticknor W., Grossman M.: Study of the usefulness of clinical and hematologic findings in the diagnosis of neonatal bacterial infections. *Clin Pediatr* 1981, 20:285–92.
14. Pittinger C.B., Eryasa Y., Adamson R.: Antibiotic-induced paralysis. *Anesth Analg* 1970, 49:487–501.
15. Mizrahi A., Barlow O., Brendon W., et al: Necrotizing enterocolitis in premature infants. *J Pediatr* 1965, 66:697–706.
16. Welborn L.G., Hannallah R.S., Higgens T.: Does anemia increase the risk of postoperative apnea in former preterm infants? *Anesthesiology* 1990, 73(3A):A1091.
17. Baum J.D.: Retrolental fibroplasia. *Dev Med Child Neurol* 1979, 21: 385–438.
18. Johnson L., Schaffer D., Boggs T.R.: Premature infants, vitamin E deficiency and retrolental fibroplasia. *Am J Clin Nutr* 1974, 29:1158–73.
19. Rescorla F.J., Grosfeld J.L.: Inguinal hernia repair in the perinatal period and early infancy: clinical considerations. *J Pediatr Surg* 1984, 19:832–8.
20. Lour H.C., Lasses P.H., Fris-Hansen B.: Impaired autoregulation of cerebral blood flow in the distressed newborn infant. *J Pediatr* 1979, 94:118–21.
21. Gregory G.A.: The baroresponses of preterm infants during halothane anesthesia. *Can Anaesth Soc J* 1982, 29:105–7.
22. Yaster M.: The dose response of fentanyl in neonatal anesthesia. *Anesthesiology* 1987, 66:433–5.
23. Emhardt J.D., Vasko M.R.: Do neonates need anesthesia? *Adv Anesth* 1990, 7:65–7.
24. Cook D.R., Davis P.J.: Anesthesia pharmacology in pediatrics, in Lake C.L. (ed): *Pediatric Cardiac Anesthesia.* Norwalk, Connecticut: Appleton & Lange, 1988: pp121–54.
25. Wellborn L.G., Rice L.J., Hannallah R.S., et al: Postoperative apnea in former preterm infants: Prospective comparison of spinal and general anesthesia. *Anesthesiology* 1990, 72:838–42.

26. Broadman L.M.: Regional anesthesia for the pediatric outpatient. *Anesthesiol Clin North Am* 1987, 5(1):53–72.
27. Gunter J.B., Dunn C.M., Bower R.J. et al: Caudal epidural anesthesia in sixty nine conscious premature and high risk infants. *Anesthesiology* 1990, 73(3S):A1096.
28. Rice L.J., Broadman L.M., Hannallah R.S.: Regional anesthesia in pediatric patients. *Adv Anesth* 1989, 6:308–9.
29. Single R.: Pediatric regional anesthesia, in Gregory G. (ed): *Pediatric Anesthesia*. New York: Churchill Livingstone, 1983, pp481–518.
30. Steward D.J.: Preterm infants are more prone to complications following minor surgery than are full term infants. *Anesthesiology* 1982, 56:304–6.
31. Kurth C.D., LeBard S.E., Downes J.J.: Association of airway obstruction, hypoxemia and postoperative apnea in preterm infants. *Anesthesiology* 1990, 73(3A):A1131.
32. Murat I., Mariette G., Blin M.C., et al: The efficacy of caffeine in the treatment of recurrent idiopathic apnea in premature infants. *J Pediatrics* 1981, 99(6):984–9.
33. Finholt D.A.: The anesthetic management of the premature nursery, in Berry F.A. (ed): *Anesthetic Management of Difficult and Routing Pediatric Patients*. New York: Churchill Livingston, 1986, pp315–48.

CHAPTER 19

The Child With Airway Obstruction

Daran W. Haber, M.D.

Case History. *An 8-year-old boy presented to the pediatric emergency room with respiratory distress and wheezing. He was sitting up with his jaw thrust forward. Although not cyanotic, he had inspiratory and expiratory wheezing, labored breathing, and stridor. He was frightened and anxious but remained alert and cooperative. A diagnosis of epiglottitis was considered. Personnel from surgery, pediatrics, otolaryngology, and anesthesiology were notified. Vital signs on admission were blood pressure, 110/70mmHg; pulse, 120/min; respirations, 30/min; temperature, 101°F orally. No attempt was made to obtain blood work or to examine the throat or epiglottis.*

Past medical history as relayed by the patient's mother was remarkable for asthma, for which he had been treated only symptomatically over several years with oral aminophylline and inhalers. His last attack had been several weeks prior to admission. Frequency between asthmatic attacks was approximately 3 to 6 months. These attacks had required emergency room visits on several occasions.

The patient's mother reported that the boy had suffered a mild upper respiratory infection the week prior to admission which had resolved spontaneously several days earlier. The episode of respiratory distress began insidiously several hours before admission and grew steadily more severe over the next few hours. This episode was reported as more severe than and different in quality from any asthma attack he had had. He had been given oral aminophylline and inhalers at the onset of the attack, but they had not helped.

While the history was being taken, the patient spontaneously regurgitated approximately 30cc of clear-pinkish fluid containing white specks (apparently aminophylline). Almost immediately, the stridor was markedly reduced, and the patient began to ventilate more easily. He had

---

Reviewed by Dr. Ingrid Hollinger, Professor, Department of Anesthesiology, Albert Einstein College of Medicine/Montefiore Medical Center.

another episode of emesis, and his symptoms continued to dissipate, although his breathing remained somewhat labored and inspiratory and expiratory wheezing remained. At that point, the patient was taken to the radiology department accompanied by his mother and the anesthesiologist, surgeon, and pediatrician. Anteroposterior and lateral x-rays of the neck and chest were taken, which were negative for significant epiglottic or aryepiglottic swelling or for foreign body or other pathology. The patient was brought back to the emergency room, and indirect laryngoscopy was scheduled. Anesthesia consult was sought.

Introduction

Airway obstruction in infants and children may be acute or chronic, congenital or acquired. When acquired, it may be infectious or traumatic in origin. It may occur anywhere along the course of the airway from the nares to the terminal bronchi. Occasionally, multiple sources of airway obstruction exist, necessitating identification of all factors for adequate treatment. An infectious or traumatic event combined with a preexisting congenital airway constriction may result in unexpectedly severe and rapidly progressive obstruction. Airway obstruction may develop insidiously and go unnoticed for days or weeks, only to suddenly reach a critical level necessitating emergent intervention.

Anatomy and Embryology of the Airway

During the 4th week of intrauterine life, the respiratory system begins to develop as an outpouching of the fetal pharynx. The facial skeleton is developed by the fusion of the first branchial arch (forming the maxilla, mandible, and zygoma) and the frontal prominence of the skull.[1] The lower face is derived from the first branchial arch. Abnormal development of structures derived from the first branchial arch may cause pathologic changes, such as Pierre Robin and Treacher Collins syndromes. The lower branchial arches form the pharynx, larynx, and tracheal cartilage. The trachea develops anteriorly and the esophagus posteriorly from a groove in the ventral foregut. Between them a tracheoesophageal septum forms. If this septum does not develop completely, a fistula or abnormal communication may remain between the esophagus and trachea.[1] The larynx develops from the proximal pharynx. The arytenoid, cricoid, cuneiform, and thyroid cartilages are formed from the fourth branchial arch. They are innervated by the superior laryngeal nerve, a branch of the vagus nerve which supplies the fourth branchial arch. The musculature of the internal larynx is innervated by the recurrent laryngeal nerve (which innervates the sixth branchial arch).[1]

The upper airway of the infant and child differs anatomically from that of the adult.[2] In the infant, the larynx is positioned at the level of C4, whereas, in the adult it is at C6. The epiglottis is omega-shaped and extends over the vocal cords at a 45° angle. This allows the infant to breathe and swallow simultaneously. Up to 2–4 months of age, infants are obligate nasal breathers; then the child's larynx descends, reaching the adult level of C6 at 4 years of age. This change in anatomy facilitates development of speech but prevents simultaneous drinking and breathing.

In the adult the narrowest part of the upper airway is at the vocal cords. In the infant and child further narrowing occurs distally at the area of the cricoid cartilage.[2] The ratio of glottic aperture to body surface area is smaller than in adults. Therefore, infants also have a disproportionately smaller larynx.[1] The diameter of the trachea in the term neonate is 4–5mm, and the length is 4–5cm. During flexion or extension of the head, the tip of an endotracheal tube may move by as much as $2cm^2$, making accidental dislodgment or endobronchial intubation likely.

Airflow Characteristics

Because of the narrow glottic and tracheal lumens of infants and children, obstructions to airflow may develop rapidly and progress quickly to complete obstruction. The passage of a gas through a partially closed tube is dependent on the Venturi principle, ie, the pressure exerted by the gas is equal in all directions except when there is linear movement along the tube. Additional pressure is created in the forward direction with a corresponding fall in lateral pressure.[3] Poiseuille's equation relating to laminar flow holds that resistance varies inversely with the fourth power of the radius (ie, resistance $[R] = 1/\text{radius}^4[r]$).[4]

A 50% reduction in airway radius increase resistance to flow by a factor of 16. In the neonate 1mm of edema will reduce airway diameter by 65%. Additionally, many airway obstructions increase turbulent flow.[4] Under these conditions, resistance becomes inversely proportional to the fifth power of the radius. A 50% reduction in airway diameter will then increase resistance by a factor of 32. Airway collapse is further increased by the softer, more distensible tissues and looser tissue planes present in the infant.[4] When air passes through a partially collapsible tube like the trachea, the Venturi effect may cause a decrease in outward pressure, causing collapse of the tracheal wall.[3]

Changes in airflow may be related to the particular type of lesion causing the obstruction. Obstructions are called "fixed" when the cross-sectional area of the lumen does not change in response to a change in transmural pressure. They are termed "variable" when the lesion causes a response to changes in transmural pressure with a corresponding change in cross-sectional area.[5] Obstructions may be intrathoracic or extrathoracic. During inspiration, intraluminal pressure is subatmospheric in the extrathoracic airway, favoring intraluminal narrowing. During expiration, intraluminal pressure is greater, leading to dilatation of the airway. Intrathoracic lesions exert a reverse effect.[5] Fixed lesions (such as tracheal stenosis) behave the same whether the result of intra- or extrathoracic pathology. Variable intrathoracic obstructions have a fixed flow throughout most of the expiratory vital capacity (eg, many tumors characteristically behave this way). Variable extrathoracic obstructions have a low constant inspiratory flow and a high midexpiratory vital capacity flow rate (eg, vocal cord lesions).[5]

Partial closure of the airway with changes in airflow may cause vibration within the tracheal lumen, producing audible sounds.[3] Stridor and wheezing may be heard, depending on the nature of the obstruction, its size, and its location. Obstructions in the supraglottic airway usually produce stridor during inspiration. Obstructions at the level of the glottis and extrathoracic trachea may produce stridor which is heard equally during inspiration and expiration. Positive pressure on the intrathoracic trachea and large bronchi may cause stridor on expiration, and classically has a "wheezy" character.[3] Generally, causes of airway obstruction in children may be congenital, infectious, or traumatic (Table 1).

Congenital Causes Of Airway Obstruction

Congenital airway obstructions in infants and children tend to progress insidiously unless compounded by a sudden infectious or traumatic process. Most often, time is available for elective diagnostic testing and treatment, without the necessity for emergent therapeutic intervention to ensure airway patency and adequate ventilatory exchange. Congenital obstructions may occur anywhere along the course of the airway.

Choanal Atresia

Choanal atresia occurs approximately once in every 8000 births.[3] It causes unilateral or bilateral nasal obstruction; when bilateral, it may cause respiratory distress immediately after birth because neonates are obligate nasal breathers. Inability to pass a suction catheter through either nostril to the pharynx confirms the diagnosis.[6] Initial treatment may involve simply placing and securing an oral nipple with an enlarged hole for the passage of air. Early surgical treatment is usually recommended. Nasopharyngeal masses, such as encephaloceles, dermoids, and teratomas may produce similar symptoms.[3]

TABLE 1. Causes of Airway Obstruction
Congenital
Choanal atresia
Craniofacial malformations (Pierre Robin, Treacher Collins syndromes)
Macroglossia (Down's, Beckwith's syndromes)
Laryngotracheomalacia
Vocal cord paralysis
Congenital subglottic stenosis
Congenital laryngeal webs
Congenital laryngeal atresia
Laryngotracheoesophageal cleft
Tracheoesophageal fistula
Subglottic hemangioma
Tracheomalacia
Tracheal stenosis
Tracheal compression by vascular rings
Infectious
Croup (laryngotracheobronchitis)
Epiglottis
Diphtheria
Retropharyngeal abscess
Traumatic
Thermal
Chemical
Foreign body aspiration

Craniofacial Malformations

Disorders of development of the first branchial arch components, such as Pierre Robin and Treacher Collins syndromes, may cause airway obstruction because of micrognathia and relative macroglossia.[6] Initial treatment may involve placing the child in the prone position during feeding to keep the tongue from falling back and obstructing the airway. Surgical advancement of the tongue is occasionally necessary. Other disorders also may be associated with macroglossia. These include Down's and Beckwith's syndromes, congenital hypothyroidism, glycogen storage disease, thyroglossal cysts, and lingual tumors.[3,6]

Laryngotracheomalacia

Laryngotracheomalacia is the most common cause of congenital stridor (about 70% of cases). It is usually a benign, self-limiting disorder characterized by high-pitched inspiratory stridor, occasionally accompanied by a soft expiratory component. It is caused by a soft laryngeal framework and shortened, redundant aryepiglottic folds which become indrawn on inspiration. Symptoms usually disappear by 18–24 months of age.[3]

Vocal Cord Paralysis

Vocal cord paralysis may be unilateral or bilateral. When bilateral, the etiology is usually central in origin and may be associated with an Arnold-Chiari malformation with compression of the vagus and cervical nerve roots by brain stem displacement and compression. Birth trauma is involved in the etiology of some bilateral cord palsies, but in this case the lesion usually resolves with age. Marked inspiratory stridor occurs with bilateral cord paralysis. Intermittent aspiration and cyanosis during feeding occur in both bilateral and unilateral cord paralysis but are more severe in bilateral paralysis. Unilateral paralysis is more common on the left side and usually involves a cardiovascular anomaly, causing a lesion of the left recurrent laryngeal nerve.[6] Radiography and endoscopy are performed for diagnosis. Nasotracheal intubation or tracheostomy is sometimes necessary.

Congenital Subglottic Stenosis

Subglottic stenosis involves a thickening of the soft tissue of the subglottic area and occasionally of the true vocal cords. Obstruction is usually greatest 2–3mm below the level of the true cords.[3] Inspiratory and expiratory stridor may occur with or without cyanosis. Often the condition may go unnoticed until compounded by infection or some other source of edema (eg, inhalation burn, endotracheal intubation). If combined with other congenital causes of subglottic narrowing, such as occurs in Down's syndrome, surgical intervention may be required.[3] Tracheostomy and/or subglottic dilatation or reconstruction may be indicated.

Congenital Laryngeal Webs and Atresia

Laryngeal webs occur often at the level of the vocal cords anteriorly. Inspiratory stridor and indrawing may be present. Direct laryngoscopy is required for diagnosis.[6] Webs may consist of a thin membrane or a thick fibrous band. Endoscopic lysis by laser may be performed electively for partial obstructions. Complete obstruction is evident immediately at birth, is incompatible with life, and must be treated immediately by forced intubation, cricothyreotomy, or tracheostomy. Laryngeal atresia is incompatible with life unless there is a distal tracheoesophageal fistula large enough to accommodate adequate ventilation. In such cases emergent tracheostomy is indicated.

Laryngotracheoesophageal Cleft and Tracheoesophageal Fistula

Failure of formation of the tracheoesophageal septum during fetal development leads to tracheoesophageal fistula when it occurs distally and laryngotracheoesophageal cleft when it occurs proximally. Infants with laryngeal clefts have respiratory distress during feeding and recurrent aspiration leading to pneumonia and death.[3] Stridor is common. Nasotracheal intubation or tracheostomy should be performed prior to surgical correction of the lesion. Tracheoesophageal fistula is most commonly associated with esophageal atresia. H-type fistulas are rare, and symptoms are less well defined.[3] Feeding problems and cyanosis may be seen soon after birth. Recurrent pneumonias are common. Barium swallow or endoscopy is required to confirm the diagnosis of tracheoesophageal fistula without esophageal atresia.

Subglottic Hemangioma

Subglottic hemangiomas most commonly present with stridor, initially inspiratory but later biphasic. The stridor is usually exacerbated by crying or by upper respiratory tract inflammation.[3] Radiologic examination may demonstrate an eccentric swelling located posteriorly in the subglottis. Endoscopy confirms the diagnosis. These lesions usually regress spontaneously, disappearing by age 2. If treatment is necessary, it is usually conservative, involving steroid therapy, with laser surgery only if necessary.[3] Tracheostomy is rarely required.

Tracheal Abnormalities

Tracheal abnormalities include tracheomalacia, tracheal stenosis, and extrinsic compression of the trachea by vascular malformations or mediastinal masses. Tracheomalacia is a disorder of the cartilaginous tracheal rings in which the rings are malformed in shape and consistency.[3] Stridor may be expiratory or biphasic. The trachea may buckle or collapse on expiration, and secretions are poorly mobilized. Diagnosis of tracheomalacia is made by bronchoscopy, and surgical correction may be warranted.[6] Tracheal stenosis is a congenital abnormality characterized by a fixed obstruction producing biphasic stridor. Diagnosis is made by bronchoscopy and soft tissue x-rays of the neck and chest. Surgical correction by tracheal dilatation or reconstruction may be required.[6] Vascular rings caused by anomalous intrathoracic vasculature may compress the trachea, as may cystic hygromas or mediastinal masses, such as tumors of the thymus or thyroid. Chest x-ray, barium swallow, or arteriography may aid in diagnosis. If the esophagus is also compressed, feeding difficulties occur. Symptoms such as biphasic stridor, respiratory indrawing, and a barking, seal-like cough are similar to those seen with other types of tracheal obstruction.[6] Nasotracheal intubation may be required until surgical correction is performed.

Infectious Causes of Airway Obstruction

Infection in the airway may act alone or combine with a preexisting congenital airway obstruction to cause a rapidly progressive and potentially life-threatening airway obstruction. Two common causes of airway infection are croup (laryngotracheobronchitis) and epiglottitis. The pathophysiology and treatment of these diseases are different, and the anesthesiologist must be able to make rapid diagnostic and treatment decisions to prevent complete airway closure, cyanosis, and death in these patients.

Croup (Laryngotracheobronchitis)

Croup and epiglottitis (and in the past, laryngeal diphtheria) account for over 80% of upper airway obstructions in infants and young children. Among these, croup is by far the most prevalent.[7,8] Croup is predominantly a viral disease, usually caused by parainfluenza myxoviruses, but it also has been caused by respiratory syncytial virus, adenovirus type 5, echoviruses and influenza viruses.[8] Rarely, Staphylococcus aureus has been implicated.[9] Both epiglottitis and croup have been more commonly reported in industrial cities, particularly those in drier northern areas, and during the autumn and spring, when greatest temperature changes occur.[8] This suggests that cold air, lack of humidity, sudden temperature change, and air pollution may play an etiologic role in predisposition to these diseases.

Croup most commonly occurs in children 6 months to 3 years of age.[9] It is characterized by a progressive inflammation of the entire tracheobronchial tree, from the larynx down to the terminal bronchi. The area of greatest obstruction occurs at the narrowest anatomic area of the upper airway, the area of the cricoid ring. Onset of the disease is insidious. Stridor is inspiratory and is accompanied by a "barking" croupy cough, tachypnea, and substernal and suprasternal retractions. Often, parents report a history of a recent prodromal upper respiratory infection. Malaise, lethargy, food intolerance, and low-grade fever may persist. Respiratory symptoms progress slowly and may result in cyanosis if untreated. In severe cases the cyanosis may persist even with supplemental oxygen therapy. Inspiratory wheezing and dimished breath sounds may be noted as the obstruction becomes more severe. Respiratory failure with severe hypoxia, bradycardia, cardiopulmonary arrest and death may ensue. Downes and Godinez have devised a scoring system to assess the clinical course of the disease and determine treatment[10] (see Table 2).

TABLE 2. Clinical Croup Score

	0	1	2
Inspiratory breath sounds	Normal	Harsh, rhonchi	Delayed
Stridor	None	Inspiratory	Inspiratory and expiratory
Cough	None	Hoarse cry	Bark
Retractions and flaring	None	Flaring and suprasternal retractions	Flaring, suprasternal, subcostal, and intercostal retractions
Cyanosis	None	In air	In 40% O_2

Best score is 0; worst is 10. A score of 4 or more warrants administration of an aerosolized vasoconstricting drug (eg, epinephrine).

Anteroposterior and lateral radiographic examination of the neck may be helpful in diagnosis. A symmetric narrowing of the subglottic airway, referred to as a "church steeple" sign on anteroposterior view, and blurring of the tracheal air shadow on lateral view are consistent with the diagnosis of croup. Leukocytosis may occur, but total white blood cell count remains below 12,000 with a relative lymphocytosis (>30%), consistent with viral infection.[8] Treatment includes oxygen administration, nebulized epinephrine solutions, observation, and if necessary, endotracheal intubation to bypass the area of most severe obstruction. Use of corticosteroids is controversial, but they may be used for cases refractory to the aforementioned treatments or for croup secondary to an acute irritation of the airway such as may occur after extubation. Specific indications for intubation include hypoxemia in spite of high inspired oxygen concentrations, persistent hypercapnia, and increasing croup score with respiratory failure in spite of nebulized epinephrine and corticosteroid therapy. Until recently, tracheostomy was preferred over endotracheal intubation because of the risk of postintubation subglottic stenosis. Tracheostomy, however, involves the risk of hemorrhage into the airway, tension pneumothorax, mediastinitis, and tracheal obstruction. Most experts prefer nasotracheal intubation, use of smaller than age-predicted endotracheal tubes, early extubation, and conversion to tracheostomy if prolonged intubation is required.[8,9]

Extubation is usually achieved within 3 to 4 days. Criteria for extubation include adequate hydration, normalization of arterial pO_2 and pCO_2, and decreasing inspired oxygen requirement. Nebulized epinephrine treatments should be continued after extubation, and the child should be monitored extensively for at least 24 hours. Bacterial superinfection may occur with croup and should be investigated by clinical examination and by chest x-ray and blood count. Early identification and treatment are essential. Negative-pressure pulmonary edema may occur and may require treatment with positive-pressure ventilation, positive end expiratory pressure, and diuretics.

Epiglottis

Epiglottitis is an acute, fulminant infection causing supraglottic inflammation involving the epiglottis, arytenoids, aryepiglottic folds, and uvula. It usually occurs in children between the ages of 2 and 7 years but may occur at any age and has been reported in adults.[11] It is rarely seen before 3 months of age because of passive maternal immunity to Hemophilus influenzae.[9] Epiglottis is usually caused by H. influenzae type b. Rarely, it may be caused by S. aureus, β-hemolytic streptococci, or Diplococcus pneumoniae.[8] As with croup, etiologic factors include dry, temperate climate; seasonal temperature change in late autumn and early spring; and high levels of air pollution. Massive supraglottic inflammation may occur which can lead to ulceration and submucosal abscess formation with necrosis of the supraglottic area. Edema may increase rapidly leading to total obstruction of the airway.

The child with epiglottitis classically presents with a high fever and sore throat, rapidly progressing to dysphagia, dysphonia, drooling, and inspiratory respiratory distress (the “4 Ds”), as well as muffled stridor at the vocal chords. In contrast to croup, onset is rapid, prodromal upper respiratory infection is usually absent, as is the “barking” croupy cough. The patient usually is acutely distressed and presents sitting upright with head extended, waist flexed, and chin thrust forward. This posture tilts the epiglottis away from the glottic aperture, helping to maintain a patent airway. Forcing the child to deviate from this position may increase the severity of the airway obstruction.[8] Because the child is unable to swallow, profuse drooling usually occurs. Cyanosis may be evident and may persist despite high inspired oxygen concentrations.

Total white blood cell count in epiglottis may reach 15,000–25,000, with a predominance of polymorphonuclear leukocytes. Lateral x-ray of the neck demonstrating a large, thickened epiglottic shadow the approximate size and shape of an adult thumb (a positive "thumb" sign) is consistent with the diagnosis. Direct pharyngoscopy demonstrating a cherry-red violaceous swelling of the epiglottis, arytenoids, false vocal cords, posterior tongue, and/or uvula confirms the diagnosis. However, this procedure performed in an unanesthetized or inadequately anesthetized patient may provoke sudden, irreversible, complete airway obstruction. It should never be performed outside the operating room or without adequate preparation and personnel experienced in intubation and tracheostomy.

Despite the obvious differences in classic symptoms of croup and epiglottitis, differentiation of these diseases may sometimes be difficult (Table 3).

TABLE 3. Differential Diagnosis of Croup and Epiglottitis

	Croup	**Epiglottitis**
Onset	Insidious	Rapid
Cough	Barking	Absent or weak
Dysphagia	Absent or weak	Severe
Dysphonia	Hoarseness, mild dysphonia	Severe dysphonia, aphonia
Drooling	Absent	Present
Protective posture	Absent	Sitting, waist flexed, chin forward
Fever	Low	High
Cyanosis	Variable	Usually present
Stridor	Present	Absent to mild
Retractions	Absent to mild	Present
Leukocytosis	Mild	Severe
Neck x-ray	"Church-steeple" sign (anteroposterior view)	"Thumb" sign (lateral view)

The most important part of the anesthetic plan is contingency planning. Assistance should always be immediately available.[12] If a diagnosis of epiglottitis is being considered, the child should not be approached or taken from its mother. Simply moving the child from its protective posture or frightening the child by attempting to obtain blood tests or to cannulate a vein may precipitate airway closure and inability to ventilate. Time should not be taken for x-rays to confirm the diagnosis unless the condition of the child is unquestionably stable. The child should be taken to the operating room by the parent, accompanied by the anesthesiologist, pediatrician, and otolaryngologist. Resuscitative, intubation, and tracheostomy equipment should accompany the patient in case sudden airway closure necessitates emergent intervention.

In the operating room, direct visualization of the epiglottis and endotracheal intubation under inhalational anesthesia may be performed. The recommended technique involves the use of halothane in oxygen. Halothane is less irritating to the airways than other inhalational agents and allows the concurrent administration of almost 100% oxygen. Once an adequately deep level of anesthesia is obtained, direct pharyngoscopy for diagnosis and orotracheal intubation may be performed. An intravenous catheter should be placed at that time, and blood should be obtained for complete blood count and arterial blood gas analyses if necessary. Cultures are taken from the epiglottis, and intravenous antibiotics are given. Later, once an airway has been established, a nasotracheal tube may be placed, again under deep anesthesia. If complete closure of the glottic aperture occurs at any time before intubation, direct laryngoscopy may disclose a swollen mass of tissue in which it is impossible to identify the larynx or even the epiglottis. Small air bubbles emerging from the glottis may be helpful in determining the correct location. If intubation is impossible, emergent cricothyrotomy (with a 14–16-gauge angiocatheter for insufflation of oxygen) or tracheostomy is indicated as a last lifesaving resort. With severe epiglottitis, however, the anatomy may be so severely distorted to make even these measures difficult or impossible.

Diaz has described an alternative technique to secure the airway by direct orotracheal or nasotracheal intubation of awake children with epiglottitis after topical anesthesia.[8] However, this technique should be reserved for those physicians who have personal expertise with it because of the propensity for sudden complete loss of airway.

Intravenous antibiotic therapy should be initiated as quickly as possible. Ampicillin (50–200mg/kg/day) is the treatment of choice. Resistant strains of H. influenzae may be treated by chloramphenicol (50–100mg/kg/day), aminoglycosides (gentamicin and tobramicin), clotrimazole, or trimethoprim-sulfamethoxazole (for cases resistant to chloramphenicol).[8] Extubation may be considered only after repeat pharyngoscopy reveals relief of obstruction and decreased swelling. Until that time the endotracheal tube should remain well secured. Arm restraint or sedation of the child may be necessary to prevent accidental extubation. Intensive monitoring of the child should continue after extubation. Rifampin prophylaxis is recommended for children and adults who have been in close contact with the patient and who have close contact with children under 4 years of age.[8]

Traumatic Causes of Airway Obstruction

Thermal and Chemical Injury

Burn injury to the face and pharynx by fire or other heat injury or by ingestion of caustic substances may cause edema leading to airway obstruction. Direct heat or chemical injury to the tracheobronchial tree is uncommon. Cooling of air through the nasal passages helps prevent distal injury, and closure of the larynx usually prevents aspiration of caustic substances.[9] Toxic products of combustion, however, may cause injury to the tracheobronchial tree and alveoli if exposure to these gases persists for prolonged periods. External signs of smoke inhalation injury may include burns to the face and neck, singeing of hair on the face and in the nares with soot deposition, and hoarseness.[9] Redness and edema of the oropharynx may be noted on laryngoscopy. Edema may slowly increase after termination of exposure, resulting in airway obstruction. Close observation during this period is essential. Endotracheal intubation may be required for severe cases.

Foreign Body Aspiration

Foreign body aspiration may occur in patients of any age but most commonly occurs in infants and children between the ages of 6 months and 4 years, most commonly in infants under 1 year of age.[9] Aspirated foreign bodies may lodge anywhere along the course of the airway and are usually radiolucent. If lodged in the trachea, a partially obstructive foreign body will produce inspiratory and expiratory stridor. If lodged low in the trachea, negative intrathoracic inspiratory pressure will dilate the airway, reducing inspiratory stridor (and possibly drawing the object deeper in the chest). On expiration, positive intrathoracic pressure will constrict the trachea, increasing expiratory stridor or closing off the airway completely and keeping the foreign body from being expelled. A foreign body in a mainstem bronchus causes airway irritation and may induce spasmodic coughing. Foreign bodies may lodge in the esophagus and cause external compression against the trachea, resulting in partial obstruction. Small foreign bodies that lodge distally in terminal bronchi may be clinically unapparent for weeks or months. At two to three weeks after aspiration a lower respiratory tract infection may develop, with chronic cough and sputum production as presenting symptoms. Localized wheezing or asymmetric breath sounds may be detected. Blood-tinged sputum may be a presenting symptom of foreign body aspiration if a sharp or irritating object is aspirated. Vegetable matter is often a cause of this symptom. History from the parents of a coughing episode during eating or during play with small objects may be helpful in making a diagnosis but foreign body aspiration is frequently unwitnessed and may go unnoticed.

Although foreign bodies are most often radiolucent, anteroposterior and lateral x-rays of the neck and chest may aid in locating anatomic deviation caused by the aspirated object. Unilateral lung or lobar hyperinflation, atelectasis, or infiltrate may be consistent with distal foreign body obstruction. Residual hyperinflation on deep expiration (noted during fluoroscopy or by "aiding" expiration with pressure on the epigastrium during end-expiratory x-ray) may help in making the diagnosis. Ventilation-perfusion scanning, thoracic computerized tomography, xerocardiography, and magnetic resonance imaging are other tests that may be useful in diagnosis. As with all types of airway obstruction, maintenance of the airway with adequate oxygenation and gas exchange is the primary concern. If the foreign body poses no immediate threat to adequate ventilation, time for gastric emptying should be allowed before any attempt is made to retrieve it.

Bronchoscopy and extraction of the foreign body is the treatment of choice. Distally lodged objects unreachable by bronchoscopy may be treated with percussion and postural drainage. Inhalation of bronchodilators also may be used. Emergency treatment for large foreign bodies causing complete airway obstruction may call for an abdominal thrust (Heimlich maneuver) or back slap. Mouth-to-mouth ventilation may be used as a last resort in an attempt to push the object into either of the mainstem bronchi, allowing for one-lung ventilation as a temporizing measure. Use of postural drainage in the treatment of large foreign bodies in the trachea or one mainstem bronchus is controversial. A foreign body freed from the initial site of obstruction may cause a complete bilateral obstruction to ventilation. Obstruction also may occur at the level of the cricoid cartilage or larynx as the object is withdrawn. Endotracheal intubation before removal of a foreign body in the upper airway may force the foreign body deeper, embedding it more firmly into the tracheal mucosa, or cause it to fragment and pass distally. For cases of distally lodged foreign bodies or for complete obstruction with inability to ventilate, endotracheal intubation should not be delayed. Muscle relaxation during bronchoscopy may ease extraction of the foreign body.

Assessment of Airway Obstruction

Initial assessment of airway obstruction involves observation for symptoms such as stridor, cyanosis, and sternal and intercostal retractions. Grunting, coughing, choking, hoarseness, tachycardia, tachypnea, hypertension, fatigue, decreased air entry, and cardiac dysrhythmias are indications for immediate intervention. Preservation of airway and oxygenation are of primary concern. Once airway stability has been established, physical examination and laboratory and x-ray testing may proceed to establish a definitive diagnosis and formulate a treatment plan. Physical examination should include a check (initially from a distance) for neck masses, tracheal location and mobility, loose or missing teeth, overbite, deviated septum, nasal polyps, size of nares, signs of external trauma, and associated cervical injury. Anteroposterior and lateral x-rays of the neck and chest may be helpful in making a diagnosis. Appropriate laboratory test include arterial blood gas analyses and complete blood count with differential. Pulse oximetry may provide a less invasive alternative or adjunct to arterial blood gas sampling.

References

1. Kingston H.G.: Airway problems in pediatric patients. *Problems in Anesthesia* 1988, 2(4):545–65.
2. Cote C.J.: How to manage the difficult pediatric airway. *Amer Soc Anes* Las Vegas, 1990, No. 262.
3. Richardson M.A., Cotton R.T.: Anatomic abnormalities of the pediatric airway. *Pediatr Clin North Am* 1984, 31(4):821–33.
4. Goldthorn J., Badgwell J.M.: Upper airway obstruction in infants and children. *Int Anesthesiol Clin* 1986, 24(1):133–44.
5. Maze A., Bloch E.: Stridor in pediatric patients. *Anesthesiology* 1979, 50:132–45.
6. Rhine E.J., Johnson G.G.: Upper airway obstruction in pediatrics. *Clinics in Anesthesiology* 1985, 3(3):721–38.
7. Badgwell J.M., McLeod E.M., Freiberg J.F.: Airway obstruction in infants and children. *Can J Anaesth* 1987, 34(1):90–8.
8. Diaz J.H.: Croup and epiglottitis in children: The anesthesiologist as diagnostician. *Anesth Analg* 1985, 64:621–33.
9. Khan M.: Common pediatric airway emergencies. *Anesthesiology Review* 1982, 9(2):29–33.
10. Downes J.J., Godinez R.I.: Acute upper airway obstruction in the child. *American Society of Anesthesiologists Refresher Courses in Anesthesiology* 1980, 8:29–47.
11. Bishop M.J.: Epiglottitis in the adult. *Anesthesiology* 1981, 55:701–2.
12. Berry F.A.: Acute airway obstruction with special emphasis on epiglottitis and croup in Berry F.A. (ed): *Anesthetic Management of Difficult and Routine Pediatric Patients.* New York: Churchill Livingstone, 1990, pp. 243–65.

CHAPTER 20

The Patient for Coronary Artery Bypass Surgery

Mark J. Badach, M.D.

Case History. *A 66-year-old white man, obese and a heavy smoker, was scheduled for a coronary artery bypass grafting (CABG) because of severe coronary artery disease (CAD). He presented to the emergency room of his local hospital complaining of sharp substernal chest pain that began shortly after he started shoveling snow from the driveway of his home. He took some nitroglycerin without any improvement. As the pain grew in intensity, he decided to seek medical attention.*

He was a known hypertensive with non-insulin-dependent diabetes well controlled with Lopressor®and Glucotrol®. He had had a myocardial infarction 2 years prior to admission, after which he developed mild congestive heart failure and was treated with digoxin and furosemide.

Physical examination demonstrated a moderately obese male, (weight 220lb, height 5'8") in acute distress: orthopneic, tachypneic, with obvious jugular venous distention.

Laboratory studies demonstrated the following: hemoglobin, 14gm/dl; hematocrit, 40%; platelets, 340,000; PT/Ptt, normal; pH 7.5; PAO_2*, 160mmHg,* $PACO_2$*, 29mmHg; on* FiO_2*, 0.5; electrolytes:* K^+*, 3.1mEq/L;* Na^+*, 140mEq/L; Cl, 98mEq/L; glucose, 240mg/dl; CPK, 330 IU; CPK–MB, 1.0 IU. ECG showed a normal sinus rhythm with elevated ST in inferolateral leads (II, III, aVF,* V_5*, and* V_6*). The ECG changes reverted to normal after infusion of nitroglycerin.*

Cardiac catheterization showed 80% left main occlusion with a 90% occlusion of the right coronary artery, poor ventricular function with an akinetic left ventricle, and an ejection fraction of 30%.

A multigated acquisition scan revealed viable muscle in the left ventricle. The patient was scheduled for urgent coronary artery bypass surgery.

Reviewed by Dr. Pierre A. Casthely, Department of Anesthesiology, St. Joseph's Hospital and Medical Center, Patterson, NJ.

It is estimated that 10 million patients have ischemic heart disease in the United States today, 4 million of whom have had a previous myocardial infarction (MI). In any one year, approximately 1.5 million patients develop a new MI, and 700,000 deaths are attributed to ischemic heart disease (IHD) per year.[1] Surgical procedures performed each year on these patients include approximately 800,000 cardiac catheterizations, with 285,000 CABGs and an equal number of coronary angloplasties. It is estimated that of the 20 million patients in the United States who undergo surgery each year, approximately 1–2 million have IHD or are at high risk for it. Thus, the problem is significant for anesthesiologists. It appears that the prognosis for patients with IHD is related to the development and severity of dysrhythmia, MI, and ventricular dysfunction.

Basic guidelines for evaluating these patients are as follows: recognize the risk factors present and attempt to optimize the patient's condition (Table 1); make an accurate diagnosis (Table 2); detect and correct the intraoperative predictors in order to optimize the patient's hemodynamic profile (Table 3).

TABLE 1. Risk Factors Predicting Outcome of a Cardiac Patient

Age	Peripheral vascular disease
Previous MI	Valvular heart disease
Angina	Cholesterol
Congestive heart failure	Cigarette smoking
Hypertension	Previous CABG surgery
Diabetes mellitus	Previous coronary angioplasty
Dysrhythmia	Cardiovascular therapy

TABLE 2. Diagnostic Testing Predictors in the Cardiac Patient

12-Lead ECG	Transesophageal echocardiography
Chest x-ray	Radionuclear imaging
Exercise stress testing	Magnetic resonance imaging
Ambulatory ECG monitoring	Magnetic resonance spectroscopy
Precordial echocardiography	Cardiac catheterization
	Routine laboratory testing

TABLE 3. Intraoperative Predictors in the Cardiac Patient
Choice of anesthetic
Surgical factors
Intraoperative cardiovascular changes
Myocardial ischemia
Ventricular dysfunction
Dysrhythmias

Known Risk Factors

Ages

The incidence of CAD increases with age. Although age does not appear to affect resting ejection fraction, left ventricular volume, and regional wall motion, it does depress cardiac response to different forms of stress, such as exercise or exogenous catecholamines.[2] Perioperative MI is now the leading cause of postoperative death in the elderly undergoing non-cardiac surgery.[3] Some studies[4] indicate a 38% incidence of ischemia, MI, or cardiac death in patients older than 70, versus 7% in those age 40–49; other authors[5] report that age is a significant predictor only when other factors are present. Thus, age may not be as important as overall physiologic status.[6]

Previous MI

Patients with prior MI are at greater risk for perioperative reinfarction 5%–8%[7–9] than are those without prior MI (0.1%–0.7%)[8,9] and have a reinfarction mortality of 36%–70%. Studies have demonstrated that the more recent the MI, the more likely is reinfarction. Within 3 months the reinfarction rate exceeds 30%; at 3–6 months, 15%;[7,8] and after 6 months, approximately 6%.[10]

One study has shown that reinfarction occurred in only 1.9% of 733 patients who had had a previous MI.[11] Perioperative reinfarction occurred in 5.7% of patients who had an MI less than 3 months previously and in 2.3% whose MI was 4–6 months earlier. From these findings it may be inferred that preoperative optimization of the patient's status, aggressive invasive monitoring and therapy, and prolonged ICU stay may significantly reduce reinfarction rates and decrease perioperative cardiac morbidity. However, whether use of these modalities lowers reinfarction rates cannot yet be determined.

Angina

Angina usually is associated with angiographically significant (>70% stenosis) changes. A typical angina (mandibular, shoulder pain) is less often (30%–65%) associated with angiographic evidence of CAD.[12]

A history of stable angina significantly increases the risk of MI and sudden death in ambulatory patients with CAD but is a controversial predictor in noncardiac surgical patients. A number of studies support angina as a predictor,[8,13] whereas other studies suggest that angina is insignificant[8] or is not a good predictor[14,15] of the risk of developing MI.

Congestive Heart Failure

Clinical or radiologic evidence of left ventricular failure is associated with a poor prognosis in patients with CAD and is one of the most important predictors of short- and longterm cardiac mortality in the patient with acute MI.

Two signs of heart failure have predictive value; a third heart sound and jugular venous distention.[8] A low preoperative ejection fraction (<0.40, as determined by radionuclear imaging or ventriculography) is predictive of perioperative MI, reinfarction, and perioperative ventricular dysfunction.[15,16]

Hypertension

The most common cardiovascular disease in the United States is hypertension, affecting more than 59 million people. Hypertension is a risk factor for ischemic heart disease, congestive heart failure, and stroke. Treatment of hypertension reduces mortality associated with stroke and heart failure but apparently not that of MI.[17]

Mild to moderate preoperative hypertension does not predict mortality.[11] However, it does predict intraoperative blood pressure stability and myocardial ischemia.

Diabetes Mellitus

Diabetes mellitus is a risk factor for CAD. It increases the risk for atherosclerotic disease two- to threefold.[18] In diabetics, MI is the leading cause of death and appears to be associated with more complications and a lower overall survival rate than in nondiabetics.[18] Myocardial infarction and myocardial ischemia tend to be silent in diabetics, more so than in other subgroups of patients. The etiology of the silent ischemic pattern may be related to altered sensory afferents.[19] Twenty percent to

40% of diabetics develop abnormal autonomic tone, which places them at a particular risk for myocardial ischemia, infarction, and cardiomyopathy.[19,20]

Dysrhythmias

Dysrhythmias are fairly common events and are usually benign in healthy patients without known heart disease. In the presence of CAD or left ventricular dysfunction, they become ominous.[21] In patients with acute MI, ventricular dysrhythmias or conduction disturbances detected in the late hospital period indicate a poor outcome. Frequent premature ventricular contractions or rhythms other than normal sinus are important findings in patients undergoing noncardiac surgery. Fascicular or bundle branch blocks, however, increase perioperative risk only when associated with more serious conditions.[22]

Peripheral Vascular Disease

Significant coronary artery stenosis (>70%) is present in 14%–78% of patients with peripheral vascular disease regardless of CAD symptoms. Such patients undergoing vascular surgery have a 15% incidence of MI, which accounts for more than 50% of their perioperative mortality.[23]

Valvular Heart Disease

The prognosis for patients with valvular heart disease depends on the disease type and its severity. Limited data indicate that aortic stenosis is associated with increased perioperative mortality.[14] Although patients with mitral valvular disease have occasionally developed congestive heart failure, only patients with aortic stenosis have an increase in mortality rate in the postoperative period.

Cholesterol

Anatomic and physiologic changes associated with atherosclerotic plaques are similar in patients who have the familial or nonfamilial form of hypercholesterolemia.[24] The Framingham study and the Multiple Risk Factor Intervention Trial demonstrated a good correlation between increased serum cholesterol and cardiovascular mortality.[25,26]

Cigarette Smoking

Smoking has acute and chronic effects on myocardial oxygen supply and demand. The acute effects on oxygen supply include increased coronary vascular resistance and increased carboxyhemoglobin levels. The chronic effects of smoking include vasoconstriction, enhanced platelet aggregation, and loss of endothelial integrity, accelerating atherosclerosis.

Previous CABG

At least 12 studies involving more than 2000 patients report a significantly lower postoperative infarction rate and cardiac mortality in prior CABG patients undergoing noncardiac surgery.[27–29] Data from these studies show the postoperative incidence of MI in these patients to be 0%–1.2% versus 1.1%–6% in patients without prior CABG surgery; mortality is 0.5%–0.9% versus 1%–2.4%. In contrast, studies of simultaneous CABG and noncardiac surgery report higher mortalities (4%–13%) attributed to the unstable nature of either the coronary or vascular disease.[30]

Previous Percutaneous Transluminal Coronary Angioplasty

Coronary angioplasty procedures, initiated approximately 8 years ago, now exceed the number of CABG surgeries performed annually. There is at present no data available describing the effects of angioplasty in patients undergoing subsequent noncardiac surgical procedures. For more complete review of preanesthetic assessment of patients for angioplasty, please see Preanesthetic Assessment 2, Chapter 4 in this series, "The Patient for Percutaneous Transluminal Coronary Angioplasty," by Dr. Alrick Brooks.

Cardiovascular Therapy

The beneficial effects of nitrates, beta blockers, and calcium channel blockers in patients with CAD are well known. Preoperative withdrawal of these therapies is associated with a higher incidence of perioperative ischemia, dysrhythmia, MI, and cardiac death.[31]

Preoperative oral beta-blocker therapy or preinduction intravenous beta-blocker administration decreases the incidence of intraoperative ischemia in both cardiac and noncardiac surgical patients. Larger-scale outcome studies are necessary to identify the patient type that will benefit from prophylactic therapy.

Diagnostic Testing

Diagnostic tests suggested for preoperative assessment of cardiac patients include electrocardiography, chest x-ray, exercise stress testing, echocardiography, radionuclear imaging, and most recently, dipyridamole thallium imaging.

12-Lead Electrocardiography

Preoperative electrocardiographic (ECG) abnormalities occur in 40%–70% of CAD patients. ST-T wave changes occur in 65%–90% of those patients, signs of left ventricular hypertrophy in 10%–20%, and Q waves in 0.5%–8%.[4,5,14]

Notwithstanding the widespread use of preoperative ECG to obtain a baseline profile of patients with suspected or known heart disease, only a few prospective studies have explored its predictive value. Abnormal preoperative ECG was the only statistically significant independent predictor of adverse cardiac outcome found in one study.[4] It was even more predictive than preoperative exercise stress test changes. In contrast, another study found that ECG abnormalities, including old Q waves, ST-T wave changes, or bundle branch blocks, had no significant predictive value.[32]

Chest Radiography

Chest radiography is routinely used in the evaluation of the cardiac patient. The presence of cardiomegaly indicates a low ejection fraction (<0.40) in more than 70% of patients with CAD and is a predictor of perioperative cardiac morbidity.[27] However, some authors dispute this finding, although they acknowledge the identification of a tortuous or calcified aorta on the preoperative chest x-ray as a significant finding.

Exercise Stress Testing

Exercise stress testing is a noninvasive test commonly used to diagnose chest pain of unknown origin and to determine prognosis for patients with known CAD. It is highly predictive of subsequent cardiac events when ST changes are (1) large (>2.5mm); (2) immediate (first 1–3 min); (3) sustained into the recovery period; (4) associated with subnormal increases in blood pressure.

Exercise stress testing has limited value for generalized screening in healthy asymptomatic patients. Preoperative exercise stress testing does not independently predict cardiac risk in noncardiac surgical patients over age 40.[4] However, preoperative exercise stress testing is a more sensitive indicator than the clinical history or preoperative ECG results, as approximately 25% of patients with a negative history and normal preoperative ECG exhibit a positive exercise stress test. Of these patients approximately 25% are at risk of developing preoperative infarction.[4]

Ambulatory ECG Monitoring

Ambulatory ECG monitoring demonstrates frequent episodes of ST depression, which are indicative of subendocardial ischemia and occur during normal daily activities.[33,34] Typically, these episodes are asymptomatic (silent) and probably unrelated to heart rate.[33,35] Of surgical patients with or at risk for CAD, more than 18%–40% have frequent ischemic episodes during the 48 hours preceding surgery.[36] However, more than 75% of these episodes appear to be clinically silent.[36] Preoperative ischemia may predict outcome.[37]

Echocardiography

Precordial echocardiography is a noninvasive, relatively inexpensive imaging technique that is used to assess ventricular and valvular function, regional wall motion and wall thickness, and pericardial tamponade and to diagnose MI left ventricular aneurysms, septal rupture, papillary muscle abnormalities and thrombus formation. Echocardiography provides information similar to radionuclear and angiographic studies but is less invasive and less expensive.

Transesophageal echocardiography, in the awake patient, is used to evaluate left atrial thrombi, valvular vegetation, and prosthetic valvular function and to assess dissecting aortic aneurysm.[38]

Radionuclear Imaging

Radionuclear imaging is used to detect MI, quantify myocardial perfusion abnormalities, and calculate ventricular performance and wall motion indices. Techniques include the use of technetium pyrophosphate (hot-spot imaging) and thallium 201 (cold-spot imaging). Technetium pyrophosphate imaging is sensitive (>90%) and moderately specific (>50%) for detection of acute MI and is most useful 2–3 days after a suspected MI.[39] Thallium imaging is highly sensitive[40] (especially during the first 24 hours after a suspected MI) but not as specific as pyrophosphate imaging.[39] The perfusion defect imaged may be infarction (fixed defect) or ischemia (transient defect). Stress thallium imaging, performed under conditions of near-maximal coronary blood flow (exercise, dipyridamole), is more sensitive than rest imaging and capable of detecting perfusion abnormalities with stenoses as low as 50%.[41]

Exercise thallium scintigraphy reportedly has 90% sensitivity in patients with multivessel disease and 60% in those with single-vessel disease.[42] Dipyridamole thallium imaging is both sensitive (93%) and specific (80%) for detection of coronary stenosis in patients selected for coronary angiography.

During lower-extremity revascularization or abdominal aortic aneurysm resection, the gated-pool—determined ejection fraction has been shown to be an independent predictor of perioperative cardiac morbidity. In one study an ejection fraction of less than 0.35 was associated with a 75%–80% incidence of perioperative MI, and an ejection fraction greater than 0.35 was associated with a 19%–20% incidence.[15] Exercise radionuclear ventriculography in older patients scheduled for elective abdominal or thoracic surgery has been found to be a better predictor of perioperative cardiac morbidity than resting ejection fraction or historical predictors.[43]

Perioperative dipyridamole thallium imaging is highly sensitive (89%–100%), reasonably specific (53%–80%), and superior to historical predictors or exercise stress testing.[44] In patients with a history of angina, prior MI, congestive heart failure, mellitus, an outcome event rate could be predicted in 37% of the cases.[44]

Magnetic Resonance

Magnetic resonance imaging is a relatively noninvasive technique that provides high-resolution tomographic and three-dimensional images. Recent studies in animals and humans indicate that magnetic resonance imaging can reliably detect acute MI, wall thinning and aneurysm formation, and subtle atrial and ventricular defects.[45]

Magnetic resonance spectroscopy can quantify intracellular pH and the levels of high-energy compounds within living cells. Since these diagnostic entities have not been studied in patients undergoing surgery, their perioperative value in patients with CAD is unknown.

Cardiac Catheterization

Cardiac catheterization is the best available tool for quantifying ventricular function and assessing coronary circulation. Ventricular function indices, such as ejection fraction, wall motion abnormalities, end-diastolic volume and change in end-diastolic pressure are predictive of perioperative ventricular dysfunction and short- and longterm outcome.[16] Angiographic findings demonstrating significant left-main or multivessel disease also are predictive of short- and longterm outcome.

There is a relatively high incidence of coronary stenosis, regardless of symptoms, in patients undergoing vascular and general surgery. Patients who had CABG surgery before vascular surgery had lower rates for early (1.5% vs 12%) and late (12% vs 26%) mortality and a higher cumulative 5-year survival rate (72% vs 43%) than patients without prior CABG.[27–30] However, the expense and morbidity associated with cardiac catheterization and the existence of alternative less costly and less risky techniques limit its application, even in high risk patients.

Routine Laboratory Testing

Routine laboratory testing usually includes hemoglobin, hematocrit, electrolytes (sodium, potassium, chloride, BUN, and creatinine), and prothrombin time and partial prothrombin time. These can be used to predict morbidity in the cardiac patient. Patients with elevated BUN and creatinine, following the dye ejection during cardiac catheterization, have an increased incidence of renal failure following open heart surgery.

Preoperative Hypokalemia

Anesthetic management of the hypokalemic patient deals with the potential risk of cardiac dysrhythmia. Clinical reports suggest that dysrhythmias are worsened by respiratory alkalosis, digitalis therapy, and myocardial ischemia or infarction in the presence of hypokalemia. Therefore, it has been traditionally taught that a surgical patient should have a serum potassium of 3.0mEq/L (undigitalized) and 3.5mEq/L (digitalized) before being subjected to anesthesia.

Clinical experience has shown that hypokalemic patients have undergone a variety of anesthetic regimens without serious cardiac dysrhythmias or instability. Malignant intraoperative dysrhythmias or inability to resuscitate patients who were hypokalemic have been reported.[46]

Hypokalemia is a far more complex problem both physiologically and clinically than has been appreciated. A few facts deserve special consideration. The most common cause of hypokalemia in the surgical patient is diuretic therapy for the treatment of systemic hypertension. A chronic hypokalemic level of just below 3.0mEq/L may be tolerated, but below 2.5mEq/L it is associated with cardiovascular dysfunction.

Clinically, hypokalemia is only a measure of intravascular potassium concentration per unit volume. Hemodilution or hemoconcentration can decrease or increase serum K^+, respectively, and serum K^+ per se may not reflect the total body potassium concentration. The electrophysiology of excitable cells is not solely dependent on transmembrane K^+ gradient but also depends on a complex relationship among the cations of potassium, sodium, calcium, and magnesium. The generalization that alkalosis induces transmembrane shift of potassium ions into the cell and acidosis induces the opposite is valid only during acute hypocapnia and hypercapnia, respectively.

Recent prospective studies in hypokalemic surgical patients suggest that hypokalemia per se is not a reason to cancel an elective case, but these patients should not be treated like normokalemic patients. Rather, supplemental potassium should be given.

Intraoperative Predictors of Perioperative Cardiac Morbidity

Choice of Anesthesia

The use of inhalational versus opioid anesthesia in patients undergoing CABG is still debated. However, most outcome studies have not demonstrated a difference in beneficial outcome between these types of anesthetics in the patient with cardiac disease. Potent inhaled anesthetics are especially useful in those with CAD and normal ventricular function. Their advantages include the ability to produce all of the objectives of anesthesia: unconsciousness, muscular relaxation, suppression of reflex responses to noxious stimulation, rapid recovery of ventilatory function, and dose-related decreases in ventricular work and oxygen consumption. Principal disadvantages are excessive cardiovascular depression under some conditions, lack of analgesia in subanesthetic concentrations during the recovery phase, and postoperative shivering and increased oxygen demand due to heat loss and peripheral vasodilation.

The potential for inducing coronary artery steal in cardiac patients has provoked arguments for restricting the use of isoflurane. A moderate coronary vasodilator, isoflurane may cause coronary steal in patients with coronary artery stenosis and steal-prone anatomy.[47] Isoflurane produces moderate coronary vasodilation of the epicardial resistance vessels, less vasodilation than adenosine but more than halothane or enflurane.[47]

Nitrous oxide tends to decrease cardiac output and increase systemic vascular resistance when given alone or in combination with opioid analgesics. It has been shown that nitrous oxide can induce ischemia in an area supplied by a critically stenotic coronary artery and thereby induce regional myocardial dysfunction. Opioid/nitrous oxide relaxant anesthesia was associated with a significantly higher incidence of myocardial reinfarction, 7.0% versus 0.5%–1.5% for other general anesthetics.[48]

The narcotic analgesics have become the anesthetic mainstay for patients with impaired cardiac performance. Currently, fentanyl and sufentanil are the drugs most commonly used as primary narcotic anesthetics. The advantages are as follows: Absence of direct effects on the heart, with no change in contractility, automaticity (but increased vagal activity), conduction, or sensitivity to catecholamines. There is no interference with autonomic or cardiovascular drug action, and blood flow autoregulation in the central nervous system, heart, and kidneys is preserved. Increased tolerance of endotracheal tube and airway manipulation is present, as is postoperative analgesia.

The disadvantages include bradycardia and hypotension during induction, muscular rigidity during induction and sometimes during emergence, and prolonged recovery time, especially to spontaneous ventilation.

There is no ideal anesthetic management for patients presenting for CABG. Techniques are based on a combination of hypnotic (lorazepam, diazepam, midazolam), opioid (fentanyl, sufentanil), and muscle relaxant (vecuronium, pancuronium), with small concentrations of inhalation anesthetics and oxygen. All anesthetic techniques aim to maintain the balance between the oxygen supply and demand of the heart. Increases in heart rate, preload, afterload, contractility, and wall tension raise the oxygen demand. The oxygen supply of the heart depends on the extent of the CAD, hemoglobin concentration, and the oxygen-hemoglobin dissociation curve.

Surgical Factors

Patients undergoing thoracic or upper abdominal surgery have a two- to threefold higher risk of perioperative cardiac complication.[7,8,11] Patients with CAD undergoing major vascular surgery are unquestionably at increased risk for perioperative MI, congestive heart failure, and cardiac death. Also, procedures lasting more than 3 hours are associated with greater mortality.[7,14] Most studies support the thesis that emergency surgery increases the risk two- to fivefold. The three major considerations with coronary bypass are prevention of MI with effective myocardial preservation; the method of improvement and storage of the vein graphs used, usually the saphenous veins; and a meticulous operative technique that constructs anastomoses with smooth internal surfaces without stenoses. Appropriate medical management and monitoring prior to surgery are important in preventing MI. One of the major advances of the 1970's was the development of hypothermic potassium cardioplegia. At present, there are both a large number of cardioplegia solutions and a wide range in the technique of application. Cardioplegia solutions are prepared by mixing the patient's blood with a previously prepared solution made of plasmalyte, mannitol, and sodium bicarbonate. To date, there is no uniform agreement as to the best technique. Myocardial temperature is carefully monitored, and the myocardium is bathed constantly with cold fluid, keeping the temperature below 15°C.

Distal anastomoses are performed first while serially injecting cold blood cardioplegia every 20–30 minutes into the ascending aorta while the aorta is clamped. The aorta clamp is then removed, and the proximal anastomoses are performed.

A left ventricular vent is ordinarily not used but particular care is taken to keep the right atrium decompressed. The aorta may be safely occluded for more than 3 hours in patients with valvular disease and for at least 2 hours in patients with CAD. Distention of the left ventricle must be scrupulously avoided. The potassium-arrested heart in diastole seems especially vulnerable to stretch injury.

The heart is most vulnerable to superfusion edema following release of the aortic clamp; it is being reperfused and rewarmed and beginning to perform cardiac work again, especially as bypass is slowed and stopped. The first hour after bypass is probably the most hazardous for the development of myocardial edema, as the heart resumes its normal workload.

A cardiac index (CI) near 2L/min or lower indicates a serious depression of cardiac output and the necessity for cardiac support, usually by insertion of a left ventricular vent and reinstitution of bypass fo 30–60 minutes until cardiac function has improved. A persistent depression of CI below 2L/min is a firm indication for use of an intraaortic balloon pump. Systolic blood pressure is maintained near 100–120mmHg. Higher pressures are avoided by use of sodium nitroprusside and nitroglycerin. If inotropic support is necessary, epinephrine and dobutamine are added to the regimen.

Intraoperative Cardiovascular Changes

Acute hypertension affects both myocardial oxygen demand and supply. During systemic hypertension, wall tension increases, which increases myocardial oxygen consumption. At any given heart rate, regional ventricular function is better when blood pressure is elevated. However, in the failing ventricle, the increases in the end-diastolic pressure may exceed the increases in the arterial diastolic pressure and decrease coronary perfusion pressure. In addition, sympathetic coronary constriction during the hypertensive episode may decrease coronary flow. Most studies suggest that fewer than 15% of ischemic episodes are associated with hypertension,[11,36] but some have shown that acute hypertensive episodes precede as many as 50% of intraoperative ischemic episodes. Thus, the predictive importance of intraoperative hypertension for perioperative cardiac morbidity is unresolved.

Hypotension reduces myocardial wall tension, decreasing oxygen demand; but as diastolic blood pressure falls below the autoregulatory limit, coronary blood flow decreases. Numerous studies have investigated the relationships between intraoperative hypotension and MI in cardiac and noncardiac patients.[36] One study reported a significantly higher reinfarction rate (15.2% vs 3.2%) among patients who developed intraoperative systolic hypotension ($\geq$30%, $\geq$10min),[7] and another study[11] found that intraoperative hypotension was the strongest dynamic predictor of perioperative MI.

Increases in heart rate deleteriously affect myocardial oxygen supply and demand. Studies in anesthetized patients undergoing cardiac and noncardiac surgery have demonstrated a causal relationship between intraoperative tachycardia and intraoperative ischemia.[36]

Intraoperative Monitoring for Ischemia

Intraoperative ECG changes consistent with MI are present in 18%–74% of noncardiac surgical patients with CAD.[15–22,24] Most changes are ST depression; ST elevation appears to be uncommon.[15,20] Most ST changes occur laterally, with 90% in leads V_4 and V_5.[15] Reversible ST changes are likely to be ischemic. In patients undergoing CABG, prebypass ischemia increases the risk of MI by two- to threefold. Segmental wall-motion and wall-thickening abnormalities are more sensitive and earlier indices of MI than ECG changes.

Most ECG changes are accompanied by transesophageal echocardiography (TEE) changes, but the converse has not been reported. TEE wall-motion and wall-thickening changes indicative of MI, even when unaccompanied by ECG changes, are predictive of perioperative cardiac morbidity.

Pulmonary artery monitoring provides useful information in assessing ventricular function. During acute coronary occlusion, exercise precipitates ECG ST changes and early and marked increases in left ventricular end-diastolic pressure. It has been shown that left ventricular end-diastolic pressure increases during ischemia because of the effects of end-atrial systolic emptying on the stiffened and ischemic left ventricle but that these increases are not reflected in the mean left ventricular diastolic pressure, the left atrial pressure, or the pulmonary capillary wedge pressure (PCWP). Although acute increases in PCWP (or development of V waves) may reflect ischemia,[49] the absence of a change in PCWP does not ensure the absence of ischemia.

Radionuclear imaging. Nuclear imaging techniques are primarily research tools for assessing ischemia as well as ventricular function. Preoperatively, they do not have widespread clinical application.

Cardiokymography is a noninvasive technique that allows analog representation of anterior wall motion. The probe is a capacitive plate placed over the chest wall; it emits a low-energy, high-frequency (10mHz) electromagnetic field. Motion within the field produces a change in capacitance and, therefore, frequency of the oscillation, which is converted to the output voltage signal. Its limitations include the inability to detract wall motion that is not anterior, the interference of noise produced by other artifactual motion, and the inability to maintain probe position during prolonged surgery or thoracic surgery. Previous studies in patients have demonstrated that cardiokymography is more sensitive and specific indicator of CAD than the ECG.[50] The relationships of cardiokymography changes to TEE wall-motion abnormalities or to adverse cardiac outcome are unknown.

Biochemical markers. Lactate production is one of the most accurate measures of MI. However, because of the complex relationship between lactate uptake and production, serum lactate measurement is an insensitive marker. Radio-labeled lactate determinations permit accurate differentiation between uptake and production.

Ventricular Dysfunction

Increased ventricular filling pressure, associated with ventricular dysfunction, deleteriously affects both myocardial oxygen supply and demand. Ischemia can precipitate ventricular dysfunction and increase end-diastolic pressure, particularly with severe coronary artery stenosis when myocardial oxygen demand is increased.

Studies using noncontinuous ECG recording techniques report that the incidence of dysrhythmias during noncardiac surgery varies from 0.9%–70%. Continuous ECG recording was used in two studies of dysrhythmias in noncardiac surgical patients and the incidence was 70%, and 28% were ventricular. The incidence was higher during general versus regional anesthesia (66% vs 52%), neurologic and thoracic surgery versus peripheral surgery (100%, 90% vs 56%), and in intubated versus nonintubated patients (72% vs 44%). Preexisting heart disease did not influence the incidence of dysrhythmias (62% vs 59%).

Other studies found no correlation between the incidence or type of dysrhythmias (other than tachycardia) and perioperative reinfarction.[11,32]

Postoperative Predictors of Perioperative Cardiac Morbidity

The postoperative period can be stressful because of the onset of pain during emergence from anesthesia, fluid shifts, temperature changes, and alteration of respiration function. Marked changes occur in plasma catecholamine concentrations, hemodynamics, ventricular function, and coagulation following noncardiac surgery, particularly in patients with preexisting cardiac disease.

Recent studies in both cardiac and noncardiac[36] surgery have shown that heart rate commonly increases postoperatively by 25%–50% over intraoperative values and that tachycardia (heart rate >100beats/min) occurs in 10%–25% of patients. Preliminary studies suggest that ischemia occurs most commonly during the postoperative period and persists for 48 hours[36] or longer. Postoperative ischemia and most infarcts appear to be silent and therefore difficult to detect.

If postoperative ischemia is proved to be an important predictor of morbidity, extended postoperative monitoring and aggressive treatment of ischemia would be indicated, thereby altering postoperative practice.

Summary

CABG for CAD is an intricacy of preoperative, intraoperative, and postoperative management and techniques. The anesthesiologist's role is an aggressive one in all aspects of care for the coronary artery bypass patient. The team approach, adequate communication, and continuous interaction among the anesthesiologist, surgeon, and cardiologist are vital to optimal-quality care of the cardiac patient.

References

1. Mangano D.T.: Assessment of the patient with ischemic heart disease, in Stanley T.H., Sperry R.J. (eds): *Anesthesiology and the Heart.* Dordrecht, The Netherlands: Kluwer Academic Publishers, 1990, pp 1–11.
2. Port S., Cobb F.R., Coleman R.E., Jones R.H.: Effect of age on the response of the left ventricular function to exercise. *N Engl J Med* 1980, 303:1133–7.
3. Djokovic J.L., Hedley-Whyte J.: Prediction of outcome of surgery and anesthesia in patients over 80. *JAMA* 1979, 242:2301–6.
4. Carliner N.H., Fisher M.L., Plotnick G.D., Gabart H., Rapoport A., Kelemen M.H., Moran G.W., Gadacz T., Peters R.W.: Routine preoperative exercise testing in patients undergoing major noncardiac surgery. *Am J Cardiol* 1985, 56:51–7.
5. Driscoll A.C., Hobika J.H., Etsten B.E., Proger S.: Clinically unrecognized myocardial infarction following surgery. *N Engl J Med* 1961, 264:633–9.
6. Mohr D.N.: Estimation of surgical risk in the elderly: A correlative review. *J Am Geriatr Soc* 1983, 31:99–102.
7. Steen P.A., Tinker J.H., Tarhan S.: Myocardial reinfarction after anesthesia and surgery. *JAMA* 1978, 239:2566–70.
8. Tarhan S., Moffitt E., Taylor W.F., Giuliani E.R.: Myocardial infarction after general anesthesia. *JAMA* 1972, 220:1451–4.
9. Plumlee J.E., Boettner R.B.: Myocardial infarction during and following anesthesia and operation. *South Med J* 1972, 65:886–9.
10. Goldman L.: Cardiac risks and complications of noncardiac surgery. *Ann Intern Med* 1983, 98:504–13.
11. Rao T.K., Jacobs K.H., El-Etr A.A.: Reinfarction following anesthesia in patients with myocardial infarction. *Anesthesiology* 1983, 59:499–505.
12. Diamond G.A., Forrester J.S.: Analysis of probability as an aid in the clinical diagnosis of coronary artery disease. *N Engl J Med* 1979, 300:1350–8.
13. Sapala J.A., Ponka J.L., Duvernoy W.F.C.: Operative and nonoperative risks in the cardiac patient. *J Am Geriatr Soc* 1975, 23:529–34.
14. Goldman L., Caldera D.L., Nussbaum S.R., Southwick F.S., Krogstad D., Murray B., Burke D.S., O'Malley T.A., Goroll A.H., Caplan C.H., Nolan J., Carabello B., Slater E.E.: Multifactorial index of cardiac risk in noncardiac surgical procedures. *N Engl J Med* 1977, 297:845–50.

15. Pasternack P.F., Imparato A.M., Bear G., Baumann F.G., Benjamin D., Sanger J., Kramer E., Wood R.P.: The value of radionuclide angiography as a predictor of perioperative myocardial infarction in patients undergoing abdominal aortic aneurysm resection. *J Vasc Surg* 1984, 1:320–5.
16. Mangano D.T.: Biventricular function after myocardial revascularization in humans: Deterioration and recovery patterns during the first 24 hours. *Anesthesiology* 1985, 62:571–7.
17. Hypertension Detection and Follow-up Program Cooperative Group: Five-year findings of the hypertension detection and follow-up program: I. Reduction in mortality of persons with high blood pressure, including mild hypertension. *JAMA* 1979, 242:2562–72.
18. Roizen M.F.: Anesthetic implications of concurrent diseases. In Miller R.D., (ed): *Anesthesia,* 3rd ed., New York: Churchill Livingstone, 1990, p 797.
19. Beard O.W., Hipp H.R., Robins M., Verzolini V.R.: Initial myocardial infarction among veterans: Ten-year survival. *Am Heart J* 1967, 73:317–21.
20. Fein F.S., Sonnenblick E.H.: Diabetic cardiomyopathy. *Prog Cardiovasc Dis* 1985, 27:255–70.
21. Olson H.G., Lyons K.P., Troope P., Butman S., Piters K.M.: The high-risk acute myocardial infarction patient at 1-year followup: Identification at hospital discharge by ambulatory electrocardiography and radionuclide ventriculography. *Am Heart J* 1984, 107:358–66.
22. Rooney S.M., Goldiner P.L., Muss E.: Relationship of right bundle-branch block and marked left axis deviation to complete heart block during general anesthesia. *Anesthesiology* 1876, 44:64–6.
23. Hertzer N.R.: Myocardial ischemia. *Surgery* 1983, 93:97–101.
24. Jensen J., Blankenhorn D.H., Kornerup V.: Coronary disease in familial hypercholesterolemia. *Circulation* 1967, 36:77–82.
25. Anderson K.M., Castelli W.P., Levy D.: Cholesterol and mortality: 30 years of follow-up from the Framingham Study. *JAMA* 1987, 257:2176–80.
26. Stamler J., Wentworth D., Neaton J.D. (for the MRFIT Research Group): Is relationship between serum cholesterol and risk premature death from coronary heart disease continuous and graded? Findings in 356,222 primary screenings of the Multiple Risk Factor Intervention Trial (MRFIT). *JAMA* 1986, 256:2823.
27. Foster E.D., David K.B., Carpenter J.A., Abele S., Fray D.: Risk of noncardiac operation in patients with defined coronary disease: The Coronary Artery Surgery Study (CASS) Registry Experience. *Ann Thorac Surg* 1986, 41:42–50.
28. Schoeppel L.S., Wilkinson C., Waters J., Meyers N.S.: Effects of myocardial infarction on perioperative cardiac complications. *Anesth Analg* 1983, 62:493–8.
29. Wells P., Kaplan J.A.: Optimal management of patients with ischemic heart disease for noncardiac surgery by complementary anesthesiologist and cardiologist interaction. *Am Heart J* 1981, 102:1029–37.

30. Reul G.J. Jr., Cooley D.A., Duncan J.M., Frazier O.H., Ott D.A., Livesay J.J., Walker W.E.: The effect of coronary bypass on the outcome of peripheral vascular operations in 1,093 patients. *J Vasc Surg* 1986, 3:788–98.
31. Engelman R.M., Hadji-Rousou I., Breyer R.H., Whittredge P., Harbison W., Chircop R.V.: Rebound vasospasm after coronary revascularization in association with calcium antagonist withdrawal. *Ann Thorac Surg* 1984, 37: 469–72.
32. Goldman L., Caldera D.L., Southwick F.S., Nussbaum S.R., Murray B., O'Malley T.A., Goroll A.H., Caplan C.H., Nolan J., Burke D.S., Krogstad D., Carabello B., Slater E.E.: Cardiac risk factors and complications in non-cardiac surgery. *Medicine* (Baltimore) 1978, 57:357–70.
33. Deanfield J.E., Maseri A., Selwyn A.P., Chierchia S., Ribeiro P., Krikler S.: Myocardial ischaemia during daily life in patients with stable angina: Its relation to symptoms and heart rate changes. *Lancet* 1983, 2:753–8.
34. Cohn P.F., Lawson W.E.: Characteristics of silent myocardial ischemia during out-of-hospital activities in asymptomatic angiographically documented coronary-artery disease. *Am J Cardiol* 1987, 59:746–9.
35. Chierchia S., Lazzari M., Freedman B., Brunelli C., Maseri A.: Impairment of myocardial perfusion and function during painless myocardial ischemia. *J Am Coll Cardiol* 1983, 1:924–30.
36. Fegert G., Hollenberg M., Browner W., Wellington Y., Levenson L., Franks M., Harris D., Mangano D.: Perioperative myocardial ischemia in the non-cardiac surgical patient (abstract). *Anesthesiology* 1988, 69:A49.
37. Raby K.E., Goldman L., Creager M.A., Cook E.F., Weisberg M.C., Whittemore A.D., Selwyn A.P.: Correlation between preoperative ischemia and major cardiac events after peripheral vascular surgery. *N Engl J Med* 1989, 321:1296–1300.
38. Steward J.B., Khandheria B.K., Oh J.K., Abel M.D., Hughes R.W., Edwards W.D., Nichols B.A., Freeman W.K., Tajik A.J.: Transesophageal echocardiography. Technique, anatomic correlations, implementation, and clinical applications. *Mayo Clin Proc* 1988, 63:649–80.
39. Stokely E.M., Buja L.M., Lewis S.E., Parkey R.W., Bonte F.J., Harris R.A. Jr., Willerson J.T.: Measurement of acute myocardial infarcts in dogs with 99 m Tc-stannous pyrophosphate scintigrams. *J Nucl Med* 1975, 17:1–5.
40. Holman B.L., Wynne J.: Infarct avid (hot-spot) myocardial scintigraphy. *Radiol Clin North Am* 1980, 18:487–99.
41. Cahalan M.K., Litt L., Botvinick E.H., Schiller N.B.: Advances in noninvasive cardiovascular imaging: Implications for the anesthesiologist. *Anesthesiology* 1987, 66:356–72.
42. Dash H., Massie B.M., Botvinick E.H., Brundage B.H.: The noninvasive identification of left main and three vessel coronary artery disease by myocardial stress perfusion scintigraphy and treadmill exercise electrocardiography. *Circulation* 1979, 60:276–84.

43. Gerson M.C., Hurst J.M., Hertzberg V.S., Doogan P.A., Cochran M.B., Lim S.P., McCall N., Adolph R.J.: Cardiac prognosis in noncardiac geriatric surgery. *Ann Intern Med* 1985, 103:832–7.
44. Eagle K.A., Singer D.E., Brewster D.C., Darling R.C., Mulley A.G., Boucher C.A.: Dipyridomole-thallium scanning in patients, undergoing vascular surgery. *JAMA* 1987, 257:2185–9.
45. McNamara M.T., Higgins C.B., Schechtmann N., Botvinick E., Lipton M.J., Chatterjee K., Amparo E.G.: Detection and characterization of acute myocardial infarction in man with use of gated magnetic resonance. *Circulation* 1985, 71:717–24.
46. Wong K.C., Shultz J.R.: Preoperative hypokalemia and the cardiac patient, in Stanley T.H., Sperry R.J. (eds): *Anesthesiology and the Heart.* Dordrecht, The Netherlands: Kluwer Academic Publishers, 1990, pp 111–8.
47. Merin R.G., Basch S.: Are the myocardial functional and metabolic effects of isoflurane really different from those of halothane and enflurane? *Anesthesiology* 1981, 55:398–408.
48. Philbin D.M., Foex P., Drummond G., et al.: Post systolic shortening of canine left ventricle supplied by a stenotic coronary artery when nitrous oxide is added in the presence of narcotics. *Anesthesiology* 1985, 62:166.
49. Kaplan J.A., Wells P.H.: Early diagnosis of myocardial ischemia using the pulmonary arterial catheter. *Anesth Analg* 1981, 60:789–93.
50. Bellows W.H., Bode R.H. Jr., Levy J.H., Foex P., Lowenstein E.: Noninvasive detection of peri-induction ischemic ventricular dysfunction by cardiokymography in humans: Preliminary experience. *Anesthesiology* 1984, 60:155–8.

Chapter 21

Multiple Gestations

Steven S. Schwalbe, M.D.

Case History. *A 23-year-old secundigravida with a known triplet gestation was admitted at 35 weeks gestation with rupture of membranes and mild uterine contractions. A previous pregnancy 2 years earlier had been delivered by cesarean section because of arrest of labor. The patient gave a history of occasional attacks of fainting during the last months of this pregnancy but could not give any details. Her physical examination and vital signs were normal; blood pressure was 126/62mmHg while sitting, heart rate was 96/min and regular, and respiratory rate was 20/min. Sonography revealed three fetuses in breech presentations with estimated weights of approximately 2000gm each. Biophysical profile and nonstress tests showed good fetal responses. However, two transient episodes of maternal hypotension to 80/50mmHg occurred during the fetal surveillance; these appeared to be related to changes in maternal position.*

A cesarean section was planned because of breech presentations of the fetuses.

Introduction

Multiple gestations are relatively infrequent occurrences in obstetrics. Twins account for only 1% of all deliveries. Triplets occur once in 9800 births and quadruplets only once in 70,000. Yet multiple gestations are associated with disproportionately high rates of maternal, fetal, and neonatal morbidity and mortality.[1]

Twin gestations may be single-ovum (monozygotic) or double-ovum (dizygotic). The rate of monozygotic twinning is approximately 0.35% of all deliveries and is unaffected by differences in ethnicity or maternal age. The rate of dizygotic twinning, on the other hand, is about 0.65%

Reviewed by Dr. Akolisa Anyaegbunam, Department of OB/GYN, Albert Einstein College of Medicine.

and increases with maternal age, parity, and a positive family history of multiple gestations. Race, ethnic background of the mother, and the use of ovulatory drugs also influence the incidence of dizygotic twinning.

Dizygotic twins will always have two separate chorions and two amnions (Figure 1 [A and B]).

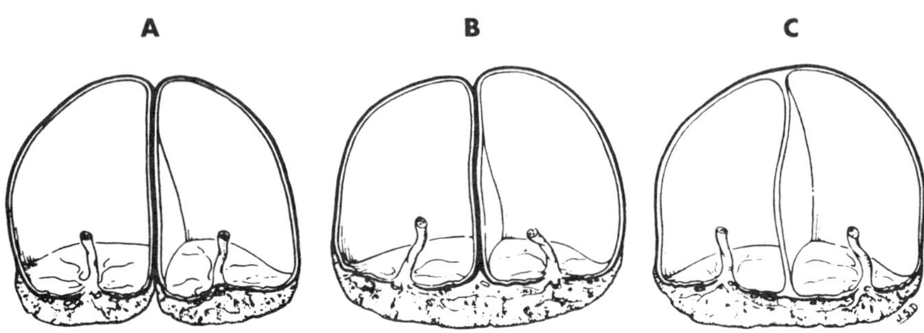

FIGURE 1. Placenta and membranes in twin pregnancies. A. Two placentas, two amnions, two chorions (from either dizygotic twins or monozygotic twins with cleavage of zygote during first 3 days after fertilization). B. Single placenta, two amnions, and two chorions (from either dizygotic twins or monozygotic twins with cleavage of zygote during first 3 days.) C. One placenta, one chorion, two amnions (monozygotic twins with cleavage of zygote from the 4th to the 8th day after fertilization). (Reprinted by permission of the publisher, from Cunningham, MacDonald and Gant: *Williams Obstetrics*. 18th Edition. Appleton and Lange 1989.)

Monozygotic twin membranes may show one of several variations, depending on when division of the fertilized ovum occurs. If division occurs within the first 72 hours after fertilization, a diamnionic, dichorionic pregnancy will result, similar in structure to that of dizygotic twinning. There may be two distinct placentas (A) or a single fused placenta (B). Between the 4th and 8th days of development, differentiation of the chorion occurs. If division occurs at this point, two amnionic sacs will develop covered by a common chorion—a diamnionic, monochorionic pregnancy (C). By the 8th day of gestation the amnion is established. Division of the zygote after this point will result in two embryos within a common amnionic sac—a monoamnionic, monochorionic twin pregnancy. When division occurs much later, after the formation of the embryonic disk, there will be incomplete cleavage, resulting in conjoined twins.

Fetal Risks

The perinatal mortality rate in multiple-gestation pregnancies is 3 to 10 times higher than that for singleton births, with severe perinatal morbidity or death ensuing in 5% to 15% of all twins.[2,3] Triplets may be at even greater risk and are five times more likely than twins to require neonatal intensive care.[4] This adverse outcome is related primarily, but not exclusively, to prematurity and intrauterine growth retardation.[3,5] Over half of these babies exhibit low birth weight (below 2500gm). One third of twins and 87% of triplets are delivered prematurely.[4,6] These preterm infants account for 90% of all neonatal deaths in multiple-gestation pregnancies. Necrotizing enterocolitis, hyaline membrane disease, and other manifestations of prematurity account for the vast majority of such deaths.

The incidence of congenital abnormalities in these premature births is three times the rate seen in singleton pregnancies and is the third leading cause of neonatal mortality.[7,8] The higher frequency of malformations is primarily seen in monozygotic (identical) twins, with the highest incidences occurring in monoamniotic twins.[9] Associated defects include macrocephaly, encephalocele, cardiac septal defects, malformations of the gastrointestinal tract, tracheoesophageal fistulas, cleft lip and palate, and cystic kidneys among others.

Monozygosity predisposes toward two other important fetal complications. Monoamniotic fetuses are at risk for intertwining and occlusion of umbilical vessels. Fetuses with monochorionic placentas (one placental disk, one chorion, one or more amnions) can develop arteriovenous connections between the fetal circulations, leading to the fetofetal transfusion or "twin-twin transfusion" syndrome. This syndrome has been diagnosed in 5.5% to 14.6% of monochorionic multiple pregnancies. It is classically characterized by a small, anemic fetus (the donor) and a large plethoric co-twin (the recipient). The donor becomes malnourished and growth retarded, and the recipient may develop congestive heart failure, polycythemia, and polyhydramnios. Overall mortality rates have been reported between 35% and 70%.[10,11]

Although dichorionic placentas are without the communications seen in the fetofetal transfusion syndrome, they are not immune to vascular abnormalities. About 7% of all twin placentas have a velamentous insertion of the umbilical cord with vasa previa.[1] Such unprotected vessels may be easily torn, resulting in fetal hemorrhage.

Other complications include abruptio placentae, placenta previa, and premature rupture of membranes with or without amnionitis. These problems are all frequently associated with infants weighing less than 1500gm.[12]

Maternal Risks

Maternal risks are increased with multiple fetal gestations. Maternal mortality is two to three times that of a singleton pregnancy. Some of the reasons have already been mentioned. Abruptio placentae and placenta previa carry a risk of maternal hemorrhage. Premature rupture of membranes may lead to chorioamnionitis. Preeclampsia-eclampsia occurs more frequently in cases of multiple gestation and potentially carries severe consequences for mother and baby. Although increases in maternal blood volume are about 40% greater in multiple-gestation pregnancies, anemia occurs more frequently and is more severe.[13] The blood loss at delivery averages twice that for singleton deliveries and manual extraction of the placenta is required twice as often. The uterus is larger, partly to accommodate the additional fetuses and amniotic cavities and partly because of the high incidence of hydramnios accompanying multiple gestations.

The mechanical effects of the enlarged uterus normally seen with pregnancy tend to be even more exaggerated. Nausea and vomiting are more frequent, and the risk of aspiration pneumonitis is greater as gastric distortion is increased.

As the enlarging uterus causes the diaphragm to move cephalad, the closing capacity may become greater than the functional residual capacity. The difference between these volumes is more than that in the singleton term parturient. Sensations of dyspnea are increased, the A-a gradient may be larger, and oxygen reserve is diminished, particularly in the supine or Trendelenburg position.[14] The larger uterus also predisposes the parturient to more aortocaval compression. Varicosities and edema in the legs tend to be greater, and the incidence of supine hypotension is more frequent. Intrapartum, one can expect a higher incidence of obstetric trauma. Postpartum, there is a greater tendency toward uterine atony and postpartum hemorrhage.

Obstetric Considerations

Proper obstetric management of such cases had in the past been complicated by the fact that in one of four cases of multiple-gestation the diagnosis was not made until either the onset of labor or delivery of the first child.[8] The increasing use of ultrasonography has dramatically changed the ability to diagnose multiple gestation in early pregnancy.[15] Ultrasound is also helpful in determining placentation, diagnosing certain complications and identifying fetal lie.[4]

Having made the diagnosis, the obstetrician may wish to conduct the antepartum management in such a way as to reduce the likelihood of preterm delivery. Unfortunately, there is at present no clearly effective way to prevent this complication. Bed rest has been advocated as a simple and relatively risk-free method to reduce the incidence of premature labor. Data regarding its efficacy are conflicting.[2,16] Nevertheless, its simplicity and safety have made prophylactic bed rest a commonly prescribed practice. Cerclage is a controversial management technique that has been used to prevent preterm delivery. However, unless there is evidence of an incompetent cervix, its use in multiple-gestation pregnancies is probably unjustified.[17] Prophylactic tocolysis with a β_2-adrenergic agent such as ritodrine or terbutaline is sometimes used to prevent preterm labor. Once again, although study data regarding the effectiveness of this method are often in conflict, some centers have reported using it successfully.[17] If tocolysis fails, steroids such as dexamethasone (6mg q6h) or betamethasone (12mg q12h) are often administered over a 24-hour period to help promote fetal lung maturity.

The death of one fetus in the second or third trimester of a multiple-gestation pregnancy may place the other fetuses as well as the mother in jeopardy. The living fetuses are at risk for either complications from the same factors that caused the demise of the first baby or from twin embolization syndrome (TES). In TES passage of thromboplastic material or placental fragments through the placenta into the circulation of the surviving twin can result in ischemic structural defects of the central nervous system, the gastrointestinal tract, and the genitourinary system.[18] The gravida is at risk of developing a disseminated intravascular coagulopathy. The degree of risk is difficult to ascertain because this is not a common event, but it appears to be related to the zygosity of the twins (monozygotic twins being at greater risk) and the gestational age at the time of the demise.[19] There is evidence to suggest that the death of a fetus in the first half of pregnancy may be less likely to affect its siblings.[20,21]

The mode of delivery chosen for multiple-gestation pregnancies varies with the number of fetuses, the fetal lie, (Table 1) the state of fetal well-being, the presence of associated conditions, the estimated fetal weight, and the experience of the obstetrician.

TABLE 1. Presentation of Fetuses in Twin Delivery

Presentation	**% Deliveries**
Both vertex	39
Vertex and breech	37
Both breech	10
Longitudinal and shoulder	8
Both shoulder	6

(Reprinted by permission of the publisher, from Shnider and Levinson: *Anesthesia for Obstetrics.* 2nd edition. Williams and Wilkins 1987.)

One recent review of 198 cases of triplets listed cesarean section as the delivery method in 94% of cases.[17] In the case of twins there is general agreement that a vaginal delivery is appropriate when both twins are vertex presentations and that cesarean section is indicated if the first twin is in a non-vertex lie.[72] For a vertex-nonvertex combination in which the estimated fetal weights are less than 2000gm (consistent with a gestational age of less than 34 weeks), it is often considered prudent to deliver by cesarean section.

But approximately one third of twin deliveries are at term or near-term with vertex-nonvertex presentations. There is considerably less agreement regarding the optimal method of delivery in these cases. Following vaginal delivery of the first child, vaginal breech delivery or external version may be attempted on the nonvertex baby. Some obstetricians prefer the use of outlet forceps to assist in this procedure. However, evidence of fetal distress, vaginal bleeding, or a prolapsed cord may require the second twin to be delivered emergently by cesarean section. Many obstetricians plan on an abdominal delivery of both twins from the outset in an effort to avoid these complications and excessive birth trauma.

Anesthetic Plan

The parturient should be evaluated as early as possible. Ideally, the first contact between the pregnant woman and the anesthesiologist should occur in the prenatal period long before the planned delivery date. This not only affords the opportunity to request any necessary consultation or laboratory studies but enables the anesthesiologist to have a frank discussion with the gravida regarding the pros and cons of the available methods of anesthesia. Unfortunately this is not always practical or possible, and many women will not be seen until after their admission to the labor unit. In such cases the anesthesiologist should make every attempt to speak with the woman as soon as possible. Anesthesia may be required emergently for a breech delivery, forceps delivery, cesarean section or version. A successful outcome requires knowledge of the maternal and fetal conditions, availability of proper equipment, availability of obstetric and anesthetic personnel, and good communication between the members of the delivery team. In addition, there must be enough equipment and skilled personnel present to handle resuscitation of all of the neonates.

The preanesthetic evaluation should take into account all of the aspects one normally considers in the labor suite, but special consideration must be paid to conditions associated with or exaggerated by a multiple-gestation pregnancy. For example, a woman with mitral stenosis is sensitive to increases in intravascular volume and cardiac output. Her clinical condition will often worsen with parturition. An exaggerated degree of physiologic impairment may occur with a multiple-gestation pregnancy. Women with obesity or airway disease whose respiratory function may be compromised with pregnancy experience correspondingly greater adverse effects with an overdistended uterus. The degree of anemia and the potential for increased blood loss in such pregnancies must be recognized.

The increased frequency and severity of aortocaval compression is often overlooked. There is a tendency to forget that if maternal blood pressure is normal, it does not necessarily follow that aortocaval compression is absent. Compensatory mechanisms, such as collatoral circulation, increased heart rate, and peripheral vasoconstriction may mask a fall in cardiac output resulting from a partial obstruction of the inferior vena cava. However, even in the absence of caval compression, uterine blood flow and fetal well-being may be severely compromised by obstruction of the lower aorta. It is common practice to relieve aortocaval compression by the use of left uterine displacement, tilting the mother's pelvis or body 15° to the left. This maneuver is not always optimal. Greater degrees of tilt or a different position may be necessary.

In singleton pregnancies, up to 12% of women may demonstrate better results in the right semilateral position than in the left semilateral.[23] In a gravida with an overdistended uterus, it can be of particular importance to assess the adequacy of the maternal position either by taking blood pressure in the lower extremities as well as in the arms or by serial noninvasive cardiac output measurements.[24] If adequate fetal monitoring is in place, the effect of a change in maternal posture on the fetal heart rates can be readily demonstrated.

Labor and Vaginal Delivery

At one time there was controversy regarding the effects of epidural block in multiple gestation. Obstetricians were concerned that the interval between deliveries of the neonates would be prolonged. Crawford, in a study of 200 twin deliveries, showed that not only was the twin-twin delivery time improved under epidural analgesia but the status of the second twin was almost always better, irrespective of the presentation at delivery.[25] He felt that this improvement was gained by eliminating the "bearing down reflex." Other studies have shown that neonatal status is not adversely affected by lumbar epidural analgesia.[26,27]

A continuous segmental lumbar epidural analgesia (either by bolus technique or continuous pump infusion) provides excellent pain relief for both stages of labor. By obviating the need for parenteral narcotics, neonatal drug depression is avoided. As noted earlier, many multiple-gestation babies are premature and therefore are particularly susceptible to narcotic depression. The lumbar epidural technique has an additional advantage in that the level of the block can easily be extended should a cesarean section become necessary or in case perineal analgesia and relaxation are desired to help with a difficult vaginal delivery.

Regional techniques offer many advantages, but they do not provide uterine relaxation. If obstetric manipulations require a relaxed uterus, general anesthesia should be given without delay. High-dose (2–3 MAC) inhalational anesthetics are administered via an endotracheal tube following a rapid sequence induction.

Cesarean Section

There are no controlled studies demonstrating a clear-cut advantage for any particular anesthetic technique when cesarean section is performed for a multiple-gestation pregnancy. The choice of anesthetic method is often based on neonatal and maternal factors other than the multiple-gesation per se. The presence of heavy vaginal bleeding, for example, may tend to favor general anesthesia, whereas the presence of asthma in the mother would favor a regional technique. However, when choosing a technique, one must bear several points in mind. First, premature infants are particularly susceptible to the respiratory and central nervous system depressant effects of narcotics and other sedatives. Second, the diminished oxygen reserve seen in parturients is decreased still further by the presence of the enlarged uterus in a multiple-gestation pregnancy. The risk of aspiration is increased. All of these factors tend to favor regional anesthesia. A recent review indicated how fine tuning of epidural analgesia may be employed to improve outcome and patient comfort.[28]

On the other hand, general anesthesia could help if uterine relaxation is desired. Whatever technique is planned, the mother should receive oxygen supplementation until all of the infants have been delivered.

Associated Conditions

As noted earlier, many of the neonates are either premature, breech presentation, or both. Proper anesthetic management requires that these situations be dealt with appropriately.

The breech presentation of the fetus is associated with a wide range of problems. The baby may experience trauma and spinal cord damage, intracranial hemorrhage, hypoxia, and prematurity. Difficulties with delivery include umbilical cord prolapse, placena previa, and abruptio placentae.

If the parturient's cooperation is desired, lumbar epidural analgesia can provide excellent analgesia for both labor and delivery, as well as maximal perineal relaxation for delivery of the head. General anesthesia should not be used routinely, because it introduces significant maternal risk. However, it may become necessary to induce general anesthesia rapidly in the event that the aftercoming head becomes trapped. Therefore, the anesthesiologist must be prepared, having checked all equipment and administered aspiration prophylaxis to the parturient.

General anesthesia is useful in the case of the entrapped head, primarily because it can provide rapid and profound perineal relaxation. Although it is true that the halogenated inhalational agents (ie, halothane, enflurane, and isoflurane) relax the body of the uterus and the lower uterine segment, they do not dilate the cervix.[29]

Whatever anesthetic method is chosen, a high inspired fraction of oxygen (FiO_2) should be administered to the mother in an effort to minimize fetal hypoxia.

If an abdominal delivery is planned, either general or regional anesthesia may be used for the delivery, and the choice is usually made on the basis of medical and obstetric considerations at hand other than the fact of a breech presentation.

Tocolytics are often administered to inhibit premature labor. If tocolysis fails, or if the intrauterine conditions are such that the children must be delivered, the anesthesiologist may be faced with the situation of administering anesthesia for an abdominal or vaginal delivery in the presence of tocolytic agents. Uterine relaxation is an effect of β_2-receptor activation. Ritodrine and terbutaline are two sympathomimetic agents commonly used for tocolysis, although at present only ritodrine is approved by the FDA for this purpose. Although predominantly β_2 active, these agents do have some β_1 activity. Heart rate and cardiac output increase, and myocardial ischemia or even cardiac failure can occur.

Pulmonary edema (either cardiogenic or noncardiogenic) is one of the more worrisome complications of prolonged β-mimetic therapy. Noncardiogenic pulmonary edema results from changes in pulmonary capillary permeability with movement of protein into the interstitium.[30] Cardiogenic pulmonary edema may be worsened by the release of antidiuretic hormone (ADH) stimulated by β-mimetics. ADH acts directly on the renal tubules to increase reabsorption of sodium. As a result, after 24 hours the patient is at risk for fluid overload, enhanced when intravenous saline, as opposed to intravenous dextrose solutions, is administered. Dextrose solutions, however, are usually avoided because of the hyperglycemia induced by β-adrenergic agents. Half normal saline is a preferred alternative. The onset of hyperglycemia is sudden, even in a nondiabetic woman. For this reason, dextrose solutions are usually considered contraindicated in women with diabetes.

β-adrenergics produce hypokalemia as intracellular transport of potassium is facilitated. No correction is necessary because serum potassium levels resolve after cessation of the tocolytic therapy.

When inducing general anesthesia, β-mimetics should be discontinued at least one-half hour prior to induction if possible. The stimulation of laryngoscopy and intubation may generate severe tachycardia and hypertension. If the urgency of the case prohibits delay, esmolol (200–500μg/kg/min as an intravenous infusion) can be a useful adjunct. Agents that cause tachycardia are best avoided. Halothane can sensitize the heart to catecholamine-induced dysrhythmias to a greater degree than do other inhalational agents and should be used with caution if at all. Hyperventilation should also be eschewed because the resulting hypocarbia further decreases serum potassium levels.

If regional anesthesia is chosen, it should be instituted cautiously. Prehydration may not be necessary and might result in pulmonary edema. Ephedrine may elicit an exaggerated response in heart rate and blood pressure. One possible approach is to increase the level of an epidural anesthetic slowly and administer small boluses of fluid as necessary.

Magnesium sulfate is another agent widely used for tocolysis. Like the β-mimimetics, it has been associated with pulmonary edema.[32] The concerns regarding prehydration following magnesium therapy are similar to those for the β-adrenergics. Considerations regarding general anesthesia and magnesium therapy center on the ability of magnesiun to potentiate both depolarizing and nondepolarizing muscle relaxants. Precurarization is not used because the magnesium itself prevents fasciculations and because its combination with even a defasciculating dose of a nondepolarizer may lead to premature paralysis.[33] As a matter of safety, the dose of succinylcholine used for intubation should not be reduced, but the degree of relaxation should be monitored by a nerve stimulator and further administration of relaxants administered accordingly.

References

1. Benirschke K., Chung K.K.: Multiple pregnancy. *N Engl J Med* 1973, 288:1276–84.
2. Hawrylyshyn P.A., Barkin M., Bernstein A., Papsin F.R.: Twin pregnancies—a continuing perinatal challenge. *Obstet Gynecol* 1982, 59:463–6.
3. McCarthy B., Sachs B.P., Layde P.: Epidemiology of neonatal mortality in twins. *Am J Obstet Gynecol* 1981, 141:252–6.
4. Sassoon D.A., Castro I.C., Davis J.L., Hobel C.J.: Perinatal outcome in triplet versus twin gestations. *Obstet Gynecol* 1990, 75:817–20.
5. Botting B.J., Davies I.M., Macfarlane A.J.: Recent trends in the incidence of multiple births and associated mortality. *Arch Dis Child* 1987, 62:941–50.

6. Ho S.K., Wu P.Y.K.: Perinatal factors and neonatal morbidity in twin pregnancy. *Am J Obstet Gynecol* 1975, 122:979–85.
7. Duncan S.B., Ginz B., Wahab H.: Use of ultrasound and hormonal assays in the diagnosis, management and outcome of twin pregnancy. *Obstet Gynecol* 1979, 53:367.
8. Keith L., Ellis R., Berger G.S., Depp R.: The Northwestern University Multihospital Twin Study. *Am J Obstet Gynecol* 1980, 138:781–91.
9. Myrianthopoulos N.C.: Congenital malformations in twins: Epidemiologic survey. *Birth Defects* 1975, 11:39–44.
10. Galea P., Scott J.M., Goel K.M.: Feto-fetal transfusion syndrome. *Arch Dis Child* 1982, 57:781–5.
11. Shah D.M., Chaffin D.: Perinatal outcome in very preterm births with twin-twin transfusion syndrome. *Am J Obstet Gynecol* 1989, 161:1111–3.
12. Medearis A.L., Jonas H.J., Stockbauer J.W.: Perinatal deaths in twin pregnancy. *Am J Obstet Gynecol* 1979, 134:413–8.
13. Rovinsky J.J., Jaffin H.: Cardiovascular hemodynamics in pregnancy: 1. Blood and plasma volumes in multiple pregnancy. *Am J Obstet Gynecol* 1965, 93:1–13.
14. James F.M.: Anesthetic considerations for breech or twin delivery. *Clin Perinatol* 1982, 9:77.
15. Benson C.B., Doubilet P.M.: Sonography of multiple gestations. *Radiol Clin North Am* 1990, 28:149–61.
16. Persson P.H., Kullander S.: Long term experience of general ultrasound screening in pregnancy. *Am J Obstet Gynecol* 1983, 146:942–6.
17. Newman R.B., Hamer C., Miller M.C.: Outpatient triplet management: A contemporary review. *Am J Obstet Gynecol* 1989, 161:547–53.
18. Patten R.M., Mack L.A., Nyberg D.A.. Filly R.A.: Twin embolization syndrome: Prenatal sonographic detection and significance. *Radiology* 1989, 173:685–9.
19. Carlson N.J., Towers C.V.: Multiple gestation complicated by the death of one fetus. *Obstet Gynecol* 1989, 73:685–9.
20. Bianchi M.T.: Triplet pregnancy with second trimester abortion and delivery of twins at 35 weeks' gestation. *Obstet Gynecol* 1984, 64:728–30.
21. Redwine F.O., Petres R.E.: Selective birth in a case of twins discordant for Tay Sachs disease. *Acta Genet Med Gemellol (Roma)* 1984, 33:35–38.
22. Chervenak F.A.: The controversy of mode of delivery in twins: The intrapartum management of twin gestation (part 2). *Semin Perinatol* 1986, 10:44–9.
23. Schwalbe S.S., Marx G.F., Meyers G.: Aortocaval compression revisited, in *Abstracts of Annual Meeting of the Society for Obstetric Anesthesia and Perinatology.* 1988, p 136.
24. Elstein I.D., Schwalbe S.S., Marx G.F.: Cardiac output measurements during and after triplet gestation. *Obstet Gynecol* 1989, 74:452–3.
25. Crawford J.S.: A prospective study of 200 consecutive twin deliveries. *Anaesthesia* 1987, 42:33–43.

26. James F.M. III, Crawford J.S., Davies P., Crawley M.: Lumbar epidural analgesia for labor and delivery of twins. *Am J Obstet Gynecol* 1976, 127: 176–80.
27. Daniels J.C., Hebre F.W.: Anesthetic considerations for complicated obstetrics: I. A retrospective study of 527 twin deliveries. *Anesth Analg* 1967, 46:527–39.
28. Leighton B.: New advances in obstetric anesthesia. 41st Annual Refresher Courses Lectures. *American Society of Anesthesiologists.* Las Vegas, NV. October 1990, No. 172.
29. Munson E.S., Embro W.J.: Enflurane, isoflurane and halothane and isolated human uterine muscle. *Anesthesiology* 1977, 46:11.
30. Wheeler A.S., Patel K.F., Spain J.: Pulmonary edema during beta-2-tocolytic therapy. *Anesth Analg* 1981, 60:695.
31. Philpsen T., Erikson P.S., Lynggard F.: Pulmonary edema following ritodrine-saline infusion in premature labor. *Obstet Gynecol* 1981, 58:304.
32. Elliott J.P.: Magnesium sulfate as a tocolytic agent. *Am J Obstet Gynecol* 1983, 147:277.
33. DeVore J.S., Asrani R.: Magnesium sulfate prevents succinylcholine induced fasciculations in toxemic parturients. *Anesthesiology* 1980, 52:76.

Chapter 22

The Patient With Pyloric Stenosis

Scott Ira Winikoff, M.D.

Case History. *A 3-week-old male was brought to the emergency room with a 3-day history of clear projectile vomiting after each feeding. He had been anuric for almost 24h and was becoming increasingly lethargic.*

Physical examination showed moderate dehydration; blood pressure, 84/50mmHg; pulse, 152/min; respiratory rate, 45/min and very shallow; temperature, 100°F; weight, 4.1kg. A small, almost midline upper abdominal mass was palpable. Laboratory findings were as follows: hemoglobin, 16.5gm/dl; hematocrit, 50%; arterial blood gases, pH 7.55, $PaCO_2$ *52,* PaO_2 *80, and* Cl^- *96mEq/L.*

He was scheduled for a semiemergency laparotomy for correction of pyloric stenosis.

Overview and History

Infantile hypertrophic pyloric stenosis is probably the most common cause of infant gastrointestinal (GI) obstruction seen in the United States today. One of the first cases of congenital pyloric stenosis was described in 1788 by Hezekiah Beardsley in a child who died at 5 years of age.[1] In 1887, Hirschsprung established the condition as a clinical entity and presented two cases.[2] In 1907, Pierre Fredet suggested that the pyloric muscle be split,[3] and in 1912, Ramstedt described the definitive procedure that is still performed today.[4] Pyloromyotomy is curative for the child with pyloric stenosis provided the management proceeds in a systematic manner, with integral input and care provided by the anesthesiologist.

Pathophysiology

Infantile pyloric stenosis usually manifests itself in the second to sixth week of life. It is caused by hypertrophy of the muscularis layer of

Reviewed by Dr. Gerard Weinberg, Associate Professor, Surgery and Pediatrics, Albert Einstein College of Medicine.

the pylorus, which swells and progressively obstructs the pyloric canal.[5] The etiology of the hypertrophy is still obscure, although many theories have been advanced to explain it. Lynn[6] has proposed a physiologic explanation that is well accepted. Milk curds moving in the stomach against the pylorus cause edema and irritation that narrow the canal. There is a work hypertrophy in response to this narrowing that further obstructs the outlet, and that obstruction is responsible for most of the symptoms.

Symptoms

The symptoms of pyloric stenosis usually begin at about the third week of life, although they have been seen as early as 36 hours after birth. Males are affected more often than females by a 4:1 ratio.[2] Also, there is a higher incidence of pyloric stenosis in children of an affected parent.[7] Symptoms begin with vomiting that occurs with one or two feedings a day progressing to projectile vomiting with every feeding. The vomitus is classically free from bile and consists of gastric juice and ingested formula. Because of the constant vomiting, the child dehydrates quickly. Other metabolic changes develop as described below.

Diagnosis

Clinical suspicion is very important in the diagnosis of pyloric stenosis. With hypertrophy of the pyloric musculature, an olivelike mass often forms. It can be palpated in the epigastrium to the right of the midline. In 75%–85% of cases, it can be palpated easily by letting the child suck a bottle of 5% dextrose or a finger wet with dextrose water. Gastric peristalsis waves that pass from the left upper quadrant to the right side may be visible during feeding. If these techniques do not help to establish the diagnosis, a barium swallow may be performed to outline the defect, a maneuver that may be dangerous because of the risk of aspiration of contrast material.

Although these techniques may help to diagnose pyloric stenosis, history and physical examination remain key. Recognizing that the child is vomiting bilious-free contents, is dehydrated, and is of the correct age is extremely important to prompt and accurate diagnosis. If the diagnosis is suggestive of pyloric stenosis but no mass is palpated, ultrasound scan is recommended.

Differential Diagnosis

Pyloric stenosis must be distinguished from other anomalies that cause GI obstruction and vomiting in the newborn, and also from sepsis and viral gastroenteritis. Other possible causes of obstruction include hiatal hernia; duodenal, ileal and jejunal atresias; achalasia of the esophagus; malrotation of the gut; Meckel's diverticulum; and intolerance of formula. However, the combination of bilious-free vomitus, visible gastric peristalsis, and a palpable "olive" mass are pathognomonic for pyloric stenosis.[8] Rapid gastric emptying of barium excludes the diagnosis of pyloric stenosis.[2]

Associated Anomalies

Associated anomalies of the renal system occur in a small percentage of patients with pyloric stenosis.[9] Jaundice may also coexist, probably resulting from decreased glucuronyl transferase activity in the liver.[10] There is an elevation of unconjugated bilirubin (indirect bilirubin) seen in about 17% of patients. The exact cause is not known, but the jaundice disappears between 5 and 10 days postoperatively and does not recur.[11]

Metabolic Derangements

Many metabolic and electrolyte changes develop in the child who has had protracted vomiting and decreased oral intake. The vomitus (mainly gastric juice) contains sodium, chloride, potassium and hydrogen ions (Table 1).

TABLE 1. Composition of External Abnormal Losses.[15]

Fluid	**Na^+ (mEq/l)**	**K^+ (mEq/l)**	**Cl^- mEq/l**	**Protein (gm)**
Gastric	20–80	5–20	100–150	—
Pancreatic	120–140	5–15	90–120	—
Small intestine	100–140	5–15	90–130	—
Bile	45–135	5–15	80–120	—
Ileostomy	10–90	3–15	20–115	—
Diarrheal	10–30	10–80	10–115	—
Sweat Normal	10–10	3–10	10–35	—
Cystic fibrosis	50–130	5–25	50–110	—
Burns	140	5	110	3–5

The children become hypochloremic, hypokalemic and alkalotic. When severe, the metabolic alkalosis produces a marked respiratory attempt at compensation.[12,13] (see Table 2). The hypoventilation that results from metabolic alkalosis can lead to apnea and atelectasis if good

TABLE 2. Metabolic Findings in the Newborn Secondary to Pyloric Stenosis

Severity of Dehydration	pH	pCO_2	CO_2	pO_2	Na^+	K^+	Cl^-	HCO_3^-
Mild	↑	↑	↑	↔	↓↔	↓	↓↓	↑
Moderate	↑↑	↑↑	↑↑	↔	↓	↓↓	↓↓↓	↑↑
Severe	↓↓	↓	↓↓	↓	↓	↓↓	↓↓↓	↓↓

pulmonary toilet is not instituted. In the patient with the mild disease, Na^+ and K^+ values are slightly decreased, and Cl^- is moderately decreased.

As the disease worsens, the most dramatic change is the decrease in serum Cl^-. The patient becomes more alkalotic, and respiratory compensation increases. Oxygenation remains close to normal. As dehydration increases, other changes develop. Severe dehydration can lead to circulatory shock, impairing renal and hepatic function. The kidneys no longer retain bicarbonate, and serum bicarbonate levels decrease. Metabolic acidosis (from the loss of HCO_3^-) produces hyperventilation and compensatory respiratory alkalosis. Arterial blood gas measurements reveal decreased pH, decreased pCO_2, and finally, a decreased pO_2.

Although the arterial blood gas and electrolyte values can vary greatly, laboratory values most often indicate hypocholoremia, hypokalemia, hyponatremia and alkalosis. Some degree of metabolic alkalosis, with a partially compensating respiratory acidosis, is usual.

The oxygen dissociation curve that measures the affinity of hemoglobin for oxygen relates to the increase in pH.[14] In adults, when the hemoglobin is 50% saturated with oxygen (P50), the PaO_2 is 27 torr. In infants, who still have up to 70% fetal hemoglobin, this number is decreased to 20–22 torr because of the increased affinity for oxygen of fetal hemoglobin. Therefore, at a given tissue pO_2, the fetal hemoglobin releases less oxygen at the tissue level (ie, the curve is shifted to the left). Other factors that shift the curve to the left include hypothermia, alkalemia and hypocapnia. So in the alkalotic patient with pyloric stenosis, the curve shifts more to the left, and less oxygen is unloaded to the tissues,[15] of special concern in the neonate because of the already low P50.

Finally, with electrolyte imbalances, there is an increased potential for seizures, especially with hypocalcemia and hyponatremia.

Timing of Surgery

Surgery for infants with pyloric stenosis should rarely be performed on an emergency basis.[16] Rather, the aim of treatment is to correct the alkalosis, dehydration, and electrolyte imbalance as quickly and as safely as possible. Any complications, such as infection, must also be corrected. Most infants can be operated on within 24 to 48 hours of admission.[2] Good medical treatment and early diagnosis have reduced mortality to nearly 0%.[4,17]

Preanesthetic Care

Assessment of the hydration status of an infant includes a detailed history, a physical examination, and laboratory verification. Whereas a history of duration of illness and other changes are often vague in adults, a detailed history is usually fairly easy to elicit in an infant. The parent often remembers with which feeding vomiting began or the time of the last oral intake. Intake and output assessment is important, as is a history of how often the baby has vomited, and when the vomiting became projectile. Inquiry must also focus on usual feeding habits with regard to quantity and quality (ie, fluids and electrolytes with formulas such as Pedialyte®or a simple sugar-water solution). Urine output usually decreases with dehydration, except in some low-birth-weight babies.[13] Normal or increased urine output with dehydration suggests diabetes mellitus or diabetes insipidus. Renal disease may also be suspected. The presence of diarrhea suggests a viral infection.

An evaluation of weight gain or loss may be difficult in an infant. These babies may just be starting to gain weight or may still be losing weight, depending on age. Therefore, weight may be an inaccurate guide to dehydration. Many other findings on physical examination describe the type and severity of the dehydration (Table 3).[12,13]

Most dehydrated infants appear ill. Mild dehydration may be indicated only by excessive thirst. If any other signs of moderate dehydration are present with thirst, the patient should be considered to have a deficit of 5%–10% of body fluid, or 50–100cc/kg.

Laboratory values should be obtained after initial evaluation and will be helpful in characterizing the type of deficit. A patient who is very dehydrated may become hemoconcentrated and show an increase in hemoglobin and hematocrit. Both BUN and serum creatinine levels may be high because of the decrease in glomerular filtration rate associated with dehydration. Also, urinalysis may show a mild to moderate proteinuria and the urine may contain casts, and both red and white blood cells. The urine will be very concentrated unless the ability to concentrate has been lost (eg, as a result of renal disease or extreme prematurity).

Fluid therapy depends on degree of dehydration. Infants with mild derangements, such as $Cl^- > 100$mEq/L, respiratory rate greater than 20 breaths per minute, no recent weight loss, and urine specific grav-

TABLE 3. Clinical Assessment of Severity of Dehydration

Type	Mild	Moderate	Severe
General appearance	Thirsty, alert, restless	Thirsty, restless or lethargic, but irritable to touch or drowsy	Drowsy, limp, cold, sweaty, comatose
Skin turgor	Poor	Very poor	Parched
Radial pulse	Normal rate and volume	Rapid and weak	Rapid and feeble
Mucous membrane and tongue	Moist to slightly dry	Dry	Very dry
Blood pressure	Normal	Hypotensive	Shock
Anterior fontannel	Flat	Depressed	Sunken
Tears	Present	Absent	Absent
Skin elasticity	Retracts immediately	Retracts slowly when pinched	Retracts very slowly (>2s) when pinched
Urine flow	Normal	Oliguric with maximum concentration	
Percentage of weight loss	4%–5%	5%–10%	>10%
Est. fluid deficit	40–50cc/Kg	60–100cc/Kg	>100cc/kg

ity >1.010, may be managed conservatively with oral glucose water, Pedialyte®, or Ringer's lactate until surgical correction can be performed.[4] Patients with moderate dehydration (Cl^- 90–100mEq/L, respiratory rate of 16 to 20 breaths per minute, and a more concentrated urine) require correction of the deficit as maintenance fluids are given.

Saline 0.45% or 0.9% is an appropriate replacement fluid. One half of the deficit is replaced over the first 8 hours and the other half over the next 16 hours. Potassium is added to the fluid once urination has begun. The maintenance fluid should consist of 5% dextrose in 0.225% saline. The rates for maintenance fluids are as follows: 1st 48 hours of life, 75cc/kg/day or 3cc/kg/h; 2 days to 1 month, 150cc/kg/day or 5cc/kg/h; 1 month and up (maximum weight 10kg), 100cc/kg/day or 4cc/kg/h.[5,12] For more severe dehydration, circulatory shock is treated first. 20cc/kg of isotonic NaCl is given over 1 hour or less. If the patient does not respond, 10cc/kg of blood or plasmanate is given over 1 to 2 hours. Thereafter, the deficit is corrected, and maintenance fluid is infused.

Pyloromyotomy

Only when the patient has been stabilized should pyloromyotomy be considered; the Ramstedt pyloromyotomy is the procedure of choice.[2] A small incision is made in the abdomen high over the liver, and the hypertrophied muscle of the pylorus is incised down to the mucosa.[2] Blood loss is minimal.

Anesthetic Plan

The stomach should be gently suctioned via a nasogastric tube prior to induction of anesthesia. In some centers, 5–7cc of sodium bicarbonate is instilled and then suctioned out after agitating the child mildly to distribute the fluid.[18] The tube is then removed, and monitors are placed. Monitors necessary for this procedure include an ECG, noninvasive blood pressure device, precordial and esophageal stethoscopes, pulse oximeter, and temperature probe, as well as the usual anesthetic machine monitors.

If atropine (0.01 – 0.02mg/kg) has not been given intramuscularly as a premedicant, it should be given intravenously. Endotracheal intubation may be performed with the patient awake, after a rapid-sequence induction or following an inhalational induction if an intravenous cannula has not yet been placed. For a rapid-sequence induction, preoxygenation, atropine, and thiopental (3–4mg/kg) with cricoid pressure are used. Once the patient is intubated, anesthesia can be maintained by almost any technique. Skeletal muscle relaxants will rarely be needed because the procedure is short. Inhalational anesthesia affords sufficient relaxation.

The patient should be awake, alert and extubatable at the end of the procedure. Oral feedings of clear fluids may be resumed in 4 to 6 hours as tolerated. The nasogastric tube is removed as soon as the infant recovers from anesthesia. If vomiting occurs frequently after feedings in the first 24 hours, washing out the stomach with saline can remove the excess mucus; thereafter, vomiting usually ceases.

The mortality for pyloric stenosis has approached 0% recently because of rapid diagnosis and preoperative medical management. Surgical correction necessitates a teamwork approach by the anesthesiologist, pediatrician and surgeon to ensure a good outcome.

References

1. Donovan E.J., Beardsley H.: Congenital hypertrophic stenosis of the pylorus. *Arch Ped* 1958, 75:359–62.
2. Benson C.D.: The stomach and the duodenum, in Mustard W.T., Ravitch M.M., Snyder W.H., et al. (eds): *Pediatric Surgery,* 2nd ed., Chicago: Yearbook Publishers, 1969, 52:818–21.

3. Dufour H., Fredet P.: Hypertrophic pyloric stenosis in newborns and surgical treatment (translated). *Surg Rev* 1908, 37:208–12.
4. Hall J.W.: Anesthesia for gastrointestinal and abdominal wall disorders, in Gregory C.A. (ed): *Pediatric Anesthesia*, New York: Churchill Livingstone, 1983, 23:707–22.
5. Lockhart C.H.: Maintenance of general anesthesia, in Gregory G.A. (ed): *Pediatric Anesthesia*, New York: Churchill Livingstone, 1983, 15:472–7.
6. Lynn H.: The mechanism of pyloric stenosis and its relationship to preoperative preparation. *Arch Surg* 1960, 81:453–9.
7. McKeown T., McMahon B.: Infantile hypertrophic pyloric stenosis in parent and child. *Arch Dis Child* 1955, 30:497–500.
8. Stevenson R.J.: Non-neonatal intestinal obstruction in children. *Surg Clin North Am* 1985, 65:1217–34.
9. Atwell J.D., Levich P.: Congenital hypertrophic pyloric stenosis and associated anomalies of the GU tract. *J Pediatr Surg* 1981, 16:1029–35.
10. Wooley M.M., Felsher B.F., Asch M.J., et al: Jaundice, hypertrophic pyloric stenosis and glucuronyl transferase. *J Pediatr Surg* 1974, 9:359–63.
11. Steward D.J.: Diseases of the gastrointestinal system, in Katz J., Steward D.J. (eds): *Anesthesia and Uncommon Pediatric Disease,* Philadelphia: WB Saunders, 1987, pp 226–35.
12. Malhotra V.: Pyloric stenosis in Yao F., Artusio J. (eds): *Anesthesiology, Problem Oriented Patient Management.* Philadelphia: JB Lippincott, 1988, pp 259–74.
13. Robeson A.M.: Parenteral fluid therapy, in Behrman R.E., Vaughan III (eds): *Nelson's Textbook of Pediatrics,* 2nd ed., Philadelphia: WB Saunders, 1985, pp 233–43.
14. Parry W.H., Adams N.R.: Acid-base homeostasis and oxygenation, in Merentein G.B., Gardner S.L. (eds): *Handbook of Neonatal Intensive Care*, St. Louis: CV Mosby, 1983, pp 117–19.
15. Bowe E.A., Klein E.J. Jr.: Acid-base, blood gas, electrolytes, in Barish P.G. (ed): *Clinical Anesthesia,* Philadelphia: JB Lippincott, 1989, pp 679–80.
16. Gregory G.A.: Pediatric anesthesia, in Miller R.D. (ed): *Anesthesia,* 2nd ed., New York: Churchill Livingstone, 1986, pp 1788–9.
17. Smith R.M.: Anesthesia in infants and children, 4th ed., St. Louis, CV Mosby, 1980, pp 324–40.
18. Berry F.: Neonatal anesthesia, in Barish P.G. (ed): *Clinical Anesthesia*, Philadelphia: JB Lippincott, 1989, pp 1253–80.

CHAPTER 23

The Patient With Multiple Sclerosis

Ya-Tseng William Lu, M.D.
Lloy E. Anderson, M.D.

Case History. *A 20-year-old Hispanic woman, gravida 2, para 0010, presented with recurrent premature cervical dilatation for emergency cervical cerclage at 20$\frac{1}{2}$ weeks' gestation with a twin pregnancy.*

Her past medical history was significant for asthma and multiple sclerosis. One year prior to admission the patient suddenly developed numbness below dermatome T8, which lasted approximately 1 month. There were no other associated signs and symptoms. Multiple sclerosis was diagnosed by a neurologist following magnetic resonance imaging (MRI) and cerebral spinal fluid (CSF) analysis. The patient had been in remission since that initial presentation.

At 12 weeks of pregnancy the patient had undergone emergency cerclage secondary to premature cervical dilatation. She was well until the day of admission, when premature cervical dilatation was found again.

Medications included multiple vitamins and ferrous sulfate.

Vital signs included temperature, 37.0°C; pulse, 76/min and regular; respiration, 20/min; blood pressure, 108/57% mmHg; airway assessment, Molampati class II; lungs, clear to auscultation bilaterally; heart, regular rate and rhythm, SI and S2, no murmurs or gallops. Laboratory findings were hemoglobin 11–0gm/dl, hematocrit 31.4% and platelet count 329,000. Weight was 78kg; height, 155cm.

The patient had eaten approximately 4 hours before coming to the operating room.

Introduction

Multiple sclerosis (MS) is the most common form of demyelinating disease of the central nervous system (CNS). Characteristics include

Reviewed by Dr. Christopher Bryan-Brown, Professor of Anesthesiology, Albert Einstein College of Medicine/Montefiore Medical Center.

recurrent attacks of focal or multifocal neurologic dysfunction of the brain, optic nerve and spinal cord. The disease occurs most commonly in early adult life between 15 and 50 years of age. Average age at the time of initial symptoms is 33 years.[1] Mild forms of MS can barely be detected initially, and symptoms may not recur for 10 to 20 years. More typically, symptoms worsen over a period of a few days to 2 to 3 weeks, then remit.

Recovery can be rapid over weeks to a few months. The extent of recovery varies from complete resolution of symptoms to some residual dysfunction before the next attack. The remission may be incomplete and may worsen progressively, with increasing permanent defect. The frequency of recurrence is greatest during the first 3 to 4 years. The prognosis is poor in patients with initial onset of symptoms after age 40, and there may be a rapid downhill course.

Etiology

The cause of MS remains unknown. Various theories are postulated: viral infection, immune-mediated response, environmental and genetic factors.[2] Some researchers suggest that the Faroe Island MS epidemic was caused by the introduction of a transmissible viral infection by British troops in World War II.[3] Others have found significant elevation of antibody titers to measles in MS patients and their siblings.[4] Elevated antibodies to other viruses (including vaccinia, rubella, varicella-zoster, herpes simplex, Epstein-Barr virus, and mumps) also have been found in both blood and CSF in MS.[4] To date no virus has been isolated.

Incidence, Prevalence and Epidemiology

The relationship between MS and geographic latitude is striking. MS is more common in temperate climatic zones in North America, Europe, Australia and New Zealand. Incidence of MS in North America, Canada and Europe is 60/100,000 population. In Denmark it is 200/100,000 population. MS is virtually unknown in Asians and African blacks, even with immigration to the temperate zones of North America.[1] Studies of migrant populations, however, reveal that the factors determining susceptibility to MS are acquired before age 15. Individuals moved from low- to high-risk zones before age 15 assume the high risk, and vice versa. The incidence of MS is also greater in urban dwellers and among affluent socioeconomic groups.[5] Average age of onset is between 30 and 40 years of age. The female-to-male ratio is 3 to 2. Average survival is greater than 30 years after onset.

Genetic factors are also implicated in MS. Blood relatives (first-degree relatives) of MS patients have an 8- to 15-fold increased risk of developing MS.[6] A study of MS in identical twins showed a 50% greater risk for MS than occurs in fraternal twins.[6] Certain histocompatibility antigens (HLA) are overrepresented in MS patients. Among whites HLA-B7 and HLA-DR2 alleles occur with increasing frequency.[7]

Precipitating Factors

Minimal temperature change, electrolyte imbalance, stress, infection or trauma can compromise conduction impulse through demyelinated nerve. Neurologic function is reduced as axonal conduction is slowed by small increases in temperature. Surgery itself does not appear to increase the risk of relapse, but postoperative pyrexia may precipitate a relapse.[8] Studies have shown that lumbar puncture and myelography are not related to relapse or progression of disability.[9,10]

Pathophysiology

The term *multiple sclerosis* is derived from the microscopic description: oligodendrocytes disappear from within plaques, and astrocytes proliferate forming scars (sclerosis).[11] Within the CNS there are multiple irregularly shaped plaques with sharp-edge lesions representing intact axon within disintegrated CNS myelin (Figures 1 & 2). Plaques occur in both gray and white matter and vary from 0.1cm to many centimeters in diameter. Most plaques occur near blood vessels. Although plaques can occur anywhere within the CNS, the paraventricular areas of the cerebrum, the optic nerve, the brain stem and the cervical spinal cord are predominant sites. The peripheral nervous system is not affected in MS patients. At autopsy approximately 20 of "silent" plaques are found.[12] Only limited myelin regeneration occurs in MS patients.

Normally, a nerve impulse is conducted down a fiber from one node of Ranvier to the next. When the internodal segment becomes demyelinated, current is shunted and the excitation of excessive nodes is slowed. Symptoms may not develop, however, until conduction fails completely. Factors contributing to total conduction failure include demyelination of several contiguous segments, changes in extracellular fluid composition caused by edema, circulating factors that block synaptic transmision, and temperature changes. For example, fever, a hot bath or exercise may cause a failure of conduction through demyelinated regions and lead to symptoms.

Because remyelination is limited in MS, remission of symptoms may be associated with factors such as the correction of transient chemical and physiologic disturbances. Evidence suggests that an increasing number

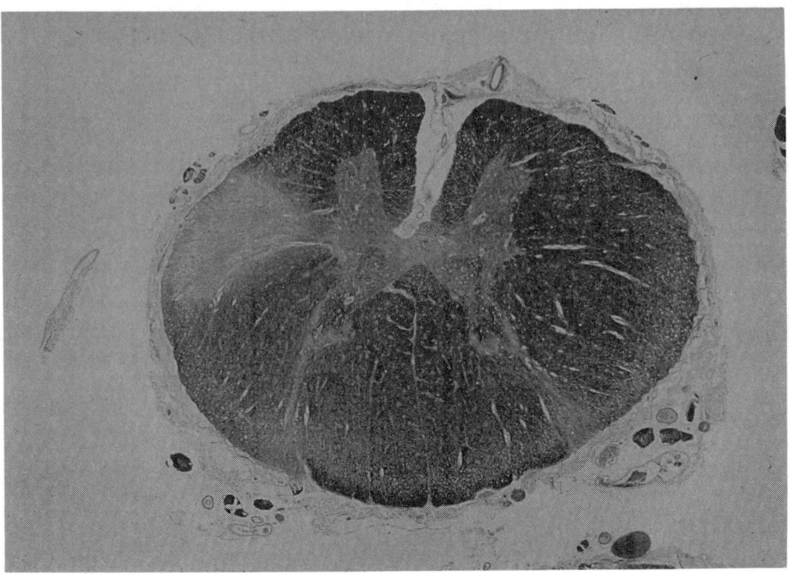

FIGURE 1. Cross-section of spinal cord (low power microscopic view) shows pale area of demyelinated section. (Courtesy of Tsuey-Ling Chen, MD, Department of Pathology, St. Peter's Medical Center, New Brunswick, NJ.)

of sodium channels may appear in demyelinated axon membranes and facilitate impulse propagation along the axon, allowing a recovery of function.[13]

Clinical Manifestation

The clinical picture of MS is associated with the location of the focal demyelination. Most initial signs are nonspecific for a particular disease, but certain signs and symptoms, such as optic neuritis, inter-nuclear ophthalmoplegia and Lhermitte's sign, are suggestive of MS.

Optic neuritis. Early signs include diminished visual acuity, central or paracentral scotoma, hyperemia and edema of the optic disk, and a defective pupillary reaction to light. Later, pallor of the temporal half of the disk develops as a result of demyelination of the papillomacular bundle.

Internuclear ophthalmoplegia. Demyelination of the medial longitudinal fasciculus (the brain stem pathway that coordinates eye movement) causes paresis of the medial rectus muscle on lateral conjugate gaze but not on convergence. Nystagmus is seen in the abducting eye.

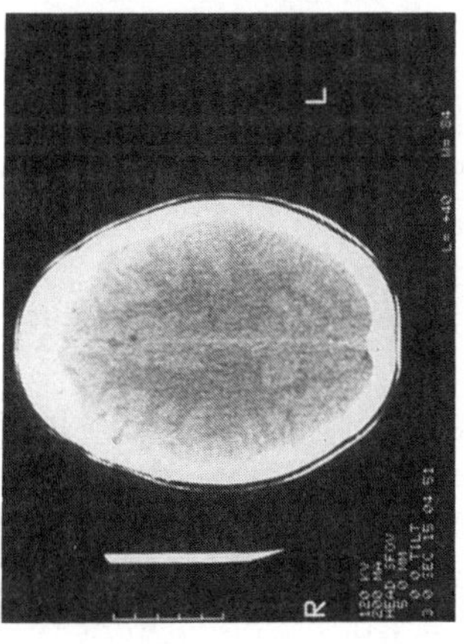

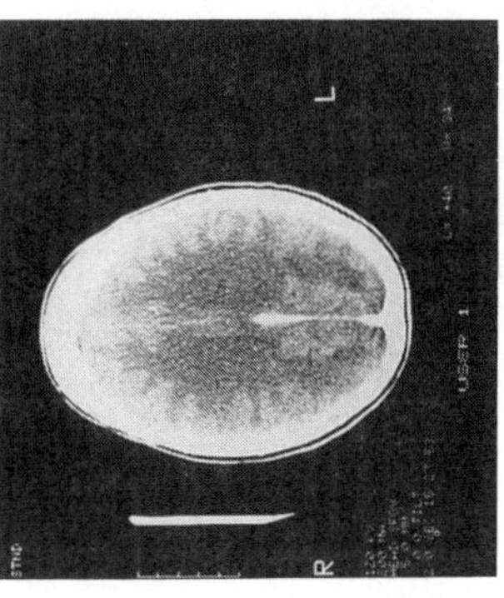

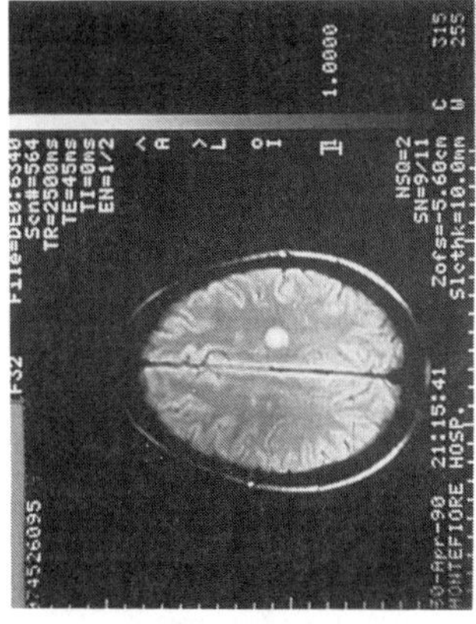

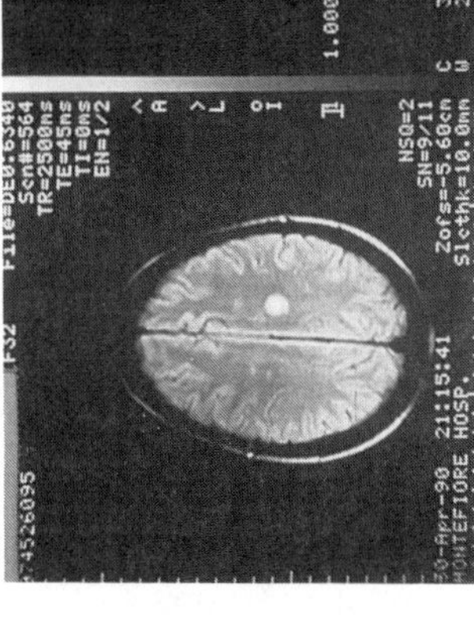

FIGURE 2. (A) CT scan without contrast and (B) CT scan with contrast revealed no plaques, whereas (C) MRI of the cerebrum showed white spots that represent demyelinated fibers or plaques (Courtesy of T. Linda Chi, MD, Department of Radiology, Montefiore Medical Center, Bronx, NY.)

Lhermitte's sign. Flexion of the neck produces a sharp sensation that is conducted down the back and into the legs. This sign suggests intramedullary disease of the cervical cord.

Other common initial symptoms include feelings of heaviness, weakness of one or more extremities, stiffness, clumsiness and spasticity. Sensory involvement includes paresthesia, girdlelike pressure and trigeminal neuralgia. Impaired two-point discrimination over the palmar surface of the fingertips suggests cervical cord lesions. Lack of coordination and loss of position and vibration sense at the ankles and toes suggest cerebellar lesions. Loss of pain and touch sensation is less common.

Other less common initial changes include urinary frequency, incontinence, hesitancy, and retention; vertigo; hearing loss; facial, extremity or truncal pain; dysarthria; changes in intellectual function; and paraparesis.

As noted, initial symptoms may be single or multiple and may be so subtle that patients ignore them. The onset of MS may be minutes to days or chronically progressive over months. Seventy percent of MS patients recover to slight or virtual disappearance of neurologic symptoms after the initial onset. In 30% of cases there is no remission, but the patient remains stable between periods of deterioration. Recovery from an acute bout is in 2 to 8 weeks.

Rarely, an acute fulminant MS is fatal in weeks to months because of intense inflammatory response within the plaques. Presenting symptoms include headache, vomiting, delirium, convulsions and coma.

The course of MS is remarkably variable. Exacerbation at unpredictable intervals over a period of several years may occur. Eventually the residual symptoms and signs persist during remission and lead to severe disability. Sometimes the disease takes a more benign course, and there is permanent remission.

Physical Disability

Multiple sclerosis is a progressive disease, and the functional impairment is unpredictable; ranging from minimal to bedridden. Approximately 50% of patients can carry on normal activity 10 years after the onset of symptoms, 25% after 20 years. Functional problems include muscle weakness, joint contractures and spasticity, visual disturbance, cognitive impairments, speech problems, dysphagia, neurogenic bladder and sexual dysfunction.

Cognitive impairments include memory loss, new learning problems, depression, denial and euphoria. Obtaining informed consent may require that relatives or friends by present.

Joint contracture may result from muscle imbalance secondary to weakness or spasticity or from limited joint motion. Sometimes spasticity causes pain and interferes with function. As the disease progresses, patients become weaker and ambulation is increasingly difficult. Falls are frequent. Positioning for establishment of regional anesthesia may be difficult.

Spasticity and incoordination of the oral, pharyngeal, laryngeal and respiratory muscles are common. The gag reflex is diminished, and swallowing is impaired. Complications include malnutrition from difficulties with mastication and oral manipulation of food as a result of delayed reflexive swallowing, impaired pharyngeal constriction, ineffective airway protection, and inadequate respiratory support for effective coughing. The ability to swallow liquids is diminished. During periods of exacerbation or febrile illness, dysphagia can worsen and lead to aspiration, pneumonia and dehydration.

End-stage MS patients are mostly bedridden, with complications such as orthostatic hypotension, thrombus formation, bacterial infections from urinary retention and decubitus ulcers, andinability to handle pulmonary secretions resulting in atelectasis and pneumonia. Death from primary pulmonary failure from a lower medullary lesion is rare.[14]

Diagnosis and Laboratory Tests

MS is a clinical diagnosis, for which laboratory tests are supportive. Clinically definite MS is diagnosed on the basis of two attacks, evidence of two lesions, and exclusion of other diseases.

CSF studies demonstrate increased CSF IgG concentration in 80% to 90% of cases with normal levels of CSF protein. The normal ratio of CSF IgG-albumin is 0.14. Agarose electrophoresis reveals an oligoclonal band in the IgG region of the CSF (but not in serum) in approximately 90% of these patients. However, neither the elevation of CSF IgG nor oligoclonal band is specific for MS. In acute exacerbations, myelin basic protein (MBP) may be detected in the CSF.

Computed tomography (CT scan) is used to rule out space-occupying lesions, and it commonly shows focal hypodensity or atrophy. Contrast-enhancing lesions are seen in more than 50% of patients during acute attacks. Magnetic resonance imaging (MRI) is more sensitive than CT scanning and shows multiple white matter abnormalities in well over 90% of patients with "clinically definite" MS. Furthermore, MRI shows abnormal results in more than half of patients at onset.

Evoked-potential studies permit non-invasive assessment of nerve fiber conduction in visual, auditory and somato-sensory pathways. The test results may reveal the presence of multiple lesions despite clinical evidence of a single abnormality.

Other demyelinating diseases that are excluded are optic neuritis, transverse myelitis, postinfectious disseminated encephalomyelitis, progressive multifocal leukoencephalopathy, adrenoleukodystrophy and central pontine myelinolysis. In practice, many diseases with scattered lesions of the nervous system may be confused with MS (eg,) systemic lupus erythematosus, polyarteritis nodosa, Behcet's disease, sarcoidosis, Chiari malformation, or subacute combined degeneration of the spinal cord.

Treatment

At present there is no specific treatment for MS. The goals of treatment include (1) amelioration of the acute episode, (2) prevention of relapses, and (3) relief of symptoms.

Currently, corticosteroids are widely used, but there is no evidence that use alters the course of the disease. ACTH may hasten recovery from an acute exacerbation, particularly optic neuritis. Other modes of treatment include administration of interferon, intensive immunosuppression, plasmapheresis, monoclonal antibodies directed against T cells, and total lymphoid irradiation.[15] Nonspecific measures recommended include avoidance of excessive fatigue, emotional stress and marked temperature changes.

Drugs used in control of spasticity or painful muscle spasms include diazepam, dantrolene and baclofen. Nerve blocks to selective muscles may be helpful in decreasing spasticity and joint contracture in localized areas. Marked hip and knee flexion posturing secondary to spasticity can be reduced by paravertebral blocks of selected nerve roots.[14]

Carbamazepine 200mg by mouth 3 times a day has been used to treat painful dysesthesias, tonic seizures and attacks of paroxysmal dysarthria and ataxia. Trials of isoniazid along with pyridoxine have been used to treat severe postural cerebellar tremor in MS patients.[16]

Pregnancy and Multiple Sclerosis

The highest risk for development of MS in women is during their reproductive years, causing great concerns about childbearing and parenthood. For the obstetrician and anesthesiologist, the concerns are management of MS patients with possible exacerbation of symptoms.

The effect of pregnancy on MS is controversial and ill defined. It is known that exacerbation may be caused by infection, fever, excessive fatigue, and emotional trauma or stress, all of which can occur during pregnancy. These complicating factors make it difficult to implicate pregnancy as the cause of exacerbation.

The relapses of MS that occur during pregnancy are commonest in the third trimester, but most exacerbations or relapses occur during the first 3 to 6 postpartum months.[17–21] Apparently, pregnancy offers some protection against exacerbations because of the associated immunosuppression of pregnancy, which is vital to the survival of the fetus. Retrospective studies by Birk et al showed a decrease in exacerbation rate and disability during pregnancy of 50% compared to nonpregnant patients, with only a 20%–40% worsening during the postpartum period.[22]

Other autoimmune diseases, such as systemic lupus erythematosus, myasthenia gravis and rheumatoid arthritis, stabilize during pregnancy and worsen postpartum. The presence of proteins such as alpha-fetoprotein (AFP), a 70 kd protein produced by the fetus, exerts a significant immunosuppressive effect on T lymphocytes, mainly helper T cells, reducing the cellular immune response. The decrease in helper T lymphocytes reduces the number and function of suppressor T cells, normally elevated in MS. The T cell helper/suppressor ratio is lowest in the third trimester and returns to normal 3 to 5 months postpartum, thus explaining the increased frequency of exacerbation postpartum and decrease of such events in the third trimester.[22] AFP also has been shown to treat and prevent experimental allergic encephalomyelitis (EAE), which clinically resembles MS.[23]

Other proteins increase immunosuppressive activity to cell-mediated immunity. For example, alpha-2 pregnancy-associated glycoprotein suppresses T cells; pregnancy-associated plasma protein, produced by the trophoblast to inhibit the activity of lymphokine produced by maternal lymphocytes, also inhibits complement-mediated lysid of trophoblast and impairs immunoadherence of macrophages at the maternal-fetal surface.[22] In addition, estrogen has been shown to suppress helper T cells.

The immunosuppressive activity of pregnancy presents an allotypic challenge, causing a reduction in disease state, which recurs postpartum when the protection is lost.

Preanesthetic Assessment

A careful history and physical examination must be completed. Essentials in the history include the time of the last MS remission and relapse, any signs of mental impairment or swallowing problems, and any previous history of aspiration, muscle weakness, joint contracture, neurogenic bladder or visual disturbances. If the patient is bedridden, a history of orthostatic hypotension, thrombus formation, pneumonia, atelectasis and decubitus ulcer must be documented. A good systems review, including cardiovascular, pulmonary and neurologic systems, is important. In women, the time of the last menstrual period and, in the case of pregnancy, the estimated date of confinement and any complications during pregnancy must be obtained.

Careful documentation of vital signs, airway assessment, testing of gag reflex, and cardiovascular, pulmonary and neurologic examination (including mental status, cranial nerves, motor and sensory pathways, and gait) is essential.

In addition, a list of medications must be made. MS patients may be taking muscle relaxants such as diazepam, dantrolene or baclofen, which may reduce the need of neuromuscular blockers for induction and maintenance of general anesthesia. In addition, case reports have shown severe bradycardia and hypotension occurring under general anesthesia in patients who received baclofen.[24] The mechanism is unknown. When anesthetizing such patients, the possibility of severe bradycardia and hypotension should be anticipated, and appropriate emergency drugs (atropine, phenylephrine hydrochloride) should be readily available. It may be necessary to reduce the baclofen dose gradually, by 5 to 10 mg per day at weekly intervals prior to surgery. Hallucinations, seizures, or both may occur following abrupt withdrawal after more than 2 months of therapy. If the patient takes steroids for more than 4 to 6 days, adrenal insufficiency must be assumed, and additional doses of steroids must be given pre-, intra- and postoperatively.

Questions regarding types of anesthesia previously received and postoperative complications and exacerbations or relapses influence the anesthetic plan. Should the patient have any mental impairment, informed consent for the anesthetic plan must be obtained in the presence of a competent next of kin or legal guardian and documented as such.

Anesthetic Considerations

Surgery and general anesthesia are stress-inducing factors that may exacerbate MS. To date no attempt to differentiate between the effects of surgery and general anesthesia on the course of MS has been made.[25]

Most general anesthetic agents appear safe for patients with MS, although thiopental has been implicated in MS relapse or exacerbation. Baskett and Armstrong reported four cases in which effects of specific anesthetic agents affected the deterioration of MS.[26] In case 1, thiopental was used, and the patient relapsed. In case 2, thiopental was associated with relapse; but on repeat operation, when thiopental was used again, no relapse occurred. In case 3, the patient had no relapse after thiopental. In all four cases, N_2O, O_2, or halothane alone or with other anesthetic agents (except thiopental) were without sequelae. Siemkowicz was unable to draw conclusions on the course of MS and general anesthesia with thiopental. In his cases, where thiopental was used and sequelae were found, the aggravated clinical status was associated with increased temperature, but fever is unlikely to be secondary to thiopental.[8]

Succinylcholine should be avoided in patients with neurologic defects and muscular atrophy because of the potential side effect of hyperkalemia. However, it can be administered to patients in remission or with mild symptoms. Non-depolarizing muscle relaxants are safe to use.[27]

Anticholinergic agents may increase temperature, a state that is poorly tolerated. It is known that any increase in temperature of 0.5 to 1.0°C may reversibly block the conduction in the marginally functioning demyelinated axons of MS, thus worsening the symptoms. Davis et al found that demyelinated axons are capable of conduction but function optimally at temperatures less than 38.0°C, and an increase in temperature may inhibit this optimal function.[28]

The use of regional anesthesia is an area of controversy. Because regional techniques are the preferred modes of anesthesia in parturients, the controversy is of particular interest to the obstetric anesthesiologist.

Epidural anesthesia is preferred over spinal anesthesia because the concentration of local anesthetic in the white matter of the spinal cord is 3 to 4 times higher following spinal administration compared to epidural (1.37μg/mg versus 0.4μg/mg).[29,30] The demyelinated neural tissue in MS predisposes the spinal cord to neurotoxicity. The blood-brain barrier is also more permeable to local anesthetic; therefore, the toxic dose may be lower than in normal patients. A limited study reported by Bamford et al revealed a 10% incidence of aggravation of MS symptoms in the post-spinal-anesthetic month, which is higher than the expected relapse rate.[25]

Epidural anesthesia is favored because it involves a peridural, rather than subarachnoid, administration of local anesthetic. Toxicity to the spinal cord is reduced.

Although epidural anesthesia for labor and delivery or for cesarean section involves deposition of a high dose of local anesthetic in the epidural space, no apparent adverse effects have been reported. Bader et al found that women who delivered vaginally or by cesarean section had a similar incidence of relapse whether or not epidural analgesia was given.[27] Types of local anesthetic used in the studies were bupivacaine 0.5% and lidocaine 2%; no reports on chloroprocaine have been documented, probably because of its alleged neurotoxicity.

It has been suggested that the associated increased emotional or physical stress, length of labor, infection and temperature change are more important contributing factors than local anesthetics. There are insufficient statistical data to justify the elimination of epidural anesthesia, especially in parturients.

Parturients who may benefit most from epidural anesthesia include those with diabetes mellitus, cardiac diseases and preeclampsia.[31,32] However, parturients should be well informed about the benefits and risks of epidural anesthesia and other options as well as postoperative sequelae.

No strong evidence links the incidence of MS exacerbation to local anesthetic administration. A study of 98 patients receiving more than 1000 local anesthetics revealed only 4 patients with symptomatology that was not consistent with a characteristic relapse of the disease.[25]

Diagnostic lumbar puncture alone apparently does not induce relapse.[9] Also, peripheral nerve block has no known association with relapses.

Anesthetic Plan

The case history presented has several important facets that influence the choice of anesthetic techniques.

The patient is in the second trimester of gestation; therefore, all physical and physiologic changes must be considered. Intubation of the trachea may be difficult secondary to capillary engorgement of the upper airway. In addition, she also has a history of asthma, and any airway irritation or emotional stress may precipitate bronchospasm. Precautions because of recent eating, increased gastric secretion, and incompetent gastroesophageal junction secondary to pregnancy must be considered. The risk of aspiration pneumonia under general anesthesia is increased.

The patient is in remission of MS with no physical or mental impairment. Finally, the safety of the fetus must be considered. A single or continuous administration of lumbar epidural anesthesia, lidocaine 2% solution, to dermatome T10 level is adequate for the cervical cerclage and carries less risk of MS relapse.

Preoperatively, informed consent with detailed discussion of the benefits and risks of general, spinal and lumbar epidural anesthesia should be obtained from the patient, who must agree to the lumbar epidural anesthesia. Neurologic examination must be included in the preoperative assessment. If time permits, metoclopramide 10mg intravenous soluset may be given to promote gastrointestinal motility and stomach emptying. A nonparticulate antacid (eg, citric acid solution) should be given preoperatively.

Prehydration with 500ml of isotonic solution prior to induction of lumbar epidural anesthesia will prevent hypotension secondary to the sympathetic block. Emergency drugs such as ephedrine, albuterol, terbutaline, and aminophylline must be available. Monitors should include

ECG, blood pressure, pulse oximeter, skin temperature, precordial stethoscope and possibly fetal heart monitor. Oxygen by nasal cannulae is advisable.

References

1. Wyngaarden J.B., Smith L.H.: The demyelinating diseases, in *Cecil Textbook of Medicine*, 18th ed., Philadelphia: WB Saunders, 1988, 2211–7.
2. McFarlin D.E., McFarlin H.F.: Multiple sclerosis (pts 1,2). *N Engl J Med* 1982, 307:1183–8, 1246–51.
3. Kurtzke J.F., Hyllested K.: Multiple sclerosis in the Faroe Island: I. Clinical and epidemiological features. *Ann Neurol* 1979, 5:6–21.
4. Rowland L.P.: Multiple sclerosis, in *Merritt's Textbook of Neurology,* 8th ed. Philadelphia: Lea & Febiger, 1989, pp 741–60.
5. Kurtzke J.F.: Epidemiologic contributions to multiple sclerosis: An overview. *Neurology (NY)* 1980, 30:61–79.
6. Currier R.D., Eldridge R.: Possible risk factors in multiple sclerosis as found in a natural twin study. *Arch Neurol* 1982, 39:140–4.
7. Ebers G.C., Paty D.: HLA typing in multiple sclerosis sibling pairs. *Lancet* 1982, 1:88–90.
8. Siemkowicz E.: Multiple sclerosis and surgery. *Anaesthesia* 1976, 31: 1211–6.
9. Schapira K.: Is lumbar puncture harmful in multiple sclerosis? *J Neurol Neurosurg Psychiatry* 1959, 22:238.
10. Wilson J.D., Braunwald E., Isselbacher K.J., et al.: Demyelinating diseases, in *Harrison's Principles of Internal Medicine,* 12th ed., New York: McGraw-Hill, 1991, pp 2039–43.
11. Robbins S.L., Cotran R.S.: Demyelinating diseases, in *Pathologic Basis of Disease,* 4th ed., Philadelphia: WB Saunders, 1989, pp 14–5.
12. Poser C., Presthus J., Horstal O.: Clinical characteristics of autopsy-proved multiple sclerosis. *Neurology* 1966, 16:791–8.
13. Waxman S.G.: Current concepts in neurology: Membranes, myelin, and pathophysiology of multiple sclerosis. *N Engl J Med* 1982, 306:1529–33.
14. Delisa J.A., Hammond M.C., Mikulic M.A., Miller R.M.: Multiple sclerosis: Common physical disabilities and rehabilitation (pts 1,2). *Am Fam Physician* 1985, 32:157–63, 127–32.
15. McFarlin D.E.: Treatment of multiple sclerosis. *N Engl J Med* 1983, 308:215–7.
16. Hallett M., Lindsey J.W., Adelstein B.D., et al.: Controlled trial of isoniazid therapy for severe postural cerebellar tremor in multiple sclerosis. *Neurology (NY)* 1985, 35:1374–7.
17. Korn-Lubetzki I., Kahana E., Cooper G., Abramsky O.: Activity of multiple sclerosis during pregnancy and puerperium. *Ann Neurol* 1984, 16:229–31.
18. Birk K., Smeltzer S.C., Rudick R.: Pregnancy and multiple sclerosis. *Semin Neurol* 1988, 8:205–13.

19. Thompson D.S., Nelson L.M., Burn A., Burks J.S., Franklin G.M.: The effects of pregnancy in multiple sclerosis: A retrospective study. *Neurology* 1986, 36:1097–9.
20. Frith J.A., McLead J.G.: Pregnancy and multiple sclerosis. *J Neurol Neurosurg Psychiatry* 1988, 51:495–8.
21. Ghezzi A., Caputo D.: Pregnancy: A factor influencing the course of multiple sclerosis? *Eur Neurol* 1981, 20:115–7.
22. Birk K., Ford C., Smeltzer S., Ryan D., Miller R., Rudick R.A.: The clinical course of multiple sclerosis during pregnancy and the puerperium. *Arch Neurol* 1990, 47:738–42.
23. Abramsky O., Brenner T., Mizrachi R., Soffer D.: Alpha-fetoprotein suppresses experimental allergic encephalomyelitis. *J Neuroimmunol* 1982, 2:1–7.
24. Sill J.C., Schumacher K., Southorn P.A., et al.: Bradycardia and hypotension associated with baclofen used during general anesthesia. *Anesthesiology* 1986, 64:255–8.
25. Bamford C., Sibley W., Laguna J.: Anesthesia in multiple sclerosis. *Can J Neurol Sci* 1978, 5:41–4.
26. Baskett P.J., Armstrong R.: Anesthetic problems in multiple sclerosis. Are certain agents contraindicated? *Anaesthesia* 1970, 25:397–401.
27. Bader A.M., Hunt C.O., Datta S., Naulty J.S., Ostheimer G.W.: Anesthesia for the obstetric patient with multiple sclerosis. *J Clin Anesth* 1988, 1:21–4.
28. Davis F.A., Michael J.A., Neer D.: Serial hyperthermia testing in multiple sclerosis: A method for monitoring subclinical fluctuation. *Acta Neurol Scand* 1973, 49:63–74.
29. Cohen E.N.: Distribution of local anesthetic agents in the neuraxis of the dog. *Anesthesiology* 1968, 29:1002–5.
30. Bromage P.R.: Mechanism of action of extradural analgesia. *J Anaesth* 1975, 47:199–211.
31. Warren T.M., Datta S., Ostheimer G.W.: Lumbar epidural anesthesia in a patient with multiple sclerosis. *Anesth Analg* 1982, 61:1022–3.
32. Jones R.N.: Lumbar epidural anesthesia in patients with multiple sclerosis [letter]. *Anesth Analg* 1983, 62:856–7.

CHAPTER 24

The Hemorrhaging Obstetric Patient

Jill Fong, M.D.

Case History. *A 34-year-old gravida 5, para 4, unregistered patient presented to the labor and delivery room with painless vaginal bleeding at 37 weeks' gestation by dates. On physical examination, the patient had a blood pressure of 90/60mmHg; pulse, 120 beats/min, and was found to be at 42 weeks gestation.*

Ultrasonography showed that the patient had placenta previa and hydramnios. The patient was contracting, and the fetal heart rate was 120–160 beats/min with intermittent variable decelarations. Laboratory values were normal, including the coagulation profile. As the medical staff was preparing the patient for cesarean section, she had a precipitous delivery.

Suddenly, she became hypotensive, and her oxygen saturation fell from 98% to 50%. The obstetrician was unable to remove the placenta completely presumably because of a partial placenta accreta. The patient was bleeding vaginally and from intravenous puncture sites. Emergency anesthetic consultation was requested as preparation was made for hysterectomy.

Introduction

Obstetric hemorrhage usually occurs either at the site of placental implantation or from disrupted blood vessels in the uterus or birth canal. Vascular integrity is required to prevent bleeding. Any interruption of vascular integrity initiates the hemostatic mechanism. There are four components of the hemostatic mechanism: vascular reaction, platelet function, coagulation cascade and clot lysis. Normally all four components respond to vascular injury simultaneously. Uterine contraction usually minimizes

---

Reviewed by Dr. Steven S. Schwalbe, Assistant Professor of Anesthesiology, Director of Obstetric Anesthesia, Albert Einstein College of Medicine/Montefiore Medical Center.

bleeding from the placental site by causing direct vascular compression. The coagulation mechanism stops bleeding if the damage to the blood vessel is small, and vascular constriction occurs to prevent intravascular pressure from displacing the platelet plug.[1]

Antepartum hemorrhage usually occurs at the placental site. Placenta previa and abruptio placentae account for about two thirds of this bleeding. Uterine rupture and fetal bleeding are less common causes.

Postpartum hemorrhage is defined as occurring within 24 hours of delivery, most frequently during the third stage of labor—the delivery of the placenta. Causes include uterine atony, lacerations, retained placenta, placenta accreta/increta/percreta, lower-uterine-segment placental implantation, uterine rupture, uterine inversion and coagulation abnormalities (Table 1).

Any obstetric patient with bleeding should have vascular access instituted and fluid resuscitation begun in addition to the rapid assessment of maternal and fetal well-being. When providing anesthesia for such patients, it is important to understand maternal physiology and placental anatomy and function, and to assess fetal well-being and the cause of the maternal bleeding.

TABLE 1. Predisposing Factors and Causes of Immediate Postpartum Hemorrhage

- **Trauma to the genital tract**
  - Large episiotomy, including extensions
  - Lacerations of perineum, vagina, or cervix
  - Ruptured uterus
- **Bleeding from placental implantation site**
  - Hypotonic myometrium—uterine atony
    - Some general anesthetics—halogenated hydrocarbons
    - Poorly perfused myometrium—hypotension
      - Hemorrhage
      - Conduction analgesia
    - Overdistended uterus—large fetus, twins, hydramnios
    - Following prolonged labor
    - Following very rapid labor
    - Following oxytocin-induced or augmented labor
    - High parity
    - Uterine atony in previous pregnancy
    - Chorioamnionitis
  - Retained placental tissue
    - Avulsed cotyledon, succenturiate lobe
    - Abnormally adherent—acreta, increta, percreta
- **Coagulation defects**
  - Intensify all of the above

Maternal Physiologic Changes

Pregnancy alters the function of every major organ system. Maternal physiologic changes of most importance to the anesthesiologist occur in the respiratory, cardiac, hematologic, gastrointestinal, renal and central nervous systems.[2] The respiratory system changes lead to a decrease in total lung capacity and functional residual capacity and an increase in respiratory rate, tidal volume and minute ventilation.[3] Increases in cardiac output and oxygen consumption and decreases in total peripheral resistance and mean arterial pressure occur.[4,5] Because there is a 10%–15% incidence of supine hypotension caused by aortocaval compression, patients of more than 20-weeks gestation should be maintained in a left lateral tilt position. Aortocaval compression can lead to maternal hypotension and a decrease in uteroplacental perfusion.[6] The pregnant patient also has an increased blood volume, with plasma volume increasing more than red blood cell volume.[7]

From the end of the first trimester and persisting for 1 to 2 weeks postpartum, gastrointestinal changes occur that lead to an increased maternal risk of aspiration of gastric content resulting from an incompetent gastroesophageal junction, decreased gastric motility and increased acid secretion.[8] Renal changes include increase in renal blood flow and glomular filtration rate.[9] Central nervous system changes include a decreased anesthetic requirement.[10,11]

Normal Placenta Anatomy and Physiology

The placenta provides the crucial link between the mother and her fetus, supplying essential nutrients to the fetus and disposing of unwanted wastes. Thus, fetal well-being depends on a properly functioning placenta. The placental implantation site is often the origin of obstetric hemorrhaging.

The placenta consists of decidua basalis, chorion, amnion and umbilical cord. The decidua basalis is of endometrial origin; the other components are of fetal origin. The chorion is directly applied to the endometrial layer of the uterus, the decidual basalis. The amnion is closest to the baby and in direct contact with the amniotic fluid. The umbilical cord brings replenished oxygenated fetal blood to the fetus via the umbilical vein and delivers this blood back to the placenta for exchange with maternal blood via the two umbilical arteries.[12]

The chorion is composed of villi of fetal origin; these villi contain fetal blood vessels that come into direct contact with intervillous maternal blood separated only by a thin fetal endothelial layer. Each maternal endometrial spiral arteriole is surrounded by a number of chorionic villi, forming units called cotyledons. Maternal arterial blood enters the basal plate of the placenta under each major cotyledon and supplies oxygenated blood to the intervillous space. This blood diffuses throughout the intervillous space and returns to maternal veins in the basal plate. Normally, at the time of placental separation the placental chorion separates from the decidua basalis.

Amniotic fluid surrounds the fetus and provides a cushion against possible injury. It helps maintain fetal temperature and yields useful information concerning the health of the fetus. Normally, the volume of amniotic fluid increases to about 1L by the 36th week of gestation and decreases thereafter. Somewhat arbitrarily, more than 2000ml of amniotic fluid is considered excessive and is referred to as hydromnios or polyhydramnios. The etiology is unclear, but it is often associated with fetal anencephaly, spina bifida fetuses, multiple gestations, or esophogeal atresia. There is a 20% incidence of associated fetal malformations. The problems with hydramnios include mechanical difficulties for the mother, such as profound aortocaval compression and an increased risk of postpartum hemorrhage or maternal abruption. Oligohydramnios refers to a volume of amniotic fluid that is less than normal.

Oligohydramnios is practically always evident when there is either fetal urinary tract obstruction or renal agenesis. Pulmonary hypoplasia can occur.[13]

The placenta has many functions. It is an organ of respiration, nutrient and drug transfer, metabolism and hormone production. Of main concern to the anesthesiologist are the respiratory and solute transferring capacities of the placenta. The most important acute fetal risk is hypoxia. Fetal oxygenation depends on uteroplacental blood flow, maternal oxygen-carrying capacity, oxygen affinity and arterial oxygen tension. Uterine blood flow can be thought of as analogous to electrical current, which is described by Ohm's Law: $I = V/r$, or Current = Driving Force/Resistance. In the case of uterine blood flow, this can be written as follows:

$$\text{Uterine blood flow} = \frac{(\text{arterial pressure} - \text{venous pressure})}{\text{uterine artery resistance}}$$

In late pregnancy 10% of the maternal cardiac output normally goes to the uterus. Of this the placenta receives 80%; the myometrium, 20%. Flow is directly proportional to pressure; the uterine vessels do not autoregulate. Any decrease in uterine artery pressure or increase in uterine vein pressure or uterine artery resistance decreases uterine blood flow. Hypotension, endogenous catecholamines, alpha-adrenergic vasopressors and aortic compression can decrease uterine artery pressure. Inferior vena caval compression and increased uterine tone can lead to increased uterine venous pressure; catecholamines increase uterine artery resistance.[12,14]

Transfer of substances such as drugs across the placenta depends in varying degrees on simple diffusion, facilitated transport and active transport. The rate of transfer of substances by simple diffusion is governed by Fick's law of diffusion, which states:

$$\text{Rate of transfer} = (\text{concentration gradient} \times \frac{\text{area} \times \text{permeability}}{\text{thickness}})$$

Permeability, in turn, is directly related to the degree of lipid solubility and inversely related to the degree of ionization, molecular weight and degree of protein binding.[13]

Intrapartum Fetal Assessment

The goal during labor is to preserve fetal well-being by the early detection and treatment of fetal distress. Fetal well-being is based on an intact uteroplacental unit. The methods used to assess the fetus take into account the important function of the uteroplacental unit and the acute risk of fetal hypoxia with its signs and symptoms. Simultaneous monitoring of fetal heart rate (FHR) and maternal contractions aids in the identification of fetal well-being. Fetal heart rate can be monitored noninvasively by means of the Doppler technique or invasively by using a fetal scalp electrode. Uterine contractions are monitored noninvasively by a tocodynamometer, which measures the tightening of the maternal abdominal muscles, or invasively by a transcervical pressure catheter connected to a strain gauge.

Normal FHR is 120 to 160 beats/min. The beat-to-beat variability is normally $>$ 6 beats/min. Abnormalities in heart rate, such as bradycardia or tachycardia, dysrhythmias, decreased or absent beat-to-beat variability, or transient decelerations, can be a sign of fetal asphyxia.

Transient decelerations in FHR fall into three categories, depending on their shape and timing with respect to maternal contractions. Early (Type 1) decelerations are symmetric, uniform decreases that coincide with the onset, peak and end of uterine contraction. These represent a vagal response to fetal head compression and are not usually associated with fetal hypoxia. Late (Type 2) decelerations are symmetric decreases that begin after the onset of a contraction and last beyond the end of the contraction; they indicate uteroplacental insufficiency and require prompt evaluation and treatment. Variable (Type 3) decelerations occur variably and usually abruptly in relationship to contractions. They result from umbilical cord compression and are usually asymmetric and often bizarre in shape. Variable decelerations are further classified by degree and duration into mild, moderate and severe categories. Mild variable decelerations usually prove insignificant. They involve a decrease in FHR to $>$ 80beats/min, lasting $<$ 30 seconds, or FHR $>$ 70 beats/min lasting $<$ 60 seconds. Moderate variables may signal mild hypoxia and are defined as a FHR $>$ 70 beats/min for 30–60 seconds. Severe variables (FHR $<$ 70 beats/min for $>$ 60 seconds) can indicate frank fetal acidosis. However, the predictive value of FHR tracing abnormalities is only about 30%–50%.

Significant fetal hypoxia leads to anaerobic metabolism and a systemic fetal acidosis, as documented by a decreased pH in the fetal scalp capillary blood sample. A normal fetal scalp pH is 7.25 to 7.45, preacidotic is 7.20 to 7.25, and acidotic is less than 7.20. If the fetal scalp sample is acidotic, the fetus should be delivered. Fetal scalp sampling has several limitations; most obviously, access to the fetal scalp is essential and is not possible in many situations, including placenta previa and breech presentation.[15]

Obstetric Hemorrhage

Antepartum Hemorrhage

Third-trimester bleeding is assumed to be a significant problem until proved otherwise.[13,16,17] (Table 2).

The bleeding patient should be admitted to the hospital, and immediate attention should be given to vascular access and availability of resuscitative fluid during maternal and fetal evaluation. Antepartum hemorrhaging is usually caused by placenta previa or aburptio placentae, accounting for one half to two thirds of all cases. Uterine rupture is another cause of antepartum, intrapartum and postpartum hemorrhage.[18]

Placenta previa occurs if the placenta is implanted very near or over the internal cervical os instead of in its normal position in the body of the uterus. A complete placenta previa covers the cervical internal os. With a partial previa, the placenta partially covers cervical os, and marginal placenta previa just overlies the margin of the internal os. A low-lying placenta refers to a placenta that is implanted in the lower uterine segment. The degree of placenta previa depends on cervical dilation at the time of examination. Placenta previa has been diagnosed in approximately 0.4% to 0.6% of pregnancies. Placental implantation on the lower uterine segment occurs once in every 200 pregnancies. Placenta previa occurs more frequently with multiparity, advancing age and previous cesarean delivery. Patients with placenta previa are also at increased risk for placenta accreta.[12,19,20]

Placenta previa rarely causes maternal death but does cause significant morbidity. Depending on the study, two thirds to 90% of patients

TABLE 2. Conditions that Predispose to or Worsen Obstetric Hemorrhage

Abnormal placentation
Placental previa
Abruptio placentae
Placenta accreta
Ectopic pregnancy
Hydatidiform mole
Trauma during labor and delivery
Complicated vaginal delivery
Cesarean section or hysterectomy
Uterine rupture; risk increased by
Previously scarred uterus
High parity
Hyperstimulation
Obstructed labor
Intrauterine manipulation
Uterine atony
Overdistended uterus
Multiple fetuses
Hydramnios
Distension with clots
Anesthesia or analgesia
Halogenated agents
Conduction analgesia with hypotension
Exhausted myometrium
Rapid labor
Prolonged labor
Oxytocin or prostaglandin stimulation
Previous uterine atony
Coagulation defects—intensifies other causes
Placental abruption
Prolonged retention of dead fetus
Amnionic fluid embolism
Saline-induced abortion
Sepsis with endotoxemia
Severe intravascular hemolysis
Massive transfusions
Severe preeclampsia and eclampsia
Congenital coagulopathies

with previa have at least one antepartum hemorrhage. In the study reporting a 90% incidence, 10%–25% of these patients developed hypovolemic shock.[21] Perinatal mortality has improved over the past 10 years, from 17%–25%[18] in the early 1970s to 4.2%–8.1%[22] in the mid-1980s.

The most characteristic event in patients with placenta previa is painless vaginal bleeding that usually appears in the third trimester of pregnancy. Because the placenta is located over the internal os, the formation of the lower uterine segment and dilation of the os inevitably results in tearing of the placental attachments and bleeding from the uterine vessels. The bleeding is augmented by the inability of the myometrial fibers of the lower uterine segment to contract and retract, compressing the uterine vessels. Coagulation problems are rarely seen with placenta previa.

The simplest, most precise and safest method of placental localization is provided by ultrasonography, which is accurate in 98% of cases.[12,23] If ultrasound or radioisotope scan is not conclusive, a definitive diagnosis can be made by direct examination of the cervical os. This procedure should be done in a delivery room only after preparations are made for cesarean section; this precaution is commonly referred to as "the double setup."

Obstetric management of the patient with placenta previa depends on many factors, including the maturity of the fetus, the incidence of active labor and the degree of hemorrhage. Prior to term the patient is usually managed with bed rest, fluid resuscitation and tocolysis if necessary, in the hope that the bleeding will stop spontaneously.[24] Fetal lung maturity is assessed by amniocentesis, and if mature, the baby is delivered. If no active labor exists, if the fetus is premature and if bleeding is minimal, there is no pressing need for delivery. If the fetus is within three weeks of term and considered to be mature, delivery is often performed.

Depending on the degree of cervical os coverage, vaginal delivery may be attempted or cesarean section selected. In a complete placenta previa, cesarean section is the accepted method of delivery, but significant blood loss can occur. The uterine incision may enter the anterior placenta and result in both maternal and fetal hemorrhage. Because the lower-uterine-segment placental implantation site does not contract as well as the normal fundal site, bleeding can occur despite good uterine contraction. If bleeding continues, the obstetrician can oversew individual bleeding sites, compress the area with an overinflated cervical Foley catheter balloon,[25] ligate the hypogastric arteries, or if necessary, perform a hysterectomy. If there is an associated placenta accreta, hysterectomy is usually performed.

Aburptio placentae refers to the separation of a normally implanted placenta after 20 weeks gestation and before the birth of the fetus. Its incidence varies from 0.2% to 2.4%.[17,18,26,27] Multiparity, pregnancy-induced hypertension, previous abruption and uterine anomalies are associated with abruptio placentae. Maternal mortality ranges from 1.8%–

11% in one series to 0%–5% in a more recent study.[18,27] Perinatal mortality varies from 19%–67% with vaginal delivery to 8%–22% with cesarean section.[27–29]

Abruptio placentae is classified according to the amount of bleeding, fetal distress and the existence of coagulopathy. In 85% of cases bleeding is apparent; however, in 15% there is occult bleeding. Abruptio placentae is considered severe if there is 1000ml of blood loss with or without fetal distress or disseminated intravascular coagulopathy (DIC). Classically, patients with severe abruptio present with vaginal bleeding, a painful hypertonic uterus, hypovolemia and absent or inaudible fetal heart tones. Mild abruptio placentae refers to < 250ml of blood loss with no fetal distress or coagulopathy.[17,30]

Uterine rupture occurs in 0.1% of deliveries in the United States.[31] It is associated with trauma, previous uterine surgery, rapid spontaneous delivery, cephalopelvic disproportion, hydromnios, cocaine abuse and multiparity. A trial of labor after a previous cesarean section via a low-lying uterine incision may be attempted but not after a classical cesarean section because of the high incidence of maternal and fetal mortality.[32]

Depending on the severity of the uterine rupture, the patient may experience severe abdominal pain, vaginal bleeding, disappearance of fetal heart tones and severe hypotension and maternal shock. Nevertheless, significant rupture may be entirely without maternal symptoms and be found only at cesarean section or postpartum tubal ligation. It is associated with approximately a 5% maternal death rate and a 50% fetal death rate.[33]

Fetal Bleeding

Antepartum bleeding is usually from the mother; fetal bleeding is unusual and difficult to identify, and it presents a much more immediate risk to the fetus. The diagnosis of fetal bleeding is made by the identification of fetal hemoglobin. Fetal bleeding can follow the disruption of fetal vessels, as can be seen with vasa previa, with velamentous insertion of the umbilical cord, or rarely, following fetal scalp blood sampling. Fetal salvage requires a high index of suspicion, rapid delivery and skillful neonatal resuscitation.[16]

Postpartum Hemorrhage

Maternal bleeding always occurs at birth. The definition of postpartum hemorrhage classically has been defined as > 500ml of blood loss. However, measurements of blood volume changes rather than blood loss, which is difficult to measure, suggest that the average blood loss is 600ml with vaginal delivery and 100ml with cesarean section.[34] Thus, the definition remains meaningless in practical terms. For this discussion, therefore, postpartum hemorrhage will include any bleeding that occurs after the delivery of the baby and that produces maternal morbidity or mortality. Postpartum hemorrhage can occur for a variety of reasons. The most common is uterine atony, followed by incidence by lacerations of the vagina and cervix. Other causes are retained placenta, placenta accreta, lower-uterine-segment placental implantation, uterine rupture, uterine inversion, and acquired or drug-induced coagulopathies.

Placenta accreta is defined by histopathology, based on the absence of deficiency of decidua basalis between the villi and myometrium. Normally, the placenta spontaneously separates from the uterine decidua basalis. If the villi actually invade the myometrium, it is referred to as a placenta increta. Placenta percreta occurs when the villi penetrate through the myometrium to involve the serosa. This abnormal adherence may involve all of the placental cotyledons and be a total placenta accreta. If a few to several cotyledons are involved, it is a partial placenta accreta. Involvement of single cotyledon is termed focal accreta.[35–37]

The true frequencies of placenta accreta, increta and percreta are unknown but appear to be on the rise. In 1980 Read and coworkers reported an incidence of clinically diagnosed placenta accreta as 1 in 2562 pregnancies (0.04%) with histologic confirmation in only 1 in 4027 pregnancies (0.02%).[38] Maternal mortality was estimated as 3.1% for these cases. A more recent study found an incidence of 0.7%.[39]

The occurrence of placenta accreta is strongly associated with placenta previa and/or a prior history of cesarean section or other uterine surgery. The risk of accreta in a woman with placenta previa who has had no prior history of cesarean section is 5%–7%. In a parturient with placenta previa and a history of prior cesarean section deliveries, the risk of accreta increases to 24%–31%; with four or more cesarean sections, it goes as high as 67%.[40,41] Antepartum hemorrhage is common but usually is a consequence of the coexisting placenta previa.

The treatment of placenta accreta is to control bleeding and restore hemodynamic stability. In deciding on the therapeutic options, the amount of bleeding, the proportion of placental surface involved, the depth of invasion and the patient's desire for further pregnancies must be taken into consideration. In small partial or focal placenta accreta, conservative treatment—manual removal of the placenta and packing of the uterus—may be sufficient. If this fails, hypogastric arterial ligation may be attempted. Total accreta, increta or percreta often require immediate hysterectomy. Conservative treatment leads to an increased incidence of maternal death when compared to immediate hysterectomy.[40]

Uterine atony occurs in 2%–5% of vaginal deliveries and is the most common cause of postpartum hemorrhage and maternal hemorrhagic death.[42,43] Normally, the uterus contracts as the placenta separates, causing the compression of spiral arteries within the myometrium and the cessation of blood flow. When this process is prevented by uterine atony, hemorrhage can be profound. There is usually painless bleeding that continues longer than a few minutes. By palpation, the uterus feels soft and has not contracted.

Uterine atony should be anticipated if the patient has any condition that can lead to inefficient myometrial contraction. Hydramnios or multiple gestation can cause uterine atony because smooth muscle fibers stretched to their full length do not contract effectively. Hypoxia or ischemia may produce metabolic interference with contraction. Pharmacologic agents such as the volatile anesthetics, tocolytics and dantrolene may increase the risk of uterine atony. Atony also occurs more commonly in patients with DIC, multiparity, and a previous history of postpartum hemorrhage.

Treatment of uterine atony consists of vigorous replacement of intravascular volume and attempts to stimulate uterine contraction. Massaging the uterus and pharmacologic therapy are the first line of treatment. Drugs that cause the uterus to contract include synthetic oxytocin, the ergot alkaloids (eg, ergonvine and methlergonovine) and the prostaglandins (eg, prostaglandin $F_2\alpha$ the 15-methyl analog and prostaglandin E_2). When pharmacologic measures fail to stimulate myometrial contraction, the obstetrician must interrupt the uterine blood supply or remove the bleeding uterus.[16]

Amniotic Fluid Embolism

The differential diagnosis of conditions causing acute cardiorespiratory failure in the pregnant patient includes pulmonary embolism caused by amniotic fluid, thrombi, or air; aspiration pneumonitis; toxicity of local anesthetics; eclampsia; intracranial hemorrhage, hemorrhagic shock; and acute heart failure. To diagnose amniotic fluid embolus, it's important to know the four cardinal features of sudden onset: (1) respiratory distress, (2) cyanosis, (3) cardiovascular collapse and (4) coma. In addition, chills, sweating, coughing, hyperreflexia and convulsions occasionally occur.

The diagnosis of amniotic fluid embolism is made by finding fetal elements in the maternal circulation. Blood should be aspirated from the right side of the heart through a central venous pressure or Swan-Ganz catheter. That pathologist should be notified, and amniotic fluid particles should be sought when examining this specimen.[44,45] However, amniotic fluid particles have been found in the blood of asymptomatic patients[46]; therefore, amniotic fluid embolism is often a diagnosis of exclusion.

The incidence of amniotic fluid embolism averages 1 in 20,000 pregnancies with various extremes reported: 1 in 1250 to 1 in 80,000 pregnancies. The cause is unknown. Some authors believe that the main route of entry of amniotic fluid is at the placental margin, as might occur with placenta previa, placental abruption, placenta accreta, retained placenta, ruptured uterus or cesarean section.[44,45]

When amniotic fluid enters the maternal circulation, pulmonary embolism, DIC and uterine atony occur. The entry of amniotic fluid into the pulmonary circulation causes cardiorespiratory collapse by increasing pulmonary artery pressure, decreasing blood return to the left heart and causing ventilation perfusion abnormalities. Uterine atony is caused by hypotension and the direct depressant effect of the embolus on uterine muscles. Disseminated intravascular coagulopathy occurs secondary to the liberation of thromboplastic substances, amniotic fluid and/or placental thromboplastin into the circulation. This thromboplastic material causes widespread clotting followed by diffuse lyses, with bleeding secondary to the consumption of coagulation factors. Laboratory tests reveal that the prothrombin time, partial thromboplastin time and thrombin time are prolonged, platelet number and fibrinogen levels are decreased, and fibrin split products are increased.[1]

Once it is determined that an amniotic fluid embolism has occurred, treatment is supportive. Cardiopulmonary resuscitation is instituted as necessary, including intubation and ventilation with 100% oxygen. Large-bore intravenous cannulas, central pressure monitoring, Foley catheter and arterial monitor are placed. Disseminated intravascular coagulopathy is treated by resolving the underlying cause. The temporary treatment consists of "feeding the fire" by giving blood products as necessary to minimize bleeding. Despite all measures, the mortality rate in amniotic fluid embolism is 86%.[45]

Anesthetic Considerations

The hemorrhaging obstetric patient should receive prompt and vigorous intravascular volume replacement, and hemodynamic stability should be maintained. Patients with antepartum bleeding often require cesarean section. If the patient is normovolemic and the fetus is not stressed, epidural or spinal anesthesia can be used. However, with excessive bleeding, the sympathetic block produced by these regional anesthetic techniques can be detrimental to both mother and fetus. General anesthesia is acceptable for all of these patients for cesarean section or hysterectomy. If general anesthesia is used, it is important to modify the technique based on the mother's hemodynamic status.

References

1. Fischbach D.P., Fogdall R.P.: *Coagulation: The Essentials.* Baltimore: Williams & Wilkins, 1981, pp 1–32.
2. Cheek T.G., Gutsche B.B.: Maternal physiologic alterations during pregnancy, in Shnider S.M., Levinson G. (eds): *Anesthesia for Obstetrics.* Baltimore: Williams & Wilkins, 1987, pp 3–13.
3. Russell I.F., Chambers W.A.: Closing volume in normal pregnancy. *Br J Anaesth* 1981, 53:1043–6.
4. Walters W.A.W., MacGegor W.G., Hill M.: Cardiac output at rest during pregnancy and the puerperium. *Clin Sci* 1966, 30:1–11.
5. Ylikorkala O., Jouppila P., Kirkinen P., Vilinikka L.: Maternal prostacyclin, thromboxane, and placental blood flow. *Am J Obstet Gynecol* 1983, 145:730–2.
6. Marx G.F.: Aortocaval compression: Incidence and prevention. *Bull NY Acad Med* 1974, 50:443–6.
7. Lund C.J., Donovan J.C.: Blood volume during pregnancy. *Am J Obstet Gynecol* 167;98:393–403.
8. Simpson K.H., Stakes A.F., Miller M.: Pregnancy delays paracetamol absorption and gastric emptying in patients undergoing surgery. *Br J Anaesth* 1988, 60:24–7.
9. Dignam W.J., Titus P., Assali N.S.: Renal function in human pregnancy: 1. Changes in glomerular filtration rate and renal plasma flow. *Proc Soc Exp Biol Med* 1958, 97:512–4.
10. Palahniuk R.J., Shnider S.M., Eger E.I., II: Pregnancy decreases the requirements for inhaled anesthetic agents. *Anesthesiology* 1974, 41:82–3.
11. Datta S., Lambert D.H., Gregus J., Gissen A.J., Covino B.G.: Differential sensitivities of mammalian nerve fibers during pregnancy. *Anesth Analg* 1983, 62:1070–2.
12. Taylor E.S.: Obstetrics and Fetal Medicine. Baltimore: Williams & Wilkins, 1977, pp 22–36, 257–70.
13. Pritchard J.A., MacDonald P.C., Grant N.F.: *Williams Obstetrics,* 17th ed. Norwalk, Conn: Appleton-Century-Crofts, 1985, pp 462–5, 389–421, 707–18.
14. Parer J.T.: Uteroplacental circulation and respiratory gas exchange, in Shnider S.M., Levinson G. (eds): *Anesthesia for Obstetrics.* Baltimore: Williams & Wilkins, 1987, pp 14–40.
15. Parer J.T.: Diagnosis and management of fetal asphyxia, in Shnider S.M., Levinson G. (eds): *Anesthesia for Obstetrics.* Baltimore: Williams & Wilkins, 1987, pp 474–88.
16. Plumer M.H.: Bleeding problems, in James F.M., Wheeler A.S., Dewan D.M. (eds): *Obstetric Anesthesia: The Complicated Patient,* 2nd ed., Philadelphia: FA Davis, 1988, pp 309–44.
17. Biehl D.R.: Antepartum and postpartum hemorrhage, in Shnider S.M., Levinson G. (eds): *Anesthesia for Obstetrics.* Baltimore: Williams & Wilkins, 1987, pp 281–9.

18. Abdul-Karim R.W., Chevil R.N.: Antepartum hemorrhage and shock. *Clin Obstet Gynecol* 1976, 19:533–59.
19. Singh P.M., Rodrigues C., Gupta A.N.: Placenta previa and previous cesarean section. *Acta Obstet Gynecol Scand* 1981, 60:367–8.
20. Brenner W.E., Edelman D.A., Hendricks C.H.: Characteristics of patients with placenta previa and results of expectant management. *Am J Obstet Gynecol* 1978, 132:180–91.
21. Hibbard L.T.: Placenta previa. *Am J Obstet Gynecol* 1969, 104:172–84.
22. McShane P.M., Heyl P.S., Epstein M.F.: Maternal and perinatal morbidity resulting from placenta previa. *Obstet Gynecol* 1985, 65:176–82.
23. Gottesfeld K.R., Thompson H.E., Holmes J.H., Taylor E.S.: Ultrasound placentography: A new method for placental localization. *Am J Obstet Gynecol* 1966, 96:538–47.
24. Tomich P.G.: Prolonged use of tocolytic agents in the expectant management of placenta previa. *J Reprod Med* 1985, 30:745–8.
25. Bowen L.W., Beeson J.H.: Use of large Foley catheter balloon to control postpartum hemorrhage resulting from a low cervical implantation. *J Reprod Med* 1985, 30:623–5.
26. Karegard M., Gennser G.: Incidence and recurrence rate of abruptio placentae in Sweden. *Obstet Gynecol* 1986, 67:523–8.
27. Green-Thompson R.W.: Antepartum haemorrhage. *Clin Obstet Gynaecol* 1982, 9:479–515.
28. Okonofua F.E., Olatunbosun O.A.: Caesarean versus vaginal delivery in abruptio placentae associated with live fetuses. *Int J Gynaecol Obstet* 1985, 23:471–4.
29. Hurd W.W., Miodovnik M., Hertzberg V., Lavin J.P.: Selective management of abruptio placentae: A prospective study. *Obstet Gynecol* 1983, 61: 467–73.
30. Pritchard J.A., Brekken A.L.: Clinical and laboratory studies on severe abruptio placentae. *Am J Obstet Gynecol* 1967, 97:681–700.
31. Peltitti D.B., Cefalo R.C., Shapiro S., Whalley P.: In-hospital maternal mortality in the United States: Time trends and relation to method of delivery. *Obstet Gynecol* 1982, 59:6–12.
32. The Cesarean Birth Task Force: National Institutes of Health consensus development statement on cesarean childbirth. *Obstet Gynecol* 1981, 57: 537–45.
33. Schrinsky D.C., Benson R.C.: Rupture of the pregnant uterus: A review. *Obstet Gynecol Surv* 1978, 33:217–32.
34. Pritchard J.A.: Changes in the blood volume during pregnancy and delivery. *Anesthesiology* 1965, 26:393–9.
35. Hutton L., Yang S.S., Bernstein J.: Placenta accreta, 26 year clinicopathologic review. *NY State J Med* 1983, 83:857–66.
36. Meyers B.: Placenta accreta. An analysis based on an unusual case. *Acta Obstet Gynecol Scand* 1955, 34:189–201.
37. Aarberg M.E., Reid M.E.: Manual removal of placenta. *Am J Obstet Gynecol* 1945, 49:368–77.

38. Read J.A., Cotton D.B., Miller F.C.: Placenta accreta: Changing clinical aspects and outcome. *Obstet Gynecol* 1980, 56:31–4.
39. Chazott C., Cohen W.R.: Catastrophic complications of previous cesarean secton. *Am J Obstet Gynecol* 1990;163:738–42.
40. Arcario T., Greene M., Ostheimer G.W., et al.: Risks of placenta previa/accreta in patients with previous cesarean deliveries. *Soc Obstet Anesth Perinatol (Astracts)*1988.
41. Clark S.L., Koonings P.P., Phelan J.P.: Placenta praevia, accreta and prior caesarean section. *Obstet Gynecol* 1985, 66:89–92.
42. Cruikshank S.H.: Management of postpartum and pelvic hemorrhage. *Clin Obstet Gynecol* 1986, 29:213–9.
43. Gibbs C.E., Locke W.E.: Maternal deaths in Texas, 1969 to 1973. *Am J Obstet Gynecol* 1976, 126:687–92.
44. Kotelko D.M.: Amniotic fluid embolism, in Shnider S.M., Levinson G. (eds): *Anesthesia for Obstetrics.* Baltimore: Williams & Wilkins, 1987, pp 274–80.
45. Morgan M.: Amniotic fluid embolism. *Anaesthesia* 1979, 34:20–32.
46. Kulhman K.A., Hidvegi D., Tamura R.K., Depp R.: Is amniotic fluid in the central circulation of peripartum patients pathologic? *Am J Perinatol* 1985, 2:295–9.

Self-Assessment Questions

Select the single letter response that most correctly answers the questions or completes the sentence.

Chapter 1

1. Common cardiac adaptations in marathon runners include:
   a. increase in left ventricle diameter
   b. disproportionate left ventricular hypertrophy
   c. normal heart rate
   d. decreased left atrial size
2. Cardiac changes in weight lifters include:
   a. left ventricular hypertrophy
   b. little or no incresae in left ventricular chamber size
   c. increased left atrial size
   d. all of the above
3. Bradycardia in athletes is due to:
   a. decreased vagal tone
   b. increased resting sympathetic tone
   c. compensation for increased stroke volume
   d. all of the above
4. Sinus dysrhythmias:
   a. are rarely found in athletes
   b. are found about as often as in the general population (ie, 2.4%)
   c. usually originate in the ventricle
   d. occur in up to 60% of athletes
5. The commonest dysrhythmia in athletes is:
   a. wandering atrial pacemaker
   b. first-degree atrioventricular block
   c. Wenckebach phenomenon
   d. bradycardia
6. Sudden cardiac death:
   a. should always be considered a major hazard in anesthetizing the athlete
   b. is usually due to hypertrophic cardiomyopathy
   c. occurs in proportion to the training undertaken
   d. is most likely associated with cystic medial necrosis and aortic rupture

7. Anemia in athletes:
   a. is due to hemolysis of older red cells from vigorously contracting muscles
   b. may be due to a dilutional fall in the hemoglobin concentration
   c. occurs commonly
   d. is due to all of the above
8. Anabolic steroids:
   a. cause a negative nitrogen balance
   b. build muscles if caloric intake is sufficient
   c. decrease athletic performance
   d. promote a better disposition
9. Laboratory abnormalities associated with ingestion of anabolic steroids are least likely to include:
   a. decresaed thyroxin-binding globulin
   b. increased total T_4 cell count
   c. elevated alkaline phosphatase
   d. decreased testosterone level
10. Cocaine:
    a. increases adrenergic activity
    b. decreases glycolysis
    c. causes hypotension under anesthesia
    d. is associated with hypothermia because of peripheral vasoconstriction

Chapter 2

1. The most common cause of spinal cord injury (SCI) is:
   a. sports injury
   b. violence
   c. motor vehicle accidents
   d. falls
2. The defect most commonly seen with high spinal cord injury is:
   a. chronic obstructive disease
   b. increase in functional residual capacity
   c. increased forced expiratory volume
   d. restrictive lung disease
3. Which of the following statement accurately describes the respiratory status SCI patients?
   a. Compliance is decreased.
   b. The work of breathing is increased.
   c. The patient has a rapid, shallow breathing pattern.
   d. All of the above are true.
4. Spinal shock is characterized by which of the following?
   a. increased vascular tone
   b. hyperactive bowel sounds

c. flaccid paralysis below the level of injury
d. all of the above

5. Methods used to treat autonomic hyperreflexia include
   a. sodium nitroprusside
   b. deepening general anesthesia
   c. phenoxybenzamine
   d. all of the above
6. Which statement concerning autonomic hyperreflexia (AH) is false?
   a. AH occurs in about 85% of patients with high spinal cord lesions above T6.
   b. Symptoms are usually slow to occur and are vague.
   c. AH occurs in pregnant SCI patients with cord lesions above T6.
   d. Orthostatic capacity is reduced in SCI patients with lesions above T6 almost 60% of the time during labor.
7. The cause of death in SCI is most frequently associated with
   a. autonomic hyperreflexia
   b. spinal shock
   c. pulmonary edema
   d. renal failure
8. Treatment of symptomatic hypercalcemia requires:
   a. hydration with glucose solutions alone
   b. diuresis with furosemide
   c. anabolic steroids in high doses
   d. administration of potassium
9. Concerning the use of muscle relaxants, pick the *incorrect* response:
   a. Succinylcholine can cause an uncontrolled release of potassium into the circulation.
   b. Nondepolarizing muscle relaxants should not be used in the SCI patient.
   c. In SCI patients the entire muscle, below the level of injury, acts as a giant endplate when exposed to succinylcholine.
   d. Succinylcholine is contraindicated in the SCI patient.
10. Concerning anesthesia for the SCI patient, pick the correct response:
   a. Regional anesthesia is contraindicated
   b. Preoperatively, SCI patients are frequently hypovolemic.
   c. The occurrence of AH under general anesthesia is not related to anesthetic depth.
   d. hyperthermia is often problematic.

Chapter 3

1. The incidence of congenital cardiac lesions is:
   a. about 2%
   b. less than 0.1%
   c. unknown because of so many variables
   d. between 0.6% and 0.8%

2. Therapy for congenital heart disease is required during the first year of life in:
   a. all infants
   b. about 50%
   c. only a small percentage, as heart strain is not apparent until the child walks
   d. children who have left-to-right shunts only
3. The commonest congenital cardiac lesion is:
   a. patent ductus arteriosus
   b. tetralogy of Fallot
   c. ventricular septal defect
   d. coarctation of the aorta
4. Pulmonary vascular resistance is increased by:
   a. hypocarbia
   b. hyperoxia
   c. hyperinflation
   d. all of the above
5. Air embolism to the systemic circulation is a common complication:
   a. irrespective of the shunt pattern
   b. only if the shunt is right to left
   c. more likely if the shunt is left to right
   d. that does not occur if pulmonary obstruction exists
6. Criteria that define pathologic murmurs include:
   a. tachycardia with gallop
   b. poor peripheral perfusion and pulses
   c. central cyanosis relieved by oxygen
   d. all of the above
7. Which of the following is not a complication of increased hematocrit?
   a. cerebral thrombosis
   b. improved oxygen transport
   c. decreased pulmonary blood flow
   d. increased cardiac work
8. Left-to-right shunting:
   a. occurs in patent ducuts arteriosus anomalies
   b. decreases pulmonary blood flow
   c. usually causes symptoms at an early age
   d. is not dependent on the relative outflow resistances of the two ventricles
9. Right-to-left shunting:
   a. occurs in ventricular septal defects without pulmonic stenosis
   b. is associated with Eisenmenger's complex
   c. is independent of the left ventricular outflow resistance
   d. rarely causes cyanosis
10. Mixing lesions:
    a. always have atrial and ventricular septal defects
    b. combine increased pulmonary blood flow with cyanosis

c. are so called because the shunt alternates with position
d. carry a good prognosis with medical therapy only

Chapter 4

1. Causes of cor pulmonale include all of the following *except:*
   a. cystic fibrosis
   b. tetralogy of Fallot
   c. COPD
   d. sleep apnea
2. Causes of increased pulmonary vascular resistance include all of the following *except:*
   a. increased hematocrit
   b. metabolic acidosis
   c. hypoxemia
   d. halogenated anesthetics
3. Hypoxic pulmonary vasoconstriction response:
   a. is significantly impaired by light levels of anesthesia
   b. reduces blood flow to collapsed alveoli
   c. may be prevented by isoflurane 1 MAC
   d. can be conveniently measured by pulse oximetry
4. The best agent for reducing pulmonary vascular resistance and maintaining systemic vascular resistance is:
   a. hydralazine
   b. nitroglycerin
   c. nitroprusside
   d. nifedipine
5. ECG findings of right ventricular hypertrophy include:
   a. right axis deviation
   b. T-wave changes in right precordial leads
   c. Incomplete right bundle branch block
   d. all of the above
6. Diagnostic components of the physical exam in patients with right heart failure include all of the following *except:*
   a. jugular venous distention
   b. peripheral edema
   c. enlarged liver
   d. a systolic murmur secondary to incompetence of the pulmonary valve
7. A sensitive and specific test for pulmonary hypertension is:
   a. ECG
   b. chest x-ray
   c. echocardiogram
   d. none of the above
8. The major problem with an extensive sympathectomy (eg, high spinal) in patients with cor pulmonale is:
   a. interference with respiratory drive

b. interference with respiratory mechanics
c. treatment of hypotension with a pressor or inotrope may exacerbate the pulmonary hypertension
d. bronchospasm associated with sympathectomy

9. Preoperative preparation of patients with COPD and cor pulmonale is likely to include all of the following *except:*
   a. antibiotics
   b. bronchodilators
   c. volume loading
   d. diuretics
10. One clear indication for digoxin in patients with cor pulmonale is
    a. bronchospasm
    b. supraventricular tachydysrhythmias
    c. right heart failure
    d. peripheral edema

Chapter 5

1. The MEN type I syndrome usually involves hyperplasia or tumors of the following three glands:
   a. pancreatic islet cells, parafollicular cells of the thyroid (calcitonin-producing), and chief cells in the stomach
   b. pancreatic islet cells, hypothalamus, and thymus
   c. parathyroid, parafollicular cells of the thyroid, and adrenal medulla
   d. parathyroid, pancreatic islet cells, and the pituitary
2. Which of the following are characteristics of MEN I patients?
   a. The disease is inherited as an autosomal recessive trait.
   b. The pancreatic islet cell tumors often secrete large amounts of antidiuretic hormone leading to hyponatremia and volume overload.
   c. The pancreatic islet cell tumors often secrete vasoactive inhibitory polypeptide (VIP), which can lead to severe secretory diarrhea with fluid and electrolyte depletion.
   d. The pancreatic islet cell tumors are characteristically unifocal but may be multifocal.
3. An agent that is considered relatively contraindicated in patients with pheochromocytomas is:
   a. atropine
   b. isoflurane
   c. thiopental
   d. diazepam
4. All of the following are present in the MEN patient who presents with hypopituitarism *except:*
   a. deficiencies of thyroid-stimulating hormone production, leading to hypothroidism

b. deficiencies of adrenocorticotropic hormone (ACTH) production, leading to hypotension and intolerance of stress
c. deficiencies of prolactin production, leading to infertility and galactorrhea
d. deficiencies in follicle-stimulating hormone (FSH) and luteinizing hormone (LH) in women, leading to infertility, amenorrhea, and decreased secondary sexual characteristics

5. All of the following statements about the MEN syndromes are true *except:*
   a. The MEN II syndrome commonly is associated with pheochromocytoma, hyperparathyroidism, and medullary carcinoma of the thyroid (MCT).
   b. The MEN III syndrome commonly is associated with pheochromocytoma, MCT, and characteristic facies with puffy lips and tongue and mucosal neuromas.
   c. The MEN II syndrome is clearly familially inherited.
   d. MEN II patients commonly present with advanced disease early in life, whereas MEN III patients are usually older.
6. Which of the following statements about MCT is *not* true?
   a. The disease is usually treated with total hyroidectomy.
   b. Besides secreting calcitonin, MCT may secrete serotonin, prostaglandins, histamine, and ACTH.
   c. MCT is fatal if not treated surgically.
   d. MCT is found in 100% of both MEN II and MEN III patients.
7. In the MEN I patient:
   a. Gastric aspiration secondary to gastrinoma may cause aspiration pneumonitis.
   b. Hypercalcemia secondary to decreased parathyroid function is common.
   c. Hyperglycemia from pancreatic tumors occurs.
   d. Postoperative hypocalcemia increases Q-T intervals.
8. In laboratory evaluation of the MEN II patient with a pheochromocytoma the least common finding is:
   a. hypoglycemia
   b. elevated hematocrit
   c. elevated white blood cell count
   d. primarily elevated norepinephrine levels compared with epinephrine levels
9. Regarding the postoperative management of MEN patients:
   a. Neuromuscular irritability after parathyroidectomy occurs in all patients.
   b. Hypoglycemia is frequent but resolves spontaneously.
   c. Anastomatic leaks after gastrinoma resection are extremely rare.
   d. Hypotension is the commonest cause of death in the first 24 hours after resection of a pheochromocytoma.
10. In patients undergoing resection of a pheochromocytoma:
    a. Propranolol should not be given in conjunction with alpha blockers.
    b. Thyroidectomy for cancer should be postponed.

c. Two weeks' preparation with phenoxybenzamine is recommended.
d. All of the above are true.

Chapter 6

1. A high incidence of trisomy 21 is most closely associated with:
   a. radiation exposure
   b. thyroid disease
   c. viral illness
   d. advanced maternal age
2. The most common congenital heart lesion present in trisomy 21 is:
   a. ASD
   b. VSD
   c. ECD
   d. TOF
3. The most common postoperative complication in patients with trisomy 21 is:
   a. inability to maintain a patent upper airway
   b. prolonged neuromuscular blockade
   c. hypertension injury
   d. respiratory infection
4. A 9-year-old boy with a history of repaired tetralogy of Fallot presents for repair of a left inguinal hernia. Which of the following regimens should he receive for endocarditis prophylaxis? (There is no history of allergy to penicillin.)
   a. no antibiotic prophylaxis required
   b. IV vancomycin and gentamicin prior to procedure
   c. IV/IM ampicillin and gentamicin prior to procedure
   d. oral penicillin V before and after the procedure
5. Sleep apnea in patients with trisomy 21:
   a. is of no concern
   b. is prevented by avoidance of barbiturate premedication
   c. is related entirely to CNS factors
   d. is caused by both mechanical and central factors.
6. Neuroendocrine abnormalities found in Down's syndrome include:
   a. decreased plasma norepinephrine
   b. decreased plasma epinephrine
   c. decreased urinary epinephrine
   d. decreased serum levels of dopa decarboxylase
7. The incidence of congenital heart disease in trisomy 21 is approximately:
   a. 60%
   b. 40%
   c. 20%
   d. 5%

8. Atropine premedication for patients with Down's syndrome:
   a. produces an abnormally rapid heart rate
   b. is contraindicated
   c. produces an exaggerated mydriatic response
   d. produces a paradoxic bradycardia
9. The preferred anesthetic technique for patients with CHD and Down's syndrome is:
   a. N_2O–narcotic–relaxant
   b. inhalational
   c. determined by the operative procedure
   d. not important; any safely administered technique is acceptable
10. Children with trisomy 21 and CHD have a greater tendency to develop pulmonary artery hypertension than do normal children with CHD. The probable reason is:
   a. the cardiac lesion is more severe than in normal counterparts
   b. there are abnormalities in the alveoli and pulmonary vascular beds
   c. the incidence of hypoxia is increased
   d. unknown

Chapter 7

1. Aprotinin works by:
   a. reducing sequestration of IgG-coated platelet by the reticuloendothelial system
   b. stimulating von Willebrand factor production
   c. clot stabilization
   d. inhibiting partial release reaction of platelets in contact with cardiopulmonary bypass circuits
2. In stimulating erythropoiesis, recombinant human erythropoietin may cause:
   a. granulocytopenia
   b. anaphylactoid reaction
   c. vasovagal response
   d. hypertension
3. Large volumes of hydroxyethyl starch may precipitate a coagulopathy with:
   a. prolonged thrombin, prothrombin, and activated partial thromboplastin time
   b. acquired capillary fragility and shortened thrombin time
   c. von Willebrand factor and platelet aggregation
   d. shortened thrombin time and acquired von Willebrand syndrome
4. A Jehovah's Witness (JW) has a hematocrit of 10%. In order to generate 1 bag of Cell Saver (225ml) with a hematocrit of 60% for autotransfusion via suction, how much blood must be collected?
   a. 55ml
   b. 775ml

   c. 1300ml
   d. 2000ml
5. The approximate number of JW in the United States is
   a. 50,000
   b. 175,000
   c. 500,000
   d. 750,000
6. JW are divided in their acceptance of:
   a. bone marrow transplants
   b. organ transplants
   c. cryoprecipitate
   d. platelets
7. The oxygen-carrying emulsion fluosol-DA 20%:
   a. is commonly used with a balanced anesthetic technique
   b. shows promise in chronic management of patients with severe obstructive lung disease
   c. has been used only as a last resort in posthemorrhage anoxic encephalopathy and only for JW
   d. is effective in the treatment of short-term anemia
8. Regarding JW, which statement is *not* true?
   a. They were founded by Brother Joseph Smith.
   b. They regard the Bible as literally true.
   c. They believe that absorption of blood will cause their excommunication.
   d. The sect has more than 3 million members worldwide.
9. A low-lying placenta has been diagnosed during pregnancy. During the preanesthetic assessment:
   a. nothing should be done, as the obstetrician can perform a cesarean section and blood will not be necessary
   b. the anesthesiologist must abide by the will of the father that all blood products be withheld
   c. a court order to transfuse in emergency situations to save the life of the baby should be sought
   d. the anesthesiologist should plan to replace blood lost with large volumes of hetastarch
10. Anemia causes:
   a. high output failure
   b. increased lactic acid production
   c. cerebral vasodilation
   d. all of the above

Chapter 8

1. Wilson's disease:
   a. is associated with deficiencies of chelating agents
   b. commonly occurs in male children

c. is inherited as an autosomal recessive disorder
d. can cause cancer

2. Regarding copper:
   a. one third of the dietary requirement is absorbed from the gut
   b. it is strongly bound to albumin
   c. metabolism in the liver is by incorporation into ceruloplasmin
   d. all of the above
3. In Wilson's disease:
   a. the amount of copper bound to albumin is decreased
   b. kidney excretion is reduced by 50%
   c. total serum copper concentration is reduced
   d. there is an increase in ceruloplasmin
4. The diagnosis of Wilson's disease is confirmed by:
   a. serum ceruloplasmin levels of less than 20mg/100ml
   b. urinary excretion of more than 25mg of copper in 24 hours
   c. elevated SGOT and SGPT levels
   d. platelet count of less than 100,000/mm^3
5. Clinical manifestations of Wilson's disease:
   a. are apparent at birth
   b. consist mainly of Kayser-Fleischer rings
   c. are seen in 50% of patients by the age of 15
   d. frequently never occur—the disease is a laboratory determination
6. Hepatocelluar involvement in Wilson's disease:
   a. rarely causes jaundice
   b. is the most frequent presentation of Wilson's disease
   c. cannot be reversed by pharmacologic means
   d. causes typical psychologic changes
7. Kayser-Fleischer rings:
   a. indicate cerebral involvement
   b. are derived from deposition of copper in Descemet's membrane in the cornea
   c. occur in all patients with Wilson's disease
   d. are readily apparent without slit-lamp examination
8. Complications of D-penicillamine therapy include:
   a. hypersensitivity reactions
   b. hepatitis
   c. bullous condition of the skin
   d. all of the above
9. The best available treatment for Wilson's disease is:
   a. decreasing copper intake
   b. 2,3-dimercaptopropanol
   c. D-penicillamine
   d. liver transplantation
10. In anesthetizing the patient with liver disease:
    a. halogenated agents should be avoided

b. spinal anesthesia is completely safe
c. local anesthesia is the technique of choice
d. the dose of muscle relaxants must be carefully titrated

Chapter 9

1. After successful resuscitation, near-drowning patients are usually all of the following except:
   a. hypothermic
   b. hypervolemic
   c. hyperrigid
   d. hyperexcited
2. According to studies by Conn and Modell, patients in categories A and B have what percent chance of normal recovery?
   a. less than 10%
   b. 50%
   c. 90%
   d. 30%
3. In considering drowning/near-drowning, which statement is *untrue*?
   a. 100,000 submersion accidents occur each year.
   b. Approximately one tenth of these patients die.
   c. Drowning is the leading cause of accidental death in children.
   d. Male victims outnumber female 5 to 1.
4. Aspiration of liquid into the lungs causes all of the following *except:*
   a. bronchospasm
   b. loss of surfactant
   c. decreased pulmonary shunting (Qs/Qt)
   d. decreased functional residual capacity
5. Aspiration/ingestion of fresh water can cause:
   a. decreased intravascular volume
   b. hypernatremia
   c. increased intravascular volume
   d. hyperchloremia
6. Hypothermia can cause all of the following *except:*
   a. decreased $CMRO_2$
   b. increased cerebral blood flow
   c. decreased cardiac output
   d. increased systemic vascular resistance
7. The mammalian diving reflex in animals is least likely to cause:
   a. apnea
   b. increased cerebral blood flow
   c. peripheral vasoconstriction
   d. decreased risk of aspiration
8. In resuscitation of the hypothermic near-drowning victim:
   a. rewarming should take priority over CPR

b. IV drugs should be used early to help restore cardiac function
c. associated injuries are infrequent
d. adequate CPR should start as soon as possible

9. Patients in category B for neurologic classification:
a. are discharged from the hospital the same day after rewarming and tolerating oral intake
b. need no extra monitoring
c. have not suffered any cerebral insufficiency
d. may develop delayed cerebral/respiratory dysfunction

10. Which of the following statements is correct?
a. ICP is usually elevated in near-drowning victims.
b. Patients with a Glasgow Coma Scale score of 3 achieve near-normal recoveries.
c. Raised ICP is indicative of severe cerebral injury.
d. Hypothermia improves immune system function.

Chapter 10

1. The muscular dystrophies are:
a. diseases of striated muscle
b. rapidly progressive
c. inherited as X-linked traits
d. characterized by pseudohypertrophy of the calves

2. Pseudohypertrophy is mostly due to:
a. abnormal glycogen metabolism
b. increased muscle protein synthesis
c. overuse of certain muscle groups
d. deposition of excess adipose and connective tissue

3. Myocardial abnormalities seen in Duchenne muscular dystrophy (DMD) may include:
a. absent atrial electrical activity
b. papillary muscle dysfunction and mitral valve prolapse
c. ventricular septal defect
d. aortic stenosis

4. Laboratory abnormalities seen in DMD may include:
a. elevated serum glucose
b. decreased ionized calcium
c. decreased serum aldolase
d. elevated CPK, SGOT, and LDH

5. Myotonic contractures may be alleviated by:
a. muscular infiltration with local anesthetics
b. succinylcholine administration
c. general anesthesia
d. spinal anesthesia

6. Myotonic contractures may be precipitated by:
   a. procainamide
   b. thiopental
   c. hyperkalemia
   d. narcotics
7. Useful therapeutic modalities in DMD are:
   a. surgical correction of scoliosis and contractures
   b. administration of procainamide
   c. chronic digitalis prophylaxis
   d. calcium channel blockade
8. A prediction of the need for mechanical ventilation postoperatively:
   a. involves vital capacity less than 45% of normal
   b. cannot reliably be obtained
   c. depends on the intactness of oropharyngeal reflexes
   d. can be based on yearly reduction of vital capacity of 5%
9. In anesthetizing patients with DMD:
   a. dantrolene should be given prophylactically
   b. susceptibility to malignant hyperthermia is common
   c. myocardial dysfunction is unaffected by inhalational agents
   d. shivering should be prevented, as it increases the oxygen debt
10. In Landouzy-Déjerine syndrome:
    a. males and females are equally affected
    b. atrial paralysis may occur
    c. ability to cough is greatly reduced
    d. all of the above

Chapter 11

1. To make the diagnosis of asthma, which of the following must be present?
   a. wheezing
   b. bronchial hyperreactivity
   c. decreased peak flow
   d. radiographic demonstration of hypertrophy
2. Diseases that may be confused with asthma include:
   a. left ventricular failure
   b. vocal cord obstruction
   c. carcinoid syndrome
   d. all of the above
3. Which of the following statements about a severe asthmatic attack is true?
   a. Functional residual capacity is decreased.
   b. Bronchoconstriction is the major cause of airway narrowing.
   c. Oxygen demand is increased.
   d. Closing volume is relativey unaffected.
4. Spirometric measurements of pulmonary function:
   a. must be performed in a pulmonary function laboratory

b. are useful for differentiating central from peripheral airways obstruction
c. record airflow as a function of time
d. are based on a vital capacity maneuver

5. Which of the following patterns best characterizes moderate obstructive lung disease?
a. $FEF_{25\%-75\%}$ about 60% of predicted
b. decreased vital capacity
c. decreased FEV_1/FVC ratio
d. bronchodilator response of at least 15%

6. All of the following ECG findings may be associated with asthma *except:*
a. right axis deviation
b. high-grade conduction block
c. ST depressions
d. premature ventricular contractions

7. In the asymptomatic asthmatic, which screening test is most useful to the anesthesiologist?
a. serum theophylline level
b. lateral and posteroanterior chest x-ray
c. peak flow measurement
d. room-air arterial blood gas analysis

8. Which of the following statements regarding anesthetic drugs is true?
a. The volatile agents (halothane, enflurane, isoflurane) stimulate bronchodilation and improve mucociliary clearance at low concentrations.
b. Ketamine is the drug of choice in induction of anesthesia in asthmatic patients as it has no side effects.
c. Barbiturates should generally be avoided both as premedication and as induction agents.
d. None of the above

9. If bronchospasm develops during general anesthesia, the appropriate initial management should be to:
a. increase the rate of aminophylline infusion (or initiate aminophylline therapy)
b. intubate the patient
c. administer metaproterenol
d. rapidly check for precipitating causes

10. Appropriate premedication of the asthmatic patient:
a. is not dependent on serum theophylline levels
b. is optimal with isoproterenol
c. should always be given at least 2h before surgery
d. requires that intravenous theophylline doses be reduced by one third

Chapter 12

1. Metrizamide is a water-soluble contrast dye that:
a. usually requires pretreatment with antihistamines and steroids prior to administration

b. requires the presence of an anesthesiologist before it is given
c. may be associated with anaphylactic reactions
d. is administered only intrathecally

2. Problems often encountered during CT scanning of small children and infants include:
a. frequent susceptibility to hypothermia
b. difficult visualization of the posterior fossa in children with hydrocephalus
c. difficulty in stabilizing the infant's head during scanning
d. all of the above

3. Prophylaxis against allergic reaction to contrast dye in patients with prior allergic history may include:
a. diphenhydramine 50mg IV, hydrocortisone 100mg IV, and epinephrine 0.25mg SC
b. diphenhydramine 50mg IV or PO before the procedure and prednisone 150mg/day for 18h before and 12h after the procedure
c. epinephrine 0.25mg SC and Alupent®0.3ml in 3ml NS via nebulized inhaler
d. narcotics such as meperidine, benzodiazapines such as diazepam and midazolam, and phenothiazines

4. Magnetic resonance imaging:
a. does not routinely require patient immobility
b. usually takes 10–15 minutes for a complete study
c. often utilizes gadolinium for contrast enhancement in the T2 pulse sequence
d. is highly sensitive in the diagnosis of white matter diseases and lesions of the posterior fossa, spine, cranial nerves, and skull base

5. Problems faced by the anesthesiologist during MRI include:
a. difficult access to the patient in the scanner
b. physical propulsion of ferromagnetic objects by the scanner and degradation of the MRI image
c. interference by the scanner with resuscitative equipment
d. all of the above

6. Myelography:
a. is rarely, if ever, used today as a diagnostic test
b. is the most sensitive test available for intradural extramedullary lesions
c. should include pretreatment with phenothiazines
d. may only be attempted via the lumbar route

7. Pneumoencephalography:
a. is currently one of the most frequently ordered neuroradiologic diagnostic tests
b. contraindicates the use of nitrous oxide as an inhalation agent
c. may include injection of air intrathecally
d. is painfree

8. Single photon emission computed tomography (SPECT):
a. and positron emission tomography (PET) are utilized for measurement

of regional cerebral blood flow, cerebral metabolism of glucose and oxygen, and cerebral blood volume
b. uses positron emitters of short half life such as carbon-11, oxygen-15, and fluorine-18
c. requires the presence of an on-site cyclotron
d. produces images of greater detail and resolution than PET

9. Interventional neuroangiography:
   a. utilizes embolic agents such as detachable balloons, silicon spheres, polyvinyl alcohol, and tissue adhesives
   b. may palliatively reduce the size of vascular lesions such as AVMs and AVFs and vascular tumors
   c. may employ the selective intra-arterial injection of sodium amytal to help map out functional and diseased brain tissue
   d. all of the above
10. Endovascular embolization:
   a. may be complicated by intracranial hemorrhage or cerebral infarction secondary to dislodgment of the embolic agent
   b. routinely requires prophylaxis with vasopressors
   c. requires the presence of an anesthesiologist only for unrestrainable patients or to give sedation
   d. is the treatment of choice for most cerebral aneurysms

Chapter 13

1. The incidence of nonobstetric surgery during pregnancy is:
   a. 5%
   b. 0.16%
   c. 10%
   d. 1.6%
2. The commonest reason for performing nonobstetric surgery during pregnancy is:
   a. acute appendicitis
   b. ruptural intracranial aneurysm
   c. heart valve replacement
   d. rapidly advancing cancer
3. Pregnant patients are prone to hypoxemia because of:
   a. decreased oxygen consumption
   b. decreased functional residual capacity
   c. decreased closing volume
   d. decreased tidal volume
4. The period of organogenesis in the developing human fetus is:
   a. 1 to 30 days
   b. 1 to 56 days
   c. 15 to 56 days
   d. 15 to 76 days

5. The anesthetic agent proved to be most teratogenic in humans is:
   a. halothane
   b. nitrous oxide
   c. morphine
   d. none of the above
6. Cardiovascular changes associated with pregnancy include:
   a. increased peripheral vascular resistance
   b. decreased cardiac output
   c. increased stroke volume
   d. decreased heart rate
7. Gastrointestinal changes associated with pregnancy include:
   a. decreased gastric motility
   b. increased gastric acidity
   c. decreased lower esophageal sphincter tone
   d. all of the above
8. The least effect on uterine venous pressure is seen with:
   a. inferior vena cava compression
   b. hypercarbia
   c. oxytocin
   d. ephedrine
9. Because of the risk of aspiration in pregnant patients:
   a. general anesthesia should be avoided
   b. extubation should be delayed until airway reflexes return
   c. 15–30ml of Maalox should be given $\frac{1}{2}$ hour prior to induction
   d. a smaller than usual endotracheal tube is indicated
10. Factors that adversely affect uterine artery blood flow, and therefore uterine placental perfusion, *least* include:
   a. maternal hyperventilation
   b. maternal hypotension
   c. ketamine <1mg/kg
   d. epinephrine

Chapter 14

1. The diagnosis of SLE is confirmed by the following pathognomonic sign:
   a. a "butterfly" rash
   b. the finding of antinuclear antibodies in the serum
   c. the presence of the lupus anticoagulant
   d. none of the above
2. The incidence of SLE in the general population is:
   a. 6 in 10,000
   b. 1 in 5000
   c. 1 in 1000
   d. 6 in 100,000

3. The disease is more prevalent in:
    a. men and dark-skinned races
    b. women
    c. Caucasians
    d. Hispanics
4. Factors that cause exacerbation of SLE are:
    a. sunlight
    b. ultraviolet light
    c. infection
    d. all of the above
5. A decrease in complement levels is often associated with the onset of:
    a. early disease
    b. arthritis
    c. mild rash
    d. glomerulitis
6. "Onion-skin" lesions are found in the:
    a. heart
    b. liver
    c. spleen
    d. lungs
7. Coagulopathy in a patient with SLE:
    a. can occur in the presence of a normal PTT
    b. is most likely a result of hypothrombinemia
    c. is most often the result of aspirin therapy
    d. will not occur if the platelet count is normal
8. An early manifestion in SLE is:
    a. cardiac involvement
    b. recurrent thrombophlebitis
    c. bilateral pleural effusion
    d. ileus
9. Pulmonary function tests show:
    a. obstructive defect with loss of lung volume
    b. restrictive defect with increased flow rates
    c. restrictive defect with normal flow rates
    d. no gross changes
10. The most sensitive test for detection of SLE is:
    a. complement fixation
    b. LE cell test
    c. immunofluorescence
    d. diffusion technique

Chapter 15

1. Which of the following is the major problem facing anesthesiologists today with respect to allergy?

   a. development of appropriate treatment plans
   b. identification of patients at risk
   c. choice of induction agents
   d. differentiating anaphylactic from anaphylactoid reactions
2. Predictable drug reactions can be described by each of the following statements *except:*
   a. They make up approximately 80% of all drug reactions.
   b. They are dose-dependent
   c. They are related to the patient's immune response
   d. They are related to known pharmacologic properties.
3. Which of the following patient groups has an incresaed propensity for severe allergy?
   a. asthmatics
   b. patients with known drug sensitivities
   c. patients with food allergies
   d. all of the above
4. All of the following statements are true of anaphylactic reactions *except:*
   a. They are IgE-mediated
   b. They require prior sensitization
   c. They release vasoactive mediators.
   d. They are clinically distinguishable from anaphylactoid reactions.
5. Type I hypersensitivity includes all of the following *except:*
   a. degranulation of mast cells and basophils
   b. immunospecific IgE
   c. sensitization
   d. T-cell lymphocytes
6. Which chemical mediator released during degranulation is necessary for anaphylaxis?
   a. histamine
   b. slow releasing substance of anaphylaxis (SRS-A)
   c. prostaglandins
   d. bradykinin
7. The severity of an allergic reaction may be influenced by:
   a. smooth muscle reactivity
   b. autonomic nervous system balance
   c. drug dosage
   d. all of the above
8. Which of the following clinical manifestations is most life-threatening?
   a. hypotension
   b. bronchospasm
   c. tachycardia
   d. dysrhythmias
9. The drug of choice for anaphylactic/anaphylactoid reactions is:
   a. corticosteroids
   b. antihistamines

c. epinephrine
d. none of the above

10. Identification of offending agents includes all of the following *except:*
    a. plasma complement C3 and C4 levels
    b. intradermal skin testing
    c. RAST
    d. radiolabeled anti-IgE antibodies

Chapter 16

1. The reason(s) for the recent surge in cocaine abuse is/are:
    a. lower street price
    b. increased purity
    c. new modes of self-administration of the drug
    d. all of the above
2. The sinus tachycardia that accompanies cocaine use is best managed by:
    a. sublingual nifedipine
    b. propranolol
    c. phenylephrine
    d. lidocaine
3. The drug of choice in management of cocaine-induced seizure activity is:
    a. digoxin
    b. phenobarbital
    c. diazepam
    d. phenytoin
4. After ingestion cocaine is detected in the urine:
    a. only for 2–4 hours
    b. hardly ever—metabolites, are excreted via the lungs
    c. for several weeks
    d. for 24–28 hours
5. The action of cocaine on nerves is best characterized by:
    a. calcium channel blockade
    b. phosphodiesterase inhibition
    c. blockade of catecholamine reuptake
    d. cholinesterase hydrolysis
6. Life-threatening side effects of cocaine include:
    a. acute hyperthermia
    b. acute myocardial infarction
    c. rhabdomyolysis
    d. all of the above
7. Intraoperative hemostasis is best obtained in the cocaine addict by:
    a. adrenaline solution 1:100,000
    b. adrenaline solution 1:400,000
    c. cautery
    d. topical

8. Myocardial vasoconstriction associated with acute cocaine intoxication can be reversed with:
   a. phentolamine
   b. reserpine
   c. propranolol
   d. digoxin
9. Free-base cocaine is:
   a. pungent
   b. readily soluble in water
   c. destroyed by heating to 90°C
   d. stable after being vaporized
10. Cocaine ingested to induce labor:
   a. decreases uteroplacental blood flow
   b. is widely used in some inner city areas
   c. may cause abruption of the placenta within hours of intranasal use
   d. all of the above

Chapter 17

1. The main structure of the steroid nucleus that is necessary for the synthesis of the hormones derived from the adrenal cortex is:
   a. glucose
   b. Δ^5-pregnenolone
   c. progesterone
   d. cholesterol
2. The principal endogenous glucocorticoid is:
   a. glucagon
   b. cortisol
   c. ACTH
   d. prednisone
3. In considering aldosterone:
   a. regulation depends on an intact renin-angiotensin system
   b. serum levels are totally dependent on ACTH secretion
   c. adequate levels are necessary for adequate sodium secretion in the distal renal tubule
   d. tubular excretion of potassium and hydrogen ions is not dependent on its presence
4. ACTH function:
   a. is independent of diurnal variation
   b. stimulates higher hormonal function in the evening
   c. depends on adequate liver metabolism
   d. controls androgen production.
5. Addison's disease may be caused by:
   a. an autoimmune disorder producing adrenal antibodies
   b. abrupt withdrawal of steroids

c. tuberculosis
d. all of the above

6. The diagnosis of Addison's disease is most reliably established by:
   a. the finding of a single bow plasma cortisol level
   b. decreased ACTH levels
   c. decreased 24-hour urine 17-hydroxicorticoid excretion after parenteral administration of ACTH
   d. increased plasma cortisol levels 30min after administration of cortisone
7. In patients critically ill with Addison's disease:
   a. immediate testing is indicated to establish the exact hormone deficiency
   b. volume depletion is essential to avoid increases in cerebral edema
   c. hydrocortisone 100mg should be given intravenously and followed by 50mg every 4 to 6 hours.
   d. hyponatermic solutions are indicated
8. In considering preoperative steroid supplementation:
   a. without additional steroids, cardiovascular problems are frequent
   b. established protocols determine the preoperative steroid requirements
   c. 500mg/day of hydrocortisone should be given to replace the maximum amount manufactured by the body
   d. deep general anesthesia can postpone the intraoperative glucocorticoid surge to the postoperative period
9. Hypoaldosteronism is characterized by:
   a. hypotension
   b. hypokalemia
   c. cardiac conduction defects
   d. hypernatremia
10. The most common cause of Cushing's syndrome is:
    a. unilateral adrenal hypoplasia
    b. bilateral adrenal hypoplasia
    c. pancreatic tumors
    d. iatrogenic administration of glucocorticoids

Chapter 18

1. Moderately premature infants are classified as:
   a. 31–36 weeks gestational age
   b. 500–1500gm weight
   c. > 2500gm weight
   d. 37–38 weeks gestational age
2. A major problem associated with prematurity is:
   a. axonal septal defect
   b. patent ducuts arteriosus
   c. aortic stenosis
   d. ventricular septal defect

3. Which of the following is characteristic of the murmur of patent ductus arteriosus?
   a. It is often continuous and at the left sternal border.
   b. It is loudest during systole.
   c. Intensity can be increased by hypoventilation.
   d. It occurs only during diastole.
4. The initial medical treatment of patent ductus arteriosus includes all of the following *except:*
   a. fluid restriction
   b. diuretics
   c. indomethacin (Indocin)
   d. digitalis
5. Temperature instability in premature infants leads to:
   a. decreased metabolic rate
   b. decreased oxygen consumption
   c. respiratory alkalosis
   d. acidosis and hyoxemia
6. Treatment of shock due to necrotizing enterocolitis requires:
   a. blood only
   b. fluids only
   c. blood and fluids
   d. colloids
7. Necrotizing enterocolitis is characterized by:
   a. abnormal distention
   b. vomiting, bloody stools
   c. shock
   d. all of the above
8. Metabolic disorders associated with prematurity include:
   a. metabolic acidosis
   b. respiratory alkalosis
   c. hypercalcemia
   d. hypobilirubinemia
9. Hypocalcemia is associated with:
   a. twitching and seizures
   b. blood calcium level of 9.5mg%
   c. hypoventilation, and respiratory acidosis
   d. shortened ST interval
10. Stage I retrolental fibroplasia is associated with:
    a. dilatation and tortuosity of retinal vessels
    b. neovascularization and peripheral clouding
    c. circumferential retinal detachment
    d. unmitigating progression

Chapter 19

1. The respiratory system:
   a. develops as an outpouch of the midbrain

  b. begins to form by the fourth week of pregnancy
  c. is derived from the 3rd branchial arch
  d. is similar to adult form by the time of delivery
2. In the infant:
  a. the larynx is positioned at the level of C6
  b. the epiglottis is directly aligned with the vocal cords
  c. nasal breathing is obligatory
  d. the narrowest part of the upper airway is just above the vocal cords
3. In the neonate, 1mm edema:
  a. reduces airway diameter by 65%
  b. will completely occlude the airway
  c. has little effect on air flow
  d. does not increase turbulent flow
4. Positive pressure on the intrathoracic trachea:
  a. classically causes stridor during inspiration and expiration
  b. is apparent as stridor on expiration
  c. is rarely associated with wheezing
  d. usually causes crepitant sounds
5. Choanal atresia:
  a. may cause respiratory distress immediately after birth
  b. occurs in 1:8000 births
  c. can be confirmed by the inability to pass a suction catheter through either nostril.
  d. includes all of the above
6. The commonest cause of congenital stridor is:
  a. Down's syndrome
  b. vocal cord paralysis
  c. laryngotracheomalacia
  d. tracheal stenosis
7. Croup:
  a. is usually caused by parainfluenza myxoviruses and respiratory syncytial viruses
  b. rarely occurs in dry and urban areas
  c. is characterized by localized inflammation of the upper airway.
  d. has a sudden, catastrophic onset
8. Epiglottitis:
  a. is most common in infants
  b. is usually associated with *S. aureus* or β-hemolytic streptococcal infection
  c. has no seasonal variation
  d. causes mainly supraglottic inflammation
9. Therapy of epiglottitis includes:
  a. prompt initiation of antibiotic therapy before cultures are obtained
  b. oxygen administration
  c. direct visualization of the epiglottis under general anesthesia
  d. all of the above

10. Foreign bodies lodged in terminal bronchi:
    a. may be unnoticed for months
    b. cause pneumonia in 2–3 days
    c. are diagnosed by detecting generalized wheezing
    d. are usually radiopaque

Chapter 20

1. The number of coronary artery bypass procedures performed each year in the United States is:
    a. 150,000–200,000
    b. 200,000–250,000
    c. 250,000 –300,000
    d. greater than 300,000
2. The leading cause of postoperative death in the elderly undergoing noncardiac surgery is:
    a. congestive heart failure
    b. diabetes
    c. CAD
    d. perioperative MI
3. The greatest risk for perioperative infarction is found in patients with:
    a. CAD
    b. congestive heart failure
    c. previous MI
    d. age >70 years
4. Reinfarction rate for a patient undergoing general anesthesia who had an MI 6 months ago is approximately:
    a. <5%
    b. 6%
    c. 8%
    d. 10%
5. The incidence of MI in patients with peripheral vascular disease undergoing vascular surgery is:
    a. <10%
    b. 15%
    c. 20%
    d. 25%
6. The valvular disease that offers the worst prognosis for patients undergoing CABG procedure is:
    a. mitral stenosis
    b. aortic insufficiency
    c. mitral insufficiency
    d. aortic stenosis
7. The chronic effects of cigarette smoking include vasoconstriction, loss of endothelial integrity, and:

a. neutropenia
b. platelet fragmentation
c. decreased platelet aggregation
d. enhanced platelet aggregation

8. Myocardial ischemia can be detected intraoperatively by:
   a. transesophageal echocardiography
   b. cardiokymography
   c. lactate levels
   d. all of the above
9. Preoperative ECG abnormalities occur in what percentage of coronary artery disease patients undergoing noncardiac surgery?
   a. 15%
   b. 30%
   c. 50%
   d. 75%
10. Exercise stress testing is highly predictive of subsequent cardiac events when ST changes are:
   a. large (>2.5mm)
   b. immediate (first 1–3 min)
   c. sustained into the recovery period
   d. all of the above

Chapter 21

1. Which of the following physiologic changes of pregnancy is exaggerated in a woman with a multiple gestation pregnancy?
   a. increased maternal blood volume
   b. decreased functional residual capacity
   c. decreased maternal hematocrit
   d. all of the above
2. The following statements about a multiple gestation pregnancy are all true *except:*
   a. Maternal mortality is 50% higher than that seen in single-gestation pregnancies.
   b. Preeclampsia occurs more frequently than in single-gestation pregnancies.
   c. The gravida is at a greater risk for developing aspiration pneumonitis than a woman with a single-fetus gestation.
   d. The risk of postpartum uterine atony and hemorrhage is greater than with a single-fetus gestation.
3. In a woman with a multiple-fetus gestation, aortocaval compression:
   a. is always relieved by a 15° tilt to the left
   b. is always relieved by a 30° tilt to the left
   c. is usually more severe than in a woman with a singleton pregnancy.
   d. is usually not problematic.

4. When used for tocolytic therapy in preterm labor, magnesium sulfate:
   a. will potentiate nondepolarizing neuromuscular blocking agents but antagonize depolarizing blockers
   b. will potentiate nondepolarizing neuromuscular blocking agents but not affect the action of depolarizing blocking agents
   c. may lead to pulmonary edema
   d. will result in none of the above
5. When used for tocolytic therapy in preterm labor, β-mimetics such as terbutaline:
   a. will potentiate nondepolarizing neuromuscular blocking agents
   b. may lead to pulmonary edema
   c. may lead to hyperkalemia
   d. do not cause tachycardia because they are β_2-selective
6. In the case of a planned vaginal delivery of twins:
   a. The second delivery may be by cesarean section even if the first delivery is vaginal.
   b. Epidural analgesia is contraindicated because it will prolong the delivery of the second twin, resulting in a poorer outcome.
   c. Spinal anesthesia will provide excellent uterine relaxation to assist with obstetric manipulations.
   d. All of the above are true.
7. Monoamniotic twins are more likely than dizygotic twins to develop which of the following complications?
   a. fetofetal transfusion syndrome
   b. intertwining and occlusion of umbilical vessels
   c. a higher incidence of malformation
   d. all of the above
8. Which of the following statements is true regarding fetal morbidity and mortality in a multiple gestation?
   a. Triplets, in general, tend to be healthier at birth than twins.
   b. Adverse outcome is primarily related to prematurity and intrauterine growth retardation.
   c. The fetofetal transfusion syndrome ensures that a smaller twin will receive an adequate blood supply from the larger twin.
   d. All of the above are true.
9. Which of the following statements is true regarding the premature delivery of twins?
   a. Fetuses of very low birth weight (less than 1000gm) should be delivered vaginally because of their small size.
   b. Prophylactic cerclage is a reliable method for preventing the premature delivery of twins.
   c. Steroids, such as dexamethasone or betamethasone, are often administered for tocolysis to prevent preterm delivery.
   d. Neonates are more likely to develop narcotic-induced respiratory depression than are infants born at term.

10. The following statements regarding the twin embolization syndrome (TES) are true *except:*
    a. TES occurs in approximately 25% of all twin pregnancies.
    b. The mother is at risk of developing a coagulopathy.
    c. Monozygotic twins are at greater risk than are dizygotic twins.
    d. The surviving twin may develop ischemic structural defects of the central nervous system.

Chapter 22

1. With recent medical and surgical advancements, the mortality of patients with pyloric stenosis is:
    a. 10%
    b. 5%–10%
    c. 3%–5%
    d. <1%
2. Symptoms of pyloric stenosis include:
    a. billous vomiting
    b. bile-free vomiting
    c. vomiting immediately after birth
    d. diarrhea
3. Diagnosis is determined by:
    a. palpable olivelike mass
    b. double bubble sign
    c. respiratory acidosis
    d. rapid gastric emptying of barium
4. Elevated bilirubin in patients with pyloric stenosis:
    a. is of the direct type
    b. disappears in 3 weeks
    c. is indirect
    d. should be ignored
5. Typical derangements in mild to moderate pyloric stenosis include:
    a. hyperchloremia
    b. metabolic alkalosis
    c. hypoxia
    d. hyperkalemia
6. The derangements in pyloric stenosis lead to:
    a. a shift in the oxyhemoglobin curves to the left
    b. a shift to the right
    c. dehydration because of hyperthermia
    d. little respiratory compensation
7. Pyloric stenosis:
    a. appears most commonly in children about 3–5 weeks of age
    b. is always a surgical emergency

c. is more common in firstborn females
d. has no inherited features

8. Which of the following pertains in a child with mild dehydration?
   a. rapid and weak radial pulse
   b. sunken anterior fontanelle
   c. estimated fluid deficit of 40–50cc/kg
   d. anuria
9. Fluids for the child undergoing pyloromyotomy must take into consideration:
   a. intraoperative losses of 15cc/kg/h
   b. maintenance of 9cc/kg/h
   c. maintenance of 4cc/kg/h
   d. dextrose 10%
10. Postoperatively:
    a. the patient can be safely extubated while still deeply anesthetized
    b. the patient must be awake and alert before extubation
    c. feeding can be restarted 2 hours postoperatively
    d. the nasogastric tube should remain for 48 hours

Chapter 23

1. Which one of the following countries has the lowest incidence of multiple sclerosis (MS)?
   a. China
   b. Denmark
   c. United States
   d. New Zealand
2. Which of the following characteristics is *least* likely to be associated with MS?
   a. loss of coordination
   b. optic neuritis
   c. demyelination of peripheral nervous system
   d. Lhermitte's sign
3. Diagnostic criteria specific for MS include:
   a. elevated CSF IgG
   b. presence of oligoclonal band in IgG region of CSF
   c. two attacks and clinical evidence of two lesions
   d. white spots on MRI
4. The highest rate of relapse occurs:
   a. during the first trimester
   b. during the third trimester
   c. 3 months postpartum
   d. 1 year postpartum
5. Which of the following events is *least* likely to cause relapse or exacerbation in an MS patient
   a. fever

b. lumbar epidural anesthesia
c. emotional upset
d. general anesthesia (O_2/N_2O/halothane/thiopental)

6. The best anesthetic choice for cesarean section in a 28-year-old patient with MS at term pregnancy would be:
a. general anesthesia
b. spinal anesthesia
c. lumbar epidural anesthesia
d. local anesthesia

7. The preanesthetic assessment for an MS patient should include:
a. history of aspiration
b. swallow evaluation
c. mental status examination
d. all of the above

8. Which of the following drugs used for MS may affect the use of neuromuscular blockers during general anesthesia?
a. ACTH
b. isoniazid
c. interferon
d. dantrolene

9. Which of the following local anesthetics should not be used for lumbar epidural anesthesia in an MS patient because of alleged neurotoxicity?
a. lidocaine
b. bupivacaine
c. chloroprocaine
d. all of the above

10. The most common cause of relapse in MS postoperatively is:
a. hyperpyrexia
b. urinary retention
c. hypoxia
d. hypoglycemia

Chapter 24

1. What is the average incidence of amniotic fluid embolism?
a. 1:5000
b. 1:100,000
c. 1:20,000
d. 1:200,000

2. The most common cause of postpartum hemorrhage is:
a. lacerations
b. uterine rupture
c. retained placenta
d. uterine atony

3. Which of the following statements is true regarding uteroplacental blood flow?
   a. Uteroplacental blood flow is decreased as uterine artery resistance falls.
   b. Uteroplacental blood flow is maintained by autoregulation as long as maternal blood pressure is maintained at 80mmHg.
   c. Uteroplacental blood flow will decrease with a fall in uterine venous pressure.
   d. Uteroplacental blood flow is directly related to the degree of uterine tone.
4. Late decelerations may signify:
   a. fetal head compression
   b. uteroplacental insufficiency
   c. umbilical cord compression
   d. fetal hyperoxia
5. An increased risk of placenta previa is associated with:
   a. oligohydramnios
   b. fetal prematurity
   c. hydramnios
   d. previous cesarean section
6. The decidua basalis is of:
   a. fetal origin
   b. myometrial origin
   c. epithelial origin
   d. endometrial origin
7. Placenta accreta is defined as:
   a. chorionic villi invading the myometrium
   b. chorionic villi penetrating through the myometrium and serosa
   c. amnion attached directly to the myometrium
   d. chorionic villi attached directly to the myometrium
8. To resolve disseminated intravascular coagulation, the most important treatment is:
   a. administration of blood products
   b. heparin
   c. epsilon aminocaproic acid
   d. to rectify the underlying cause
9. The risk of placenta accreta is increased by:
   a. abruptio placentae
   b. placenta previa
   c. hydramnios
   d. fetal malformation
10. If an amniotic fluid embolism is promptly treated, the maternal mortality rate is:
    a. 86%
    b. 8.6%
    c. 50%
    d. 5.0%

Answers to Self-Assessment Questions

Chapter 1	*Chapter 2*	*Chapter 3*	*Chapter 4*
1.a	1.c	1.d	1.b
2.d	2.d	2.b	2.d
3.c	3.d	3.c	3.b
4.d	4.c	4.c	4.b
5.d	5.d	5.a	5.d
6.b	6.b	6.d	6.d
7.d	7.d	7.b	7.d
8.b	8.b	8.a	8.c
9.b	9.b	9.b	9.c
10.a	10.b	10.b	10.b

Chapter 5	*Chapter 6*	*Chapter 7*	*Chapter 8*
1.d	1.d	1.d	1.c
2.c	2.c	2.d	2.d
3.a	3.a	3.d	3.c
4.c	4.c	4.c	4.a
5.d	5.d	5.d	5.c
6.d	6.c	6.c	6.b
7.a	7.b	7.d	7.b
8.a	8.c	8.a	8.d
9.d	9.c	9.c	9.c
10.d	10.d	10.d	10.d

Chapter 9	*Chapter 10*	*Chapter 11*	*Chapter 12*
1.a	1.a	1.b	1.c
2.c	2.d	2.d	2.d
3.c	3.b	3.c	3.b
4.c	4.d	4.d	4.d
5.c	5.a	5.a	5.d
6.b	6.c	6.b	6.b
7.b	7.a	7.c	7.b
8.d	8.a	8.d	8.a
9.d	9.b	9.d	9.d
10.c	10.d	10.d	10.a

Chapter 13	*Chapter 14*	*Chapter 15*	*Chapter 16*
1.d	1.d	1.b	1.d
2.a	2.c	2.c	2.b
3.b	3.b	3.d	3.c
4.c	4.d	4.d	4.d
5.d	5.d	5.d	5.c
6.c	6.c	6.a	6.d
7.d	7.a	7.d	7.c
8.d	8.b	8.b	8.a
9.b	9.c	9.c	9.d
10.c	10.c	10.a	10.d

Chapter 17	*Chapter 18*	*Chapter 19*	*Chapter 20*
1.d	1.a	1.b	1.c
2.b	2.b	2.c	2.d
3.a	3.a	3.a	3.c
4.d	4.d	4.b	4.b
5.d	5.d	5.d	5.b
6.c	6.c	6.c	6.d
7.c	7.d	7.a	7.d
8.d	8.a	8.d	8.d
9.c	9.a	9.d	9.c
10.d	10.a	10.a	10.d

Chapter 21	*Chapter 22*	*Chapter 23*	*Chapter 24*
1.d	1.d	1.a	1.c
2.a	2.b	2.c	2.d
3.c	3.a	3.c	3.a
4.c	4.c	4.c	4.b
5.b	5.b	5.b	5.d
6.a	6.a	6.c	6.d
7.d	7.a	7.d	7.b
8.b	8.c	8.d	8.d
9.d	9.c	9.c	9.b
10.a	10.b	10.a	10.a

Index

The chapters in these volumes are revised and updated articles that originally appeared as lessons in consecutive issues of *Anesthesiology News.*

Preanesthetic Assessment 1

ISBN: 0-8176-3376-6
ISBN: 0-8176-3376-6

Lesson	Chapter	Title
1	1	*The Patient with Cerebrovascular Disease,* by Michael Domurat
2	2	*The Trauma Victim,* by Jeffrey Berman
3	3	*The Intravenous Drug Abuse Patient,* by Clifford Gewirtz
4	4	*The Child Undergoing Outpatient Surgery,* by Nirmala Balan
5	5	*The Patient Undergoing Renal Transplantation,* by Rhoda Levine
6	6	*The Patient with Pheochromocytoma,* by Richard B. Patt
7	7	*The Patient with Known or Suspected Malignant Hyperthermia,* by Anna Stanec
8	8	*The Spinal Cord Injury Patient,* by Michael Domurat
9	9	*The Patient with Juvenile Laryngeal Papillomas,* by Ashok Kumar
10	10	*The Patient with Aortic Valvular Disease,* by Irene Osborn
11	11	*The Patient in Diabetic Ketoacidosis,* by Richard B. Patt
12	12	*The Morbidly Obese Patient,* by Werner Pfisterer
13	13	*The Patient with Myasthenia Gravis,* by John Prinscott
14	14	*The Head-Injured Patient,* by Elizabeth A.M. Frost

Lesson	Chapter	Title
15	15	*The Patient with Pregnancy-Induced Hypertension*, by Richard B. Patt
16	16	*The Patient with Neurofibromatosis*, by David Amar
17	17	*The Patient with Trigeminal Neuralgia*, by Zenaida Aurellano-Hoh
18	18	*The Patient with Acquired Immune Deficiency Syndrome*, by Elizabeth A.M. Frost
19	19	*The Patient with Porphyria*, by Gary Hartstein
20	20	*The Patient with Snakebite*, by Robert Hirsh
21	21	*The Patient with Sleep Apnea*, by Ingrid Hollinger
22	22	*The Patient with Mitral Valve Prolapse*, by Lydia Alcantara
23	23	*The Elderly Patient with a Hip Fracture*, by Jonathan Halevy
24	24	*The Patient with a Bleeding Disorder*, by Gerald M. Scheinman
25	25	*The Patient with a Ruptured Intracranial Aneurysm*, by Catherine Gulati
26	26	*The Patient with Scoliosis*, by James B. Mueller

Preanesthetic Assessment 2

ISBN: 0-8176-3430-4
ISBN: 0-8176-3430-4

Lesson	Chapter	Title
27	1	*The Patient with Abdominal Aortic Aneurysm*, by Raghubar P. Badola
28	2	*The Patient with Pituitary Disease*, by Irene Osborn

46	20	*The Patient with Chickenpox*, by Anil De Silva
47	21	*The Patient with Rheumatoid Arthritis*, by Werner Pfisterer
48	22	*The Patient for Outpatient Cataract Surgery*, by C.J. Eagle and Jan M. Davies
49	23	*The Patient with Peripheral Vascular Disease*, by Raghubar P. Badola
50	24	*The Patient with Parkinson's Disease*, by Michael S. Ackerman

Preanesthetic Assessment 3

ISBN: 0-8176-3430-4
ISBN: 0-8176-3430-4

Lesson	Chapter	Title
51	1	*The Athlete*, by Stephen G. Solomon
52	2	*The Patient with Chronic Spinal Cord Injury*, by James B. Mueller
53	3	*The Child with Congenital Heart Disease for Noncardiac Surgery*, by Ingrid Hollinger
54	4	*The Patient with Cor Pulmonale*, by James E. Tobin
55	5	*The Patient with Multiple Endocrine Neoplasia*, by Robert Adams
56	6	*The Patient with Down's Syndrome*, by David R. Sofair
57	7	*The Patient who is a Jehovah's Witness*, by Jon D. Samuels
58	8	*The Patient with Wilson's Disease*, by Elizabeth A.M. Frost
59	9	*The Drowning/Near-Drowning Patient*, by James B. Mueller